Lecture Notes in Mathematics

For information about Vols. 1–758, please contact your bookseller or Springer-Verlag.

Vol. 759: R. L. Epstein, Degrees of Unsolvability: Structure and Theory. XIV, 216 pages. 1979.

Vol. 760: H.-O. Georgii, Canonical Gibbs Measures. VIII, 190 pages. 1979.

Vol. 761: K. Johannson, Homotopy Equivalences of 3-Manifolds with Boundaries. 2, 303 pages. 1979.

Vol. 762: D. H. Sattinger, Group Theoretic Methods in Bifurcation Theory. V, 241 pages. 1979.

Vol. 763: Algebraic Topology, Aarhus 1978. Proceedings, 1978. Edited by J. L. Dupont and H. Madsen. VI, 695 pages. 1979.

Vol. 764: B. Srinivasan, Representations of Finite Chevalley Groups. XI, 177 pages. 1979.

Vol. 765: Padé Approximation and its Applications. Proceedings, 1979. Edited by L. Wuytack. VI, 392 pages. 1979.

Vol. 766: T. tom Dieck, Transformation Groups and Representation Theory. VIII, 309 pages. 1979.

Vol. 767: M. Namba, Families of Meromorphic Functions on Compact Riemann Surfaces. XII, 284 pages. 1979.

Vol. 768: R. S. Doran and J. Wichmann, Approximate Identities and Factorization in Banach Modules. X, 305 pages. 1979.

Vol. 769: J. Flum, M. Ziegler, Topological Model Theory. X, 151 pages. 1980.

Vol. 770: Séminaire Bourbaki vol. 1978/79 Exposés 525–542. IV, 341 pages. 1980.

Vol. 771: Approximation Methods for Navier-Stokes Problems. Proceedings, 1979. Edited by R. Rautmann. XVI, 581 pages. 1980.

Vol. 772: J. P. Levine, Algebraic Structure of Knot Modules. XI, 104 pages. 1980.

Vol. 773: Numerical Analysis. Proceedings, 1979. Edited by G. A. Watson. X, 184 pages. 1980.

Vol. 774: R. Azencott, Y. Guivarc'h, R. F. Gundy, Ecole d'Eté de Probabilités de Saint-Flour VIII-1978. Edited by P. L. Hennequin. XIII, 334 pages. 1980.

Vol. 775: Geometric Methods in Mathematical Physics. Proceedings, 1979. Edited by G. Kaiser and J. E. Marsden. VII, 257 pages. 1980.

Vol. 776: B. Gross, Arithmetic on Elliptic Curves with Complex Multiplication. V, 95 pages. 1980.

Vol. 777: Séminaire sur les Singularités des Surfaces. Proceedings, 1976-1977. Edited by M. Demazure, H. Pinkham and B. Teissier. IX, 339 pages. 1980.

Vol. 778: SK1 von Schiefkörpern. Proceedings, 1976. Edited by P. Draxl and M. Kneser. II, 124 pages. 1980.

Vol. 779: Euclidean Harmonic Analysis. Proceedings, 1979. Edited by J. J. Benedetto. III, 177 pages. 1980.

Vol. 780: L. Schwartz, Semi-Martingales sur des Variétés, et Martingales Conformes sur des Variétés Analytiques Complexes. XV, 132 pages. 1980.

Vol. 781: Harmonic Analysis Iraklion 1978. Proceedings 1978. Edited by N. Petridis, S. K. Pichorides and N. Varopoulos. V, 213 pages. 1980.

Vol. 782: Bifurcation and Nonlinear Eigenvalue Problems. Proceedings, 1978. Edited by C. Bardos, J. M. Lasry and M. Schatzman. VIII, 296 pages. 1980.

Vol. 783: A. Dinghas, Wertverteilung meromorpher Funktionen in ein- und mehrfach zusammenhängenden Gebieten. Edited by R. Nevanlinna and C. Andreian Cazacu. XIII, 145 pages. 1980.

Vol. 784: Séminaire de Probabilités XIV. Proceedings, 1978/79. Edited by J. Azéma and M. Yor. VIII, 546 pages. 1980.

Vol. 785: W. M. Schmidt, Diophantine Approximation. X, 299 pages. 1980.

Vol. 786: I. J. Maddox, Infinite Matrices of Operators. V, 122 pages. 1980.

Vol. 787: Potential Theory, Copenhagen 1979. Proceedings, 1979. Edited by C. Berg, G. Forst and B. Fuglede. VIII, 319 pages. 1980.

Vol. 788: Topology Symposium, Siegen 1979. Proceedings, 1979. Edited by U. Koschorke and W. D. Neumann. VIII, 495 pages. 1980.

Vol. 789: J. E. Humphreys, Arithmetic Groups. VII, 158 pages. 1980.

Vol. 790: W. Dicks, Groups, Trees and Projective Modules. IX, 127 pages. 1980.

Vol. 791: K. W. Bauer and S. Ruscheweyh, Differential Operators for Partial Differential Equations and Function Theoretic Applications. V, 258 pages. 1980.

Vol. 792: Geometry and Differential Geometry. Proceedings, 1979. Edited by R. Artzy and I. Vaisman. VI, 443 pages. 1980.

Vol. 793: J. Renault, A Groupoid Approach to C*-Algebras. III, 160 pages. 1980.

Vol. 794: Measure Theory, Oberwolfach 1979. Proceedings 1979. Edited by D. Kölzow. XV, 573 pages. 1980.

Vol. 795: Séminaire d'Algèbre Paul Dubreil et Marie-Paule Malliavin. Proceedings 1979. Edited by M. P. Malliavin. V, 433 pages. 1980.

Vol. 796: C. Constantinescu, Duality in Measure Theory. IV, 197 pages. 1980.

Vol. 797: S. Mäki, The Determination of Units in Real Cyclic Sextic Fields. III, 198 pages. 1980.

Vol. 798: Analytic Functions, Kozubnik 1979. Proceedings. Edited by J. Ławrynowicz. X, 476 pages. 1980.

Vol. 799: Functional Differential Equations and Bifurcation. Proceedings 1979. Edited by A. F. Izé. XXII, 409 pages. 1980.

Vol. 800: M.-F. Vignéras, Arithmétique des Algèbres de Quaternions. VII, 169 pages. 1980.

Vol. 801: K. Floret, Weakly Compact Sets. VII, 123 pages. 1980.

Vol. 802: J. Bair, R. Fourneau, Etude Géometrique des Espaces Vectoriels II. VII, 283 pages. 1980.

Vol. 803: F.-Y. Maeda, Dirichlet Integrals on Harmonic Spaces. X, 180 pages. 1980.

Vol. 804: M. Matsuda, First Order Algebraic Differential Equations. VII, 111 pages. 1980.

Vol. 805: O. Kowalski, Generalized Symmetric Spaces. XII, 187 pages. 1980.

Vol. 806: Burnside Groups. Proceedings, 1977. Edited by J. L. Mennicke. V, 274 pages. 1980.

Vol. 807: Fonctions de Plusieurs Variables Complexes IV. Proceedings, 1979. Edited by F. Norguet. IX, 198 pages. 1980.

Vol. 808: G. Maury et J. Raynaud, Ordres Maximaux au Sens de K. Asano. VIII, 192 pages. 1980.

Vol. 809: I. Gumowski and Ch. Mira, Recurences and Discrete Dynamic Systems. VI, 272 pages. 1980.

Vol. 810: Geometrical Approaches to Differential Equations. Proceedings 1979. Edited by R. Martini. VII, 339 pages. 1980.

Vol. 811: D. Normann, Recursion on the Countable Functionals. VIII, 191 pages. 1980.

Vol. 812: Y. Namikawa, Toroidal Compactification of Siegel Spaces. VIII, 162 pages. 1980.

Vol. 813: A. Campillo, Algebroid Curves in Positive Characteristic. V, 168 pages. 1980.

Vol. 814: Séminaire de Théorie du Potentiel, Paris, No. 5. Proceedings. Edited by F. Hirsch et G. Mokobodzki. IV, 239 pages. 1980.

Vol. 815: P. J. Slodowy, Simple Singularities and Simple Algebraic Groups. XI, 175 pages. 1980.

Vol. 816: L. Stoica, Local Operators and Markov Processes. VIII, 104 pages. 1980.

continuation on page 295

Lecture Notes in Mathematics

Edited by A. Dold and B. Eckmann

973

Matrix Pencils

Proceedings of a Conference
Held at Pite Havsbad, Sweden, March 22–24, 1982

Edited by B. Kågström and A. Ruhe

Springer-Verlag
Berlin Heidelberg New York 1983

Editors
Bo Kågström
Axel Ruhe
Institute of Information Processing
Department of Numerical Analysis, University of Umeå
901 87 Umeå, Sweden

AMS Subject Classifications 1980: 15 A 18, 15 A 42, 65-06, 65 F 15, 65 F 20, 65 F 25, 65 F 35, 65 L 02

ISBN 978-3-540-11983-8 Springer-Verlag Berlin Heidelberg New York
ISBN 978-0-387-11983-0 Springer-Verlag New York Heidelberg Berlin

2146/3140-543210

Preface

A conference devoted to Matrix Pencils was held March 22-24, 1982 at Hotel Larus, Pite Havsbad in Northern Sweden. It was organized jointly by the Institute of Information Processing, Numerical Analysis department at the University of Umeå, and the Swedish Institute of Applied Mathematics (ITM).

The choice of topic reflects the research interest of the group in Umeå since more than a decade, and the aim of the conference was to introduce and discuss recent mathematical and numerical research in matrix computations and related areas, as well as the use of matrix computations in applied problems. More than fifty people from fourteen nations participated at the conference; twenty-three talks were presented and six short contributions. The present volume contains a selection of the contributions to the conference, and here we give a short survey of the contents.

The first group of papers deal with general $A-\lambda B$ pencils. Though the mathematical theory for such pencils has been well understood for a long time, we have only recently got any reliable numerical methods. Pioneering work has been performed by Vera Nikolaevna Kublanovskaya from Leningrad, and her talk at this conference gives an introduction to that work. She presents the AB-algorithm and its modification as one approach for handling spectral problems of linear pencils. Paul Van Dooren treats essentially the same problem, with the emphasis on transformations to block triangular form and finding reducing subspaces. This new concept extends the notion of deflating subspaces for regular pencils, as introduced by G.W. Stewart, to the singular case. He also points out the relevance of those concepts in the theory of dynamic systems. Bo Kågström extends his works on algorithms for the Jordan Normal Form (JNF) of a matrix, to the pencil case, and shows how the JNF software can be used for regular pencils. He also formulates a modification of the AB-algorithm for deflating zero and infinite eigenvalues of a regular pencil, in terms of the Singular Value Decomposition. James Demmel reports some results on the condition of block diagonalization, i.e. the part of an algorithm that preceeds the computation of reducing subspaces. The bounds on the condition number of the transformation matrix are expressed in terms of the projection matrices determined by the partitioning.

The general matrix pencil problem can be very hard to deal with numerically. One reason is that the JNF is not a continuous function of the matrix elements. It is therefore necessary to see which numerical results that are relevant for the applied problem that gives rise to the matrix pencil. One important class of such applied problems is systems of ordinary differential equations $F(t,y,y')=0$. If these contain algebraic equations, and $\partial F/\partial y'$ is singular, many difficulties arise. C.W. (Bill) Gear and Linda Petzold deal with such differential algebraic equation (DAE) systems, and point out that they are considerably different from ordinary differential systems, as we are used to know them. Linear DAE systems with constant coefficients can be completely characterized via the Kronecker canonical form of the matrix pencil $A-\lambda B$. However the linear nonconstant-coefficient case is not a simple extension of the constant coefficient case when the nilpotency of $A-\lambda B$ exceeds one.

In a special topic paper Kam-Moon Liu and Eduardo Ortiz are presenting numerical results from the application of the Tau method to the solution of eigenvalue problems defined by ordinary differential equations.

The symmetric case (A symmetric and B positive definite) is well understood, and here the emphasis is on development of efficient algorithms for large sparse problems, and their use in applications. Finite element computations is big business, and the papers by Liv Aasland and Petter Bjørstad, Eric Carnoy and M. Geradin, and Thomas Ericsson deal mainly with usage of the Lanczos algorithm in such computations. The Lanczos algorithm is now on the verge of replacing simultaneous iteration in commercial finite element packages. The symmetric $(A-\lambda B)$-problem with B positive definite is also discussed in the papers by D.J. Evans and Alan Jennings. Their topics are preconditioned iterative methods for the determination of extreme eigenpairs and bounds for the eigenvalues when a Cholesky factorization of B is available, respectively.

Generalizations of the Lanczos algorithm for nonsymmetric problems are discussed by Saad and Ruhe. Youcef Saad gives a unified view, in terms of projections of the most commonly used algorithms for large sparse problems, including the nonsymmetric Lanczos algorithm, the Arnoldi method and the subspace iteration. He also presents some à priori error bounds for these methods, in terms of the distance from the exact eigenvector to the approximating subspace. In the paper by Axel Ruhe the two-sided Arnoldi method is presented. The algorithm computes both left and right eigenvector approximations, by applying the Arnoldi method both to the matrix and its transpose. For nonsymmetric problems much work remains to be done, and it is not yet clear which of the generalizations of Lanczos that is the winner.

The other major group of papers deal with singular value problems, and applications to statistical data analysis. Gene Golub gave a review of statistical computations and generalized eigenvalue problems. He discussed estimates from linear models and compared the variance of a weighted least squares estimate, with that of a minimum variance estimate. This is important when using weighting to obtain better behaved estimates, and gives guidelines on how to choose weights and compute estimates.

Charles van Loan was the first to introduce the generalized singular value decomposition (GSVD). Here he uses the GSVD to analyse some weighting methods for least squares problems with equality constraints. So far it has not existed any stable algorithm to actually compute the GSVD. G.W. (Pete) Stewart describes how the problem of computing the GSVD can be reduced to that of computing a CS-decomposition (Cosine-Sinus) of a matrix with orthonormal columns, and suggests an algorithm, that is made stable by a reorthogonalization technique.

Two papers deal with theoretical aspects. Jiguang Sun from Beijing gives perturbation results both for the generalized eigenvalue and the generalized singular value problems. As an example he generalizes some classical perturbation theorems like the Bauer-Fike, Hoffman-Wielandt, Weyl-Lidskii and the Davis-Kahan theorems. Per Åke Wedin studies angles between subspaces. By utilizing the singular value decomposition and orthogonal projections, he shows how angles between two subspaces of C^n are related to the principal angles between certain invariant two dimensional subspaces. Further these concepts are used to point out that whenever relevant perturbation identities are known, it is easy to use an angle function to get perturbation bounds. Svante Wold, in the last local contribution, deals with applications to chemical data analysis.

Finally we would like to take this opportunity of thanking all speakers, chairmen and participants for their contributions. In particular we are indebted to Christina Holmström and Inga-Lena Olsson for their careful typing of various documents associated with the conference, and some of the typing in this volume.

The financial support from ITM is gratefully acknowledged.

Umeå, September 1982

Bo Kågström Axel Ruhe

Contents

In addition the following talks were presented at the conference:

Germund Dahlquist, Royal Institute of Technology, Stockholm, Sweden.
Some reflections on transformation of time dependent matrices to block diagonal form.

Emeric Deutsch, Polytechnic Institute of New York, New York, U.S.A.
Lower bounds for the Perron root of a non-negative matrix.

Gene Golub, Stanford University, Stanford, California, U.S.A.
The generalized eigenvalue problem and statistical computations.

Kresimir Veselic, Fernuniversität Hagen, Hagen, Germany
A Jacobi-like algorithm for the problem $Ax=\lambda Bx$ with symmetric indefinite A and B.

Olof Widlund, Courant Institute, New York, U.S.A.
Iterative solution of large generalized eigenvalue problems.

ADDRESSES OF THE AUTHORS

Liv Aasland, Petter Bjørstad:	Det Norske Veritas FDIV 40 Postboks 300 N-1322 HØVIK, NORGE
Eric Carnoy, M. Geradin:	L.T.A.S. Dynamique des constructions mécaniques Université de Liège Rue Ernest Solvay, 21 B-4000 Liège BELGIQUE
James Demmel:	Computer Science Division/EECS Dept University of California Berkely, CA 94720 U.S.A.
Thomas Ericsson:	Institute of Information Processing Department of Numerical Analysis University of Umeå S-901 87 UMEÅ, SWEDEN
David Evans:	Department of Computer Studies University of Technology Loughborough, Leicestershire LE 11 3 TU UNITED KINGDOM
C.W. Gear:	Department of Computer Science University of Illinois 1304 W Springfield URBANA IL 61801 U.S.A.
Alan Jennings:	Civil Engineering Department Queens University Belfast, BTG 5PG UNITED KINGDOM
Vera Kublanovskaya:	USSR Leningrad D-II Fontanka, 25 Inst of Mathematics Acad. of Sciences
Bo Kågström:	Institute of Information Processing Department of Numerical Analysis University of Umeå S-901 87 UMEÅ, SWEDEN
Eduardo Ortiz, K-M. Liu:	Imperial College of Science and Tech-nology Department of Mathematics Huxley Building Queen's Gate, London SW7 2BZ UNITED KINGDOM

Linda Petzold: Sandia National Laboratories
Applied Mathematics Division 8331
Livermore, CA 94550,
U.S.A.

Axel Ruhe: Institute of Information Processing
Department of Numerical Analysis
University of Umeå
S-901 87 UMEÅ, SWEDEN

Youcef Saad: Yale University
Computer Science Department
10, Hillhouse Ave, Box 2158,
Yale Station
New Haven CT 06520, U.S.A.

G.W. Stewart: Department of Computer Science
University of Maryland
College Park MD 20742
U.S.A.

Ji-Guang Sun: Computing Center
Academia Sinica
Peking
PR China

Paul van Dooren: Philips Research Laboratory
Av. van Becelaere, 2 Box 8
B-1170 Brussels
BELGIUM

Charles van Loan: Department of Computer Science
405 Upson Hall
Cornell University
Ithaca, New York 14853
U.S.A.

Per Åke Wedin: Institute of Information Processing
Department of Numerical Analysis
University of Umeå
S-901 87 UMEÅ, SWEDEN

Swante Wold: Department of Organic Chemistry
University of Umeå
S-901 87 UMEÅ, SWEDEN

SECTION A.1

OF

GENERAL $(A-\lambda B)$-PENCILS

CANONICAL REDUCTIONS - THEORY AND ALGORITHMS

The Condition Number of Equivalence Transformations that Block Diagonalize Matrix Pencils

James Demmel

Computer Science Division
University of California
Berkeley, CA, 94720, USA

ABSTRACT

How ill-conditioned must a matrix S be if its columns are constrained to span certain subspaces? We answer this question in order to find nearly best conditioned matrices S_R and S_L that block diagonalize a given matrix pencil $T = A+\lambda B$, i.e. $S_L^{-1}TS_R = \Theta$ is block diagonal. We show that the best conditioned S_R has a condition number approximately equal to the cosecant of the smallest angle between right subspaces belonging to different diagonal blocks of Θ. Thus, the more nearly the right subspaces overlap the more ill-conditioned S_R must be. The same is true of S_L and the left subspaces. For the standard eigenproblem ($T = A-\lambda I$), $S_L = S_R$ and the cosecant of the angle between subspaces turns out equal to an earlier estimate of the smallest condition number, namely the norm of the projection matrix associated with one of the subspaces. We apply this result to bound the error in an algorithm to compute analytic functions of matrices, for instance $\exp(T)$.

1. Introduction

Consider the problem of finding the eigenvalues of a matrix T. Two measures of the ill-conditioning of this problem have appeared frequently in the literature. One is the condition number of a matrix S which (block) diagonalizes T under similarity (i.e. $S^{-1}TS$ is block diagonal), and the other is the norm of the projection matrix P_i belonging to the spectrum of the i-th diagonal block of $S^{-1}TS$ (if the i-th block is 1 by 1, the norm of P_i is usually denoted $1/|s_i|$ [19]). Many authors have shown that the larger the condition number of S, or the larger the norm of P_i, the more sensitive to perturbations are at least some of the eigenvalues of T. Bauer and Fike [3], Kato [11], Kahan [9], Ruhe [12], Wilkinson [19,20] and others have all contributed theorems stating this result in different ways. Recently Sun [16] has extended many of these results to regular matrix pencils.

Our goal in this paper is to show that these two measures of ill-conditioning are nearly equivalent. We state our result in terms of angles between subspaces because this makes sense for pencils $T=A+\lambda B$ as well as the standard eigenproblem $T=A-\lambda I$: the condition number of the best S which displays the block structure is within a small constant factor of the cosecant of the smallest angle between a subspace belonging to one diagonal block and the subspace spanned by all the other subspaces together. In the case of the standard eigenproblem this cosecant turns out equal to the largest of the norms of the projections P_i.

We exhibit a best S for decomposing T into two blocks and compute its condition number exactly in terms of the norm of a projection (see part 2 below). This result was obtained independently by Bart et. al. [0] and improves an earlier estimate of Kahan [9]. Wilkinson [19, p 89] and Bauer [2] relate the two measures when $S^{-1}TS$ is completely diagonal; we generalize their results to diagonal blocks of arbitrary sizes in theorems 3 and 3a below.

For our results, $\|\cdot\|$ will denote the 2-norm for vectors and also the matrix norm induced by the vector norm:

$$\|S\| = \max_{x\neq 0}\|Sx\| / \|x\| \quad .$$

$\kappa(S)$ will denote the condition number of S with respect to $\|\cdot\|$:

$$\kappa(S) \equiv \|S\| \; \|S^{-1}\| \quad .$$

The angle between subspaces is defined as the smallest possible angle between a vector u in one subspace $\mathbf{S}^1$ and a vector v in another subspace $\mathbf{S}^2$:

$$\vartheta(\mathbf{S}^1, \mathbf{S}^2) = \min \{\arccos |u^*v| \text{ when } u\in\mathbf{S}^1, v\in\mathbf{S}^2, \|u\| = \|v\| = 1\} \tag{1.1}$$

(ϑ will be discussed more fully later).

If $\mathbf{S}^1, \ldots, \mathbf{S}^b$ is a collection of subspaces, the space spanned by their union is denoted $\text{span}\{\mathbf{S}^1, \ldots, \mathbf{S}^b\}$.

With this preparation, let us consider the subspaces associated with the block diagonal matrix $S_L^{-1}TS_R = \Theta = \text{diag}(\Theta_1, \ldots, \Theta_b)$, where Θ_i is r_i by c_i; r_i and c_i must be equal unless $T = A+\lambda B$ is a singular pencil [7]. From $S_L^{-1}TS_R=\Theta$ follows $TS_R = S_L\Theta$ which implies that T maps the space $\mathbf{S}_R^1$ spanned by the first c_1 columns of S_R into a space $\mathbf{S}_L^1$ spanned by the first r_1 columns of S_L. Similarly, columns $c_1+\cdots+c_{i-1}+1$ to $c_1+\cdots+c_i$ of S_R span a space $\mathbf{S}_R^i$ that T maps into a space $\mathbf{S}_L^i$ spanned by columns $r_1+\cdots+r_{i-1}+1$ to $r_1+\cdots+r_i$ of S_L. Stewart [15] calls the pairs $\mathbf{S}_R^i$, $\mathbf{S}_L^i$ deflating pairs since they deflate T to block diagonal form. For the standard eigenproblem $T=A-\lambda I$ we have $\mathbf{S}_R^i = \mathbf{S}_L^i$ [7] in which case they are denoted by $\mathbf{S}^i$ and called invariant subspaces and then no generality is lost by assuming $S_R = S_L$. Henceforth we drop the subscripts R and L of S since they are unnecessary for the standard eigenvalue problem and since our results apply to each case separately for the general problem $T = A+\lambda B$.

Our problem is to choose the columns of S to minimize $\kappa(S)$ subject to the condition that the columns span the subspaces $\mathbf{S}^i$. (It is not important for the proofs of our results that the $\mathbf{S}^i$ be defined by an eigenvalue problem; we ask only that the $\mathbf{S}^i$ be linearly independent and together span all of euclidean space. Thus our results may be interpreted as results on one-sided block diagonal scaling of matrices.) Our first result will be that by choosing the columns spanning each subspace to be orthonormal, we will have an S whose condition number is within a factor $\sqrt{b}$ of optimal, where b is the number of diagonal blocks of Θ:

$$\kappa(S_{ORTHO}) \le \sqrt{b}\ \kappa(S_{OPTIMAL}) \ . \tag{1.2}$$

S_{ORTHO} denotes any matrix S whose columns are orthonormal in groups as described above, and $S_{OPTIMAL}$ denotes any matrix S whose condition number is as small as possible. This extends a result of Van der Sluis [13] where all subspaces $\mathbf{S}^i$ are one-dimensional. Van Dooren and Dewilde [17] have also shown the choice of S_{ORTHO} is nearly best, and in fact optimal if the subspaces $\mathbf{S}^i$ are orthogonal.

Furthermore, we shall bound $\kappa(S_{ORTHO})$ above and below in terms of the angles between the subspaces $\mathbf{S}^i$ spanned by its columns. Let ϑ_i denote the smallest angle between $\mathbf{S}^i$ and the subspace spanned by all the other subspaces together:

$$\vartheta_i = \vartheta(\mathbf{S}^i \ , \ \operatorname*{span}_{j \ne i}\{\mathbf{S}^j\}) \ . \tag{1.3}$$

We shall show

$$\max_i\ (\csc \vartheta_i + \sqrt{\csc^2 \vartheta_i - 1}) \le \kappa(S_{OPTIMAL}) \le \kappa(S_{ORTHO}) \le \sqrt{b}\ \sqrt{\sum_{i=1}^{b} \csc^2 \vartheta_i} \tag{1.4}$$

When $b=2$ (i.e. we have only 2 diagonal blocks) S_{ORTHO} is in fact optimal, and

$$\kappa(S_{ORTHO}) = \kappa(S_{OPTIMAL}) = \csc \vartheta + \sqrt{\csc^2 \vartheta - 1} = \cot\ \vartheta/2 \ . \tag{1.5}$$

For the standard eigenproblem $\csc \vartheta_i = \| P_i \|$, where P_i is the projection associated with subspace i. It follows from (1.4) that the two measures of ill-conditioning $\kappa(S_{OPTIMAL})$ and $\max_i \| P_i \|$ we wanted to show nearly equivalent can differ by no more than a constant factor:

$$\max_i \| P_i \| \le \kappa(S_{OPTIMAL}) \le b \cdot \max_i \| P_i \| \ . \tag{1.6}$$

The rest of this paper is organized as follows. Part 2 shows the choice S_{ORTHO} is optimal for $b=2$ diagonal blocks. Part 3 discusses breaking T into more than 2 blocks. Part 4 applies the results to an error bound for computing a function of a matrix $f(T)$. Part 5 has the proof of a technical result used in part 2 and some related results.

2. How to Decompose T into 2 blocks

In this section we show that the best conditioned S whose first c columns span a given subspace $\mathbf{S}^1$ and whose remaining $n-c$ columns span another given complementary subspace $\mathbf{S}^2$ has condition number

$$\kappa(S_{OPTIMAL}) = \csc \vartheta + \sqrt{\csc^2 \vartheta - 1} = \cot \vartheta/2 \tag{2.1}$$

where $\vartheta = \vartheta(\mathbf{S}^1, \mathbf{S}^2)$. Note that we assume $\mathbf{S}^1$ and $\mathbf{S}^2$ are linearly independent, for otherwise S would be singular.

To prove (2.1) we will need a technical result, Theorem 1, that bounds the norms of submatrices of a positive definite matrix in terms of its condition number. Theorem 1 is a slight generalization of an inequality of Wielandt [4] and the proof technique used here yields several other inequalities (Theorem 4) one of which (5.21) is an inequality of Bauer [1].

Let

$$H = \begin{bmatrix} A & B \\ B^* & C \end{bmatrix}$$

be a Hermitian positive definite matrix, partitioned so that A is n by n, B is n by m, and C is m by m. Let $\kappa = \|H\| \; \|H^{-1}\|$ be the condition number of H. Let $X^{-1/2}$ denote any matrix such that $X^{-1/2}(X^{-1/2})^* = X^{-1}$.

Theorem 1: If H and κ are defined as above, then

$$\|(A^{-1/2})^* B C^{-1/2}\| \le \frac{\kappa - 1}{\kappa + 1} \tag{2.2}$$

or, equivalently,

$$\kappa \ge \frac{1 + \|(A^{-1/2})^* B C^{-1/2}\|}{1 - \|(A^{-1/2})^* B C^{-1/2}\|} . \tag{2.3}$$

Furthermore, this bound is sharp. In fact, given any n by m matrix Z such that $\|Z\| < 1$, both sides of inequality (2.2) are equal for the matrix

$$H = \begin{bmatrix} I & Z \\ Z^* & I \end{bmatrix} .$$

This theorem will be proved in Part 5.

We also need another definition of the (smallest) angle ϑ between subspaces that is more useful than the one stated in the introduction. As stated there, ϑ is the smallest possible angle between a vector in one subspace and a vector in the other subspace (the largest possible angle may be much larger than the smallest if the subspaces are not one dimensional). If S_1 is an n by c matrix of orthonormal columns which form a basis of $\mathbf{S}^1$ and S_2 is an n by $n-c$ orthonormal basis of the second space $\mathbf{S}^2$, then ϑ may also be expressed as [5]

$$\vartheta(\mathbf{S}^1, \mathbf{S}^2) = \arccos \|S^*{}_1 S_2\| = \arccos \sup_{x,y} |y^* S^*{}_1 S_2 x| \tag{2.4}$$

$$= \inf_{u,v} \arccos |u^* v|$$

where the sup is over arbitrary unit vectors x and y, and where the inf is over unit vectors u in $\mathbf{S}^1$ and v in $\mathbf{S}^2$.

Now consider a candidate matrix S:

$$S_{ORTHO} = [S_1 \mid S_2] \tag{2.5}$$

where S_1 and S_2 are orthonormal bases of $\mathbf{S}^1$ and $\mathbf{S}^2$ respectively. We may describe every other possible S whose columns span $\mathbf{S}^1$ and $\mathbf{S}^2$ in terms of S_{ORTHO}:

$$S_D = S_{ORTHO}\, D = S_{ORTHO}\, \mathrm{diag}(D_1\,,\, D_2) = [S_1 D_1 \mid S_2 D_2]\ , \tag{2.6}$$

where D_1 is a nonsingular c by c matrix and D_2 is a nonsingular $n-c$ by $n-c$ matrix. (2.6) states simply that any basis of $\mathbf{S}^i$ can be expressed as a nonsingular linear combination $S_i D_i$ of the columns of one basis S_i. We want to know which D minimizes $\kappa(S_D)$. We compute

$$\kappa^2(S_D) = \kappa(S_D{}^* S_D) \tag{2.7}$$

$$= \kappa \begin{bmatrix} D_1{}^* D_1 & D_1{}^* S_1{}^* S_2 D_2 \\ D_2{}^* S_2{}^* S_1 D_1 & D_2{}^* D_2 \end{bmatrix} .$$

We may now invoke Theorem 1 with $A^{-1/2} = D_1^{-1}$, $B = D_1{}^* S_1{}^* S_2 D_2$, and $C^{-1/2} = D_2^{-1}$ to find

$$\kappa^2(S_D) \geq \frac{1 + \| S_1{}^* S_2 \|}{1 - \| S_1{}^* S_2 \|}$$

$$= \frac{1 + \cos\vartheta}{1 - \cos\vartheta} \qquad (\,0 < \vartheta \leq \pi/2\,)$$

$$= \cot^2(\vartheta/2)$$

or

$$\kappa(S_D) \geq \cot\ \vartheta/2\ . \tag{2.8}$$

If D_1 and D_2 are unitary, it is easy to verify that we have equality in (2.8), proving (2.1) with $S_{OPTIMAL} = S_{ORTHO}$ as desired. Note, however, that $S_{OPTIMAL}$ is far from unique, since there are many orthonormal bases for a given space. It is also possible to find an $S_{OPTIMAL}$ with nonorthonormal S_1 and S_2 [6].

It remains to show $\csc\vartheta = \| P \|$ where P is the projection onto $\mathbf{S}^1$ parallel to $\mathbf{S}^2$. We may express P several ways. For example, let $S = [S_1 \mid S_2]$ be any matrix where S_i spans $\mathbf{S}^i$. Write S^{-1} as

$$S^{-1} = \begin{bmatrix} (S^{-1})^{(1)} \\ (S^{-1})^{(2)} \end{bmatrix} \tag{2.9}$$

where $(S^{-1})^{(i)}$ contains as many rows as S_i contains columns. Then P may be written

$$P = S_1 \cdot (S^{-1})^{(1)}\ .$$

Instead of using this representation of P, we use a second one which shows directly how P is associated with the standard eigenproblem $T = A - \lambda I$.

By Schur's Theorem [8], we may reduce T to upper triangular form by a unitary matrix Q

$$Q^* T Q = \begin{bmatrix} E' & G \\ 0 & F' \end{bmatrix} \tag{2.10}$$

Since the transformation by Q in (2.10) is just a unitary change of coordinates it does not change any angles between subspaces so we may assume without loss of generality that T is initially upper triangular.

We seek a matrix S such that

$$S^{-1} \begin{bmatrix} E' & G \\ 0 & F' \end{bmatrix} S = \begin{bmatrix} E & 0 \\ 0 & F \end{bmatrix} \tag{2.11}$$

(where E' is similar to E and F' to F) and a corresponding projection P which projects onto the invariant subspace belonging to E.

The matrices S and P can be exhibited as follows. Define R by solving $G = RF' - E'R$; this equation can be rearranged to form a triangular system of linear equations whose solution is R with its entries rearranged to form a vector. If E and F have disjoint spectra, this triangular system is nonsingular so that a solution R must exist. Otherwise G, E' and F' must be so related that the equations $G = RF' - E'R$ are consistent (R will exist, for example, if T has all linear elementary divisors).

Then one may verify that S must be of the form:

$$S = \begin{bmatrix} 1 & R \\ 0 & 1 \end{bmatrix} \begin{bmatrix} D_1 & 0 \\ 0 & D_2 \end{bmatrix} . \tag{2.12}$$

where D_1 and D_2 are arbitrary nonsingular matrices of the same dimensions as E' and F' respectively as discussed above.

Now observe that

$$P = \begin{bmatrix} 1 & -R \\ 0 & 0 \end{bmatrix} = P^2 \tag{2.13}$$

so that P is a projection. Since

$$PT = TP = \begin{bmatrix} E' & -E'R \\ 0 & 0 \end{bmatrix} .$$

P projects onto the invariant subspace corresponding to E'. Note that $||P||^2 = 1 + ||R||^2$. Also, since $||P|| = ||I-P||$ [10], where $I-P$ is the projection corresponding to F', it does not matter which projection we use.

By choosing

$$D_1 = I \quad \text{and} \quad D_2 = (I + R^* R)^{-1/2} \tag{2.14}$$

where D_2 can be any matrix such that $D_2 D_2^* = (I + R^*R)^{-1}$ we obtain an

$$S = \begin{bmatrix} I & R \\ & I \end{bmatrix} \begin{bmatrix} D_1 & \\ & D_2 \end{bmatrix} = \begin{bmatrix} I & R(I + R^*R)^{-1/2} \\ & (I + R^*R)^{-1/2} \end{bmatrix}$$

$$= [S_1 \mid S_2]$$

where S_1 and S_2 contain orthonormal vectors.

Thus

$$\vartheta = \arccos \| S_1^* S_2 \| \qquad (0 < \vartheta \le \pi/2) \tag{2.15}$$

$$= \arccos \| R(I + R^*R)^{-1/2} \|$$

so

$$\csc \vartheta = (1 - \cos^2 \vartheta)^{-1/2} \tag{2.16}$$

$$= \sqrt{1 + \|R\|^2}$$

$$= \|P\|$$

as desired.

3. How to Decompose T into b Blocks when $b > 2$

In this section we first consider partitioned matrices

$$S = [S_1 \mid \cdots \mid S_b] \tag{3.1}$$

where each submatrix S_i must span a given subspace $\mathbf{S}^i$ and show that S is nearly best conditioned when each S_i's columns are orthonormal. Next we bound the condition number of the best such S above and below in terms of $\max_i \csc \vartheta_i$, where

$$\vartheta_i = \vartheta(\mathbf{S}^i, \operatorname{span}_{j \neq i}\{\mathbf{S}^j\}) . \tag{3.2}$$

Finally we will discuss a different choice of S (also discussed in the literature [14,18]) which is harder to compute and has slightly different bounds on its condition number.

Theorem 2: Let S be

$$S = [S_1 \mid \cdots \mid S_b] \tag{3.3}$$

where S_i contains c_i columns.

If we choose the columns constituting S_i to be any orthonormal basis of the subspace $\mathbf{S}^i$, then S will have a condition number no larger than $\sqrt{b}$ times the smallest possible:

$$\kappa(S) \le \sqrt{b} \cdot \kappa(S_{OPTIMAL}) . \tag{3.4}$$

Said another way, choose S so that S^*S has identity matrices (of sizes c_i by c_i) as diagonal blocks.

Proof: This proof is a simple generalization of the proof that by diagonally scaling an n by n positive definite matrix to have unit diagonal, its condition number is within a factor of n of the lowest condition number achievable by diagonal scaling [13]. We generalize diagonal scaling for unit diagonal to be block diagonal scaling for block unit diagonal, i.e. to have identity matrices (of various sizes) on the diagonal. We show that a block diagonal scaling with b blocks produces a matrix whose condition number is within a factor b of the lowest possible condition number.

Assume S_i forms an orthonormal basis of $\mathbf{S}^i$ and let D be a block diagonal nonsingular matrix whose blocks D_i are c_i by c_i. Then any S' whose columns S'_i span $\mathbf{S}^i$ can be written $S' = SD$ for some D. Now

$$\sqrt{b}\,\kappa(SD) = \sqrt{b}\,\frac{\max\limits_{w\neq 0}\dfrac{\| Sw \|}{\| D^{-1}w \|}}{\min\limits_{z\neq 0}\dfrac{\| Sz \|}{\| D^{-1}z \|}} \geq \frac{\| D^{-1}z_o \|}{\| D^{-1}w_o \|}\cdot\frac{\sqrt{b}\,\| Sw_o \|}{\sigma_{\min}(S)} , \tag{3.5}$$

where z_o is chosen so that $\| z_o \| = 1$ and $\| Sz_o \| = \sigma_{\min}(S)$ = the smallest singular value of S, and w_o is chosen so $\| w_o \| = 1$ and $\| D^{-1}w_o \| = \sigma_{\min}(D^{-1})$. With this choice of w_o the factor $\| D^{-1}z_o \| / \| D^{-1}w_o \|$ is at least one. Since D is block diagonal, w_o can be chosen to have nonzero components corresponding to only one block of D. Thus, $\| Sw_o \|^2 = \| w_o{}^*S^*Sw_o \| = \| w_o{}^*w_o \| = 1$. Since the largest singular value $\sigma_{\max}(S)$ satisfies

$$\sigma_{\max}(S) = \| S \| \leq \sqrt{\sum_{i=1}^{b} \| S_i \|^2} = \sqrt{\sum_{i=1}^{b} 1} = \sqrt{b} ,$$

we get

$$\sqrt{b}\,\kappa(SD) \geq \frac{\sigma_{\max}(S)}{\sigma_{\min}(S)} = \kappa(S) . \tag{3.6}$$

Since (3.6) is true for any D, it is true in particular when $SD = S_{OPTIMAL}$. Q.E.D.

Van Dooren and Dewilde [17] have improved the factor $\sqrt{b}$ and shown, in particular, that if the subspaces are themselves orthogonal, then the above choice of S is in fact optimal.

In the case $b=2$ we expressed $\kappa(S_{OPTIMAL})$ in terms of $\csc\vartheta$, where ϑ was the smallest angle between $\mathbf{S}^1$ and $\mathbf{S}^2$. We can also bound $\kappa(S)$ here in terms of the $\csc\vartheta_i$, where ϑ_i is the angle between $\mathbf{S}^i$ and its complement $\operatorname{span}_{j\neq i}\{\mathbf{S}^j\}$:

Theorem 3: Let T, S and $\csc\vartheta_i$ be defined as above Then

$$\max_i\,(\csc\vartheta_i + \sqrt{\csc^2\vartheta_i - 1}) \leq \kappa(S) \leq \sqrt{b}\cdot\sqrt{\sum_{i=1}^{b}\csc^2\vartheta_i} , \tag{3.7}$$

or weakened slightly,

$$\max_i \csc \vartheta_i \le \kappa(S) \le b \cdot \max_i \csc \vartheta_i \quad . \tag{3.8}$$

Proof: This proof is based on a similar result of Wilkinson's [19, p. 89] when all invariant subspaces are one dimensional. First we will prove the lower bound and then the upper bound.

From (2.8) we know that any S (not just the one defined above) which has one group of columns spanning $\mathbf{S}^i$ has a condition number bounded from below:

$$\kappa(S) \ge \cot \vartheta_i / 2 = \csc \vartheta_i + \sqrt{\csc^2 \vartheta_i - 1} \quad . \tag{3.9}$$

Since (3.9) is true for all i, the lower bound follows easily.

We compute the upper bound as follows:

$$\kappa(S) = \| S \| \ \| S^{-1} \| \le \sqrt{b} \ \| S^{-1} \| \tag{3.10}$$

since $\| S \| \le \sqrt{b}$ (as mentioned in the proof of Theorem 2). Using notation analogous to (3.3) and (2.9) define the matrix P_i

$$P_i = S_i \ (S^{-1})^{(i)} \tag{3.11}$$

(which would be the matrix projection onto $\mathbf{S}^i$ for the standard eigenproblem). Since S_i consists of orthonormal columns, (3.11) and then (2.16) yield

$$\| (S^{-1})^{(i)} \| = \| P_i \| = \csc \vartheta_i \tag{3.12}$$

Thus

$$\| S^{-1} \| \le \sqrt{\sum_{i=1}^{b} \| (S^{-1})^{(i)} \|^2} = \sqrt{\sum_{i=1}^{b} \csc^2 \vartheta_i} \tag{3.13}$$

and the upper bound follows. Q.E.D.

The lower bound in Theorem 3 has been proven by Bauer [2] in the case when all invariant subspaces are one-dimensional .

The other choice of S discussed in the literature is scaled so that the i-th diagonal block of S^*S is $\csc \vartheta_i$ times an identity matrix of size c_i by c_i. With this choice of S the i-th diagonal block of $(S^*S)^{-1}$ has the same norm as the corresponding block of S^*S, namely $\csc \vartheta_i$. Smith [14] showed in the case when all invariant subspaces are one-dimensional that this choice of S is optimally scaled with respect to the condition number

$$\kappa_F(S) \equiv \| S \|_F \ \| S^{-1} \|_F$$

where $\| \cdot \|_F$ is the Frobenius norm:

$$\| S \|_F \equiv \sqrt{\sum_{i=1}^{n} \sum_{j=1}^{n} | S_{ij} |^2} \quad .$$

More generally, with this choice of S, Theorem 2 is weakened slightly to become:

Theorem 2a: With S chosen so that the i-th diagonal block of S^*S is $\csc \vartheta_i$ times an identity matrix, we have

$$\kappa(S) \le b \cdot \kappa(S_{OPTIMAL}) \ . \tag{3.14}$$

Proof: Similar to Theorem 2.

Theorem 3, on the other hand, becomes slightly stronger:

Theorem 3a: With S chosen as in Theorem 2a, we can bound $\kappa(S)$ as follows:

$$\max_i \left(\csc \vartheta_i + \sqrt{\csc^2 \vartheta_i - 1}\right) \le \kappa(S) \le \sum_{i=1}^{b} \csc \vartheta_i \ . \tag{3.15}$$

Proof: Similar to Theorem 3.

The upper bound of Theorem 3a generalizes a result of Wilkinson [19, p 89] for one dimensional invariant subspaces. Note that the "spectral condition numbers" $1/|s_i|$ used by Wilkinson and others [14,19] are just $\csc \vartheta_i$ (or $\|P_i\|$) when the invariant subspaces are one-dimensional. When $\sum_{i=1}^{b} \csc \vartheta_i$ is large the upper bound in (3.15) is comparable with the upper bound on $\kappa(S_{OPTIMAL})$ given by Bauer [2, Theorem VII] in the case of one-dimensional invariant subspaces.

This choice of S is more difficult to compute than the S of Theorems 2 and 3 because of the need to compute the $\csc \vartheta_i$, though not much more difficult if the subspaces are all one or two dimensional.

4. Computing a Function of a Matrix

In this section we want to show why a well conditioned block diagonalizing matrix S is better than an ill-conditioned one for computing a function of a matrix T. Assuming $f(T)$ is an analytic function of T, we compute $f(T)$ as follows:

$$f(T) = f(S\Theta S^{-1}) = Sf(\Theta)S^{-1} = S \begin{bmatrix} f(\Theta_1) & & \\ & \cdot & \\ & & f(\Theta_m) \end{bmatrix} S^{-1} \ . \tag{4.1}$$

The presumption is that it is easier to compute f of the small blocks Θ_i than of all of T. We will not ask about the error in computing $f(\Theta_i)$ but rather the error in computing $\Theta = S^{-1}TS$ and $f(T) = Sf(\Theta)S^{-1}$. In general, we are interested in the error in computing the similarity transformation $X = SYS^{-1}$.

We assume for this analysis that we compute with single precision floating point with relative precision ε. That is, when $\bullet$ is one of the operations $+, -, *$ or $/$, the relative error in computing $fl(a \bullet b)$ is bounded by ε:

$$fl(a \bullet b) = (a \bullet b)(1+e) \quad \text{where } |e| \le \varepsilon \tag{4.2}$$

Using (4.2) it is easy to show

Lemma 1: Let A and B be real n by n matrices, where $n\varepsilon < .1$. Let $|A|$ denote the matrix of absolute entries of A: $|A|_{ij} = |A_{ij}|$. Then to first order in ε the error in computing the matrix product AB is bounded as follows:

$$|fl(AB) - AB| \le n\varepsilon |A|\ |B| \ . \tag{4.3}$$

Proof: See [18].

Computing $X = SYS^{-1}$ requires two matrix products: $Z = fl(SY)$ and $X = fl(ZS^{-1})$, where we assume S and S^{-1} are known exactly. Applying Lemma 1 to these two products yields

Lemma 2: If $n\varepsilon < 1/10$, then to first order in ε

$$\| fl(SYS^{-1}) - SYS^{-1} \| \le 3n^{5/2}\kappa(S) \| Y \| \varepsilon \ . \tag{4.4}$$

Proof: Straightforward.

Assuming this bound is realistic, it is clear that picking S to keep $\kappa(S)$ small is advantageous. The error in computing similarity transformations of matrices is discussed in more detail in Wilkinson [19, chap 3].

5. Proof of Theorem 1

This theorem was stated in Part 2.

Unit vectors $x \in \mathbf{C}^m$ and $y \in \mathbf{C}^n$ satisfying

$$y^*(A^{-1/2})^* B C^{-1/2} x = \| (A^{-1/2})^* B C^{-1/2} \| \tag{5.1}$$

must exist. Use them to construct the unit vectors

$$z = A^{-1/2} y / \| A^{-1/2} y \| \quad , \quad w = C^{-1/2} x / \| C^{-1/2} x \| \ , \tag{5.2}$$

and

$$s(\vartheta) = \begin{bmatrix} z \sin\vartheta \\ w \cos\vartheta \end{bmatrix} . \tag{5.3}$$

We want to consider H acting on the 2-dimensional subspace in which $s(\vartheta)$ lies. Now

$$s^*(\vartheta) H s(\vartheta) \le \Lambda \tag{5.4}$$

implies

$$[z^* \sin\vartheta, w^* \cos\vartheta] \begin{bmatrix} A & B \\ B^* & C \end{bmatrix} \begin{bmatrix} z \sin\vartheta \\ w \cos\vartheta \end{bmatrix} \le \Lambda \ , \tag{5.5}$$

or

$$\sin^2\vartheta \cdot z^* A z + \cos^2\vartheta \cdot w^* C w + \sin\vartheta\cos\vartheta\, (w^* B^* z + z^* B w) \le \Lambda \ . \tag{5.6}$$

To simplify notation, let $a \equiv z^*Az$ and $c \equiv w^*Cw$.

From (5.1) and (5.2) we know that

$$z^*Bw = \| (A^{-1/2})BC^{-1/2} \| / (\| A^{-1/2}y \| \cdot \| C^{-1/2}x \|) \tag{5.7}$$

$$= \| (A^{-1/2})BC^{-1/2} \| \cdot \| A^{1/2}z \| \cdot \| C^{1/2}w \| \quad .$$

Since $(C^{1/2})^*C^{1/2} = C$, we get $c \equiv w^*Cw = \| w^*(C^{1/2})^*C^{1/2}w \| = \| C^{1/2}w \|^2$. Similarly, $a \equiv z^*Az = \| A^{1/2}z \|^2$, so (5.7) becomes

$$z^*Bw = \| (A^{-1/2})^*BC^{-1/2} \| \cdot \sqrt{ac} \quad . \tag{5.8}$$

Substituting (5.8) into (5.6) and rearranging, we obtain

$$\left(\frac{c+a}{2}\right) + \left(\frac{c-a}{2}\right) \cos 2\vartheta + \sqrt{ac} \; \| (A^{-1/2})^*BC^{-1/2} \| \; \sin 2\vartheta \le \Lambda \tag{5.9}$$

Since ϑ was arbitrary, we can maximize the L.H.S. of (5.9) over ϑ yielding

$$\left(\frac{c+a}{2}\right) + \sqrt{\left(\frac{c-a}{2}\right)^2 + ac \; \| (A^{-1/2})^*BC^{-1/2} \|^2} \le \Lambda \; , \tag{5.10}$$

or

$$\| (A^{-1/2})^*BC^{-1/2} \| \le \frac{\sqrt{(\Lambda - (c+a)/2)^2 - ((c-a)/2)^2}}{\sqrt{ac}} \tag{5.11}$$

$$= \frac{\sqrt{(\Lambda - a)\,(\Lambda - c)}}{\sqrt{ac}} \; .$$

Similarly, the inequality

$$\lambda \le s^*(\vartheta) H s(\vartheta) \tag{5.12}$$

implies

$$\lambda \le \left(\frac{c+a}{2}\right) + \left(\frac{c-a}{2}\right) \cos 2\vartheta + \| (A^{-1/2})^*BC^{-1/2} \| \; \| A^{1/2}z \| \; \| C^{1/2}w \| \; \sin 2\vartheta \; . \tag{5.13}$$

Minimizing the R.H.S. of (5.13) over ϑ we obtain

$$\lambda \le \left(\frac{c+a}{2}\right) - \sqrt{\left(\frac{c-a}{2}\right)^2 + ac \; \| (A^{-1/2})^*BC^{-1/2} \|^2} \tag{5.14}$$

or, rearranging,

$$\| (A^{-1/2})^*BC^{-1/2} \| \le \frac{\sqrt{(a - \lambda)\,(c - \lambda)}}{\sqrt{ac}} \; . \tag{5.15}$$

Combining (5.11) and (5.15) yields

$$\| (A^{-1/2})^*BC^{-1/2} \| \le \min\left[\sqrt{(a-\lambda)\,(c-\lambda)/(ac)} \; , \; \sqrt{(\Lambda - a)\,(\Lambda - c)/(ac)}\right] \; .$$

All we know about $z^*Az \equiv a$ is that $\lambda \le a \le \Lambda$, and similarly $\lambda \le c \equiv w^*Cw \le \Lambda$. Thus

$$\|(A^{-1/2})^* BC^{-1/2}\| \le \max_{\lambda \le \alpha, \gamma \le \Lambda} \min\left[\sqrt{(\alpha-\lambda)(\gamma-\lambda)/(\gamma\alpha)}\,,\ \sqrt{(\Lambda-\alpha)(\Lambda-\gamma)/(\gamma\alpha)}\right]. \tag{5.16}$$

Since $(\alpha - \lambda)/\alpha$ is an increasing function of α and $(\Lambda - \alpha)/\alpha$ is a decreasing function of α in the range $\lambda \le \alpha \le \Lambda$, we see the max in the last inequality occurs when the two arguments of the min are equal. This equality implies

$$(\alpha - \lambda)(\gamma - \lambda) = (\Lambda - \alpha)(\Lambda - \gamma) \tag{5.17}$$

or

$$\alpha + \gamma = \Lambda + \lambda\ . \tag{5.18}$$

Substituting (5.18) into (5.16) yields

$$\|(A^{-1/2})^* BC^{-1/2}\| \le \max_{\lambda \le \gamma \le \Lambda} \sqrt{(\gamma-\lambda)(\Lambda-\gamma)} / \sqrt{\gamma(\Lambda+\lambda-\gamma)} \tag{5.19}$$

$$= \frac{\Lambda - \lambda}{\Lambda + \lambda}$$

$$= \frac{\kappa - 1}{\kappa + 1}\ .$$

as desired.

Any 2 by 2 positive definite matrix whose diagonal entries are equal shows the the inequality of Theorem 1 is sharp.

We now show that given κ and $Z = (A^{-1/2})^* BC^{-1/2}$ such that $\|Z\| < 1$ and the inequality of the theorem is sharp, it is possible to construct an H with the given constraints. Simply choose

$$A = I \quad , \quad C = I \quad \text{and} \quad B = Z \tag{5.20}$$

corresponding to the (arbitrary) choice $\Lambda = 1 + \|Z\|$ and $\lambda = 1 - \|Z\|$. It is easy to verify that every inequality in the proof is sharp for this choice of A, B, and C. Q.E.D.

Theorem 4: Let H, Λ, λ, and κ be as above. Define $X^{-1/2}$ such that $X^{-1/2}(X^{-1/2})^* = X^{-1}$. Then the following inequalities are sharp:

$$\|BC^{-1}\| \le \frac{1}{2}(\sqrt{\kappa} - 1/\sqrt{\kappa}) \tag{5.21}$$

$$\|A^{-1}B\| \le \frac{1}{2}(\sqrt{\kappa} - 1/\sqrt{\kappa}) \tag{5.22}$$

$$\|B\| \le \frac{1}{2}(\Lambda - \lambda) \tag{5.23}$$

$$\|(A^{-1/2})^* B\| \le \sqrt{\Lambda} - \sqrt{\lambda} \tag{5.24}$$

$$\|BC^{-1/2}\| \le \sqrt{\Lambda} - \sqrt{\lambda} \tag{5.25}$$

Proof: All the proofs are analogous to the proof of Theorem 1. To prove (5.21), for example (also proved in [1]), choose z and y unit vectors such that

$$z^* BC^{-1} y = \| BC^{-1} \|$$

and let

$$x = C^{-1} y / \| C^{-1} y \|$$

Consider H restricted to the two dimensional subspace in which

$$s(\vartheta) = \begin{bmatrix} z \sin\vartheta \\ x \cos\vartheta \end{bmatrix}$$

lies. The rest of the proof follows similarly to that of Theorem 1.

We can also show that given κ and arbitrary $R = BC^{-1}$ such that (5.21) is sharp, it is possible to construct an H with the given constraints. Simply choose

$$C = I \quad , \quad A = \left(\frac{\kappa^2 + 1}{2\kappa}\right) I \quad \text{and} \quad B = R \tag{5.26}$$

corresponding to the (arbitrary) choice $\Lambda = (\kappa + 1)/2$ and $\lambda = (\kappa + 1)/2\kappa$. It is easy to verify that every inequality in the proof is sharp for this choice of A, B, and C.

Note that Theorems 1 and 4 are still true when A, B, and C are conforming submatrices extracted from a larger H (or $Q^* HQ$ with Q unitary) since the bounds are monotonic in κ (or Λ and λ). In particular, if A, B, and C are scalar Theorem 1 becomes an inequality of Wielandt [4].

Acknowledgement

I thank Prof. W. Kahan for suggesting the application of the theorem to the problem of finding best conditioned similarities, and both him and Prof. B. Parlett for much constructive criticism regarding the presentation of the results. I also thank Paul Van Dooren for pointing out his improvement of Theorem 2 [17] and the referee for pointing out reference [0]. I also acknowledge the financial support of the U. S. Department of Energy, Contract DE-AM03-76SF00034, Project Agreement DE-AS03-79ER10358, and the Office of Naval Research, Contract N00014-76-C-0013.

References

[0] H. BART, I. GOHBERG, M. A. KAASHOEK, P. VAN DOOREN, Factorizations of Transfer Functions, SIAM J. Control, vol 18, no 6, November 1980, pp 675-696

[1] F. L. BAUER, A further generalization of the Kantorovic inequality, *Numer. Math.*, 3, pp 117-119, 1961

[2] F. L. BAUER, Optimally scaled matrices, *Numer. Math.*, 5, pp 73-87, 1963

[3] F. L. BAUER, C. T. FIKE, Norms and Exclusion Theorems, *Numer. Math.*, 2, pp 137-141, 1960

[4] F. L. BAUER, A. S. HOUSEHOLDER, Some inequalities involving the euclidean condition of a matrix, *Numer. Math.*, 2, pp 308-311, 1960

[5] C. DAVIS, W. KAHAN, Some new bounds on perturbations of subspaces, Bull. A. M. S., 75, 1969, pp 863-8

[6] J. DEMMEL, The Condition Number of Similarities that Diagonalize Matrices, Electronics Research Laboratory Memorandum, University of California, Berkeley, 1982

[7] F. R. GANTMACHER, *The Theory of Matrices*, trans. K. A. Hirsch, Chelsea, 1959, vol. 2

[8] E. ISAACSON, H. B. KELLER, *Analysis of Numerical Methods*, Wiley, 1966

[9] W. KAHAN, Conserving Confluence Curbs Ill-Condition, Technical Report 6, Computer Science Dept., University of California, Berkeley, August 4, 1972

[10] T. KATO, Estimation of Iterated Matrices, with Application to the von Neumann Condition, *Numer. Math.*, 2, pp 22-29, 1960

[11] T. KATO, *Perturbation Theory for Linear Operators*, Springer-Verlag, 1966

[12] A. RUHE, Properties of a Matrix with a Very Ill-conditioned Eigenproblem, *Numer. Math.*, 15, pp 57-60, 1970

[13] A. VAN DER SLUIS, Condition Numbers and Equilibration of Matrices, *Numer. Math.*, 14, pp 14-23, 1969

[14] R. A. SMITH, The Condition Numbers of the Matrix Eigenvalue Problem, *Numer. Math.*, 10, pp 232-240, 1967

[15] G. W. STEWART, Error and Perturbation Bounds for Subspaces Associated with Certain Eigenvalue Problems, *SIAM Review*, vol. 15, no. 4, Oct 1973, p 752

[16] J. G. SUN, The Perturbation Bounds for Eigenspaces of Definite Matrix Pairs, to appear

[17] P. VAN DOOREN, P. DEWILDE, Minimal Cascade Factorization of Real and Complex Rational Transfer Matrices, *IEEE Trans. on Circuits and Systems*, vol. CAS-28, no. 5, May 1981, p 395.

[18] J. H. WILKINSON, *Rounding Errors in Algebraic Processes*, Prentice Hall, 1963

[19] J. H. WILKINSON, *The Algebraic Eigenvalue Problem*, Oxford University Press, 1965

[20] J. H. WILKINSON, Note on Matrices with a Very Ill-Conditioned Eigenproblem, *Numer. Math.*, 19, pp 176-178, 1972

An approach to solving the spectral problem of $A-\lambda B$

by

V.N. Kublanovskaya
Mathematical Institute of the Academy of Sciences
Leningrad, USSR

1. Introduction and summary

We consider the spectral problem for a linear pencil of matrices

(1.1) $A-\lambda B$.

We will treat both the regular and singular case. We start with how to find the eigenvalues, eigenvectors and principal vectors of a pencil with nonsingular A and B. For that we use the AB algorithm, which is described in section 2. The AB algorithm is a natural generalization of the QR algorithm to pencils (1.1), and we will state the conditions for the convergence of the iterants to block triangular from. See [2], [3]. The AB algorithm is related to, but different from, the QZ algorithm by Moler and Stewart [6].

The inclusion of shifts is also described, as well as two implementation variants, one based on orthogonal and the other on elimination matrices. We continue in section 3, by describing how to treat singular A (zero eigenvalues) and singular B (infinite eigenvalues). A finite sequence of range nullspace separations is necessary to eliminate such singularities. In section 4, we describe how to compute Jordan vector chains when the eigenvalues have been computed. The algorithm is related to the algorithm to compute the Jordan Normal Form of a matrix eigenvalue problem [1], [5]. Finally in section 5 we treat the case of a singular pencil of possibly rectangular matrices. See also [4] for these matters.

We will not discuss the effects of rounding errors. The problem of replacing zero submatrices by matrices of a small norm is a non-trivial one, see e.g. [5].

2. Reduction of an arbitrary regular pencil to block triangular form.

Let $A-\lambda B$ be a regular pencil. We will now describe an algorithm to compute a sequence of pencils,

$$(2.1) \qquad A_k - \lambda B_k, \quad k = 1,2,..$$

such that $A_k - \lambda B_k$ tends to a pencil of block triangular form. The algorithm starts by taking,

$$A_0 = A, \quad B_0 = B,$$

and then continues by computing a sequence (2.1) satisfying,

$$(2.2) \qquad A_k B_{k+1} = B_k A_{k+1}, \quad k = 0,1,2,..,$$

$$A_k \text{ or } B_k \text{ upper triangular for } k = 1,2,..$$

The condition (2.2) does not specify the sequence uniquely, but we note that:

1. All pencils $A_k - \lambda B_k$ are equivalent to $A-\lambda B$.

If B is nonsingular we see that,

$$(2.3) \qquad A_{k+1} = B_k^{-1} A_k B_{k+1},$$

$$B_{k+1} = B_k^{-1} B_k B_{k+1},$$

and there is a similar expression for the equivalence if A is nonsingular. In the general case a linear combination of A and B has to be taken. From (2.3) we also get the relation between successive eigenvectors. If u_i^k denote an eigenvector to $A_k - \lambda B_k$, we see that

$$u_i^0 = V_k u_i^k, \quad V_k = B_1 B_2 \ldots B_k.$$

2. The sequence $A_k - \lambda B_k$ converges to block triangular form, with equal absolute value of the eigenvalues in each diagonal block, under the following conditions:

i) If $P^{-1}JP = A^{-1}B$ is a transformation into Jordan Normal Form, the factorization $P = LR$ into lower and upper triangular factors exists.

ii) The eigenvalues λ_i of $A-\lambda B$ satisfy

$$0 < |\lambda_1| = \ldots = |\lambda_{\delta_1}| < |\lambda_{\delta_1+1}| = \ldots = |\lambda_{\delta_2}| < \ldots = |\lambda_{\delta_n}| < \infty.$$

iii) The sequences $||V_k||$ and $||V_k^{-1}||$ are bounded.

The size of the diagonal blocks will be $\sigma_i \times \sigma_i$ with $\sigma_i = \delta_i - \delta_{i-1}$.

In the case of a simple pencil i.e. when J is diagonal, (i), (ii), (iii) are satisfied and furthermore the diagonal elements of A_k and B_k satisfy:

$$(2.4) \qquad a_{ii}^k / b_{ii}^k = \lambda_i \{1 + O\left(\left|\frac{\lambda_{\delta_i}}{\lambda_{\delta_i+1}}\right|^k\right) + O\left(\left|\frac{\lambda_{\delta_{i-1}}}{\lambda_{\delta_i}}\right|^k\right)\}.$$

For a proof see [2].

3. The sequence $A_k - \lambda B_k$ can be computed by a sequence of range nullspace separations.

Introduce the rectangular matrices

$$(2.5) \qquad \mathbb{A}_k := [A_k, B_k]$$

$$Q_k := \begin{bmatrix} -B_{k+1} \\ A_{k+1} \end{bmatrix}$$

Now (2.2) implies that the columns of Q_k span the nullspace of $\mathbb{A}_k$. Different range-nullspace separations give rise to different algorithms. We note e.g. that the QR and LR algorithms for the standard eigenvalue problem satisfy (2.2).

The QR-algorithm for $A-\lambda I$ is,

$$A_k = L_k Q_k^T, \quad A_{k+1} = Q_k^T L_k = L_{k+1} Q_{k+1}^T,$$

which implies (2.2) since,

$$L_k Q_{k+1} = Q_k L_{k+1}.$$

The LR-algorithm is described analogously.

In the case of a general regular pencil the AB-algorithm performs the range nullspace separation by means of a QR factorization, and the AB-1 algorithm by means of a LR factorization.

In the AB algorithm we compute an orthogonal T_k satisfying

(2.6) $$\mathbb{A}_k T_k = S_k = [L_k | 0].$$

We then get A_{k+1}, B_{k+1} from the last n columns of T_k,

$$T_k = \left[* \left| \begin{array}{c} -B_{k+1} \\ A_{k+1} \end{array} \right. \right].$$

If A_0 is Hessenberg and B_0 is triangular all A_k, B_k retain that form, and we can build up T_k as a sequence of plane rotations in the coordinate planes chosen in the sequence:

$$(i,j) = (n,2n-1),(n,2n),(n-1,2n-2),(n-1,2n-1),(n-1,2n),\ldots,$$

$$(1,n+1),(1,n+2),\ldots,(1,2n).$$

In the AB-1 algorithm we compute an upper triangular R_k satisfying

$$\mathbb{A}_k R_k = S_k = [L_k | 0].$$

Now R_k is built up as a product of elementary elimination matrices,

$$R_{ij} = I + r_{ij} e_i e_j^T,$$

differing from the unit matrix only in position (i,j). To retain triangular Hessenberg form we now first annihilate the lower codiagonal of B_k, by choosing

$$(i,j) = (n+2,1),(n+3,2),\ldots,(2n,n-1)$$

and

$$r_{ij} = -\alpha_{i-n,j}/\alpha_{i-n,i}$$

where α_{ij} are the elements of the successively transformed $\mathbb{A}_k$. Then we annihilate B_k row by row from the bottom up by taking

$$\begin{aligned}(i,j) = {} & (n,2n) \\ & (n-1,2n),(n-1,2n-1) \\ & \ldots\ldots\ldots\ldots \\ & (1,2n),(1,2n-1),\ldots,(1,n+1)\end{aligned}$$

and we see that finally R_k will have the form

$R_k =$ 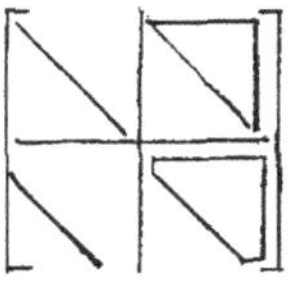

giving A_{k+1} Hessenberg, and B_{k+1} upper triangular.

The AB and AB-1 algorithm solve the same problem, but we note that the AB-1 algorithm needs only 1/4 th of the number of arithmetic operations of the AB algorithm, and can also be made to retain band form of A and B. On the other hand, the AB algorithm can be expected to be more stable numerically.

Introduction of shifts. We can, precisely as for the LR or QR algorithms, introduce a sequence of shifts $t_1,t_2,\ldots,t_k$. We then take instead of (2.5)

$$(2.7) \qquad \mathbb{A}_k = [A_k - (t_k - t_{k-1})B_k, B_k]$$

and proceed as above, with separating range and nullspaces.

The asymptotical relation is now replaced by the following theorems (See [2] and [4] for proofs).

Theorem 1. Let the shifts t_k be chosen so that $t_k \to \sigma$ $(k \to \infty)$, and assume that $B^{-1}A$ is nondefective

$$B^{-1}A = P^{-1}\Delta P, \quad \Delta = \mathrm{diag}(\lambda_i).$$

If for any r it holds

1) $|\lambda_i - \sigma| \leq \ldots \leq |\lambda_r - \sigma| < |\lambda_{r+1} - \sigma| \leq \ldots \leq |\lambda_n - \sigma|$.

2) $P_{n-r,n-r}$ is nonsingular if

$$P = \begin{bmatrix} P_{r,r} & P_{r,n-r} \\ P_{n-r,r} & P_{n-r,n-r} \end{bmatrix} \quad \text{with } P_{r,r} \text{ and } r x r \text{ matrix.}$$

3) PV_k has its last $n-r$ rows linearly independent.

4) $||V_k||$ are bounded.

Then the following asymptotical equalities hold

$$\det((A_k)_{n-r,n-r})\det((B_k^{-1})_{n-r,n-r}) =$$

$$= (\lambda_{r+1}-t_k)\ldots(\lambda_n-t_k)\left\{1 + 0\left(\frac{\varphi_k(\lambda_r)}{\varphi_k(\lambda_{r+1})}\right)\right\}$$

Here φ_k is the polynomial

$$\varphi_k(\lambda) = (\lambda-t_1)\cdot\ldots\cdot(\lambda-t_k).$$

<u>Theorem 2</u>. Let the conditions of Theorem 1 hold for $r=1,2,..,n-1$. Then

$$\frac{a_{ii}^k}{b_{ii}^k} = (\lambda_i-t_k)\left\{1 + 0\left(\frac{\varphi_k(\lambda_i)}{\varphi_k(\lambda_{i+1})}\right) + 0\left(\frac{\varphi_k(\lambda_{i-1})}{\varphi_k(\lambda_i)}\right)\right\}, \quad i=1,2,..,n.$$

Moreover, if we choose shifts in the obvious fashion, we will get quadratical convergence:

<u>Theorem 3</u>. Let the condition of Theorem 2 hold and assume that

$$\left|(\lambda_1-t_k) - \frac{a_{11}^k}{b_{11}^k}\right| < \varepsilon.$$

If we choose

$$t_{k+1} = \frac{a_{11}^k}{b_{11}^k} + t_k$$

we will get

$$\left| (\lambda_1 - t_{k+1}) - \frac{a_{11}^{k+1}}{b_{11}^{k+1}} \right| < \mu\varepsilon^2, \quad \mu \text{ constant.}$$

3. Deflating zero and infinite eigenvalues.

In this section, let

(3.1) $D(\lambda) = A - \lambda B$

be a regular $n \times n$ pencil, but A or B or both are singular. We will describe how to treat the case when A is singular, i.e. a zero eigenvalue; the case of an infinite eigenvalue is treated analogously replacing A by B. We consider three different algorithms.

The first algorithm makes a normalized QR-decomposition of A and solves a block linear system of equations, yielding an $r \times r$ matrix eigenvalue problem (r is the rank of A). See [4] for more details.

The second algorithm transforms $D(\lambda)$ into a block upper triangular pencil by means of an orthogonal equivalence. Like the first algorithm, it starts with a normalized QR-factorization;

(3.2) $\theta A T = [L_0 | 0]$

θ permutation

T orthogonal

L_0 $n \times r$ lower trapezium with

$$|\ell_{ii}| \geq |\ell_{jk}|, \quad j \geq i, \quad k \geq i,$$

and partition,

$$T = [Q_r \; Q_{n-r}]$$

where Q_{n-r} is a basis of the nullspace of A. If we now permute the columns, we get

$$Q = [Q_{n-r}\ Q_r].$$

This is the right hand transformation of the equivalence, to get the left hand one; we perform a similar QR factorization of $(BQ_{n-r})^T$, yielding

$$T_1 = [P_{n-r}\ P_r] \equiv P.$$

We then get

$$D_1(\lambda) = P^T D(\lambda) Q = \begin{bmatrix} D_{11}(\lambda) & D_{12}(\lambda) \\ 0 & D_{22}(\lambda) \end{bmatrix}$$

where $D_{22}(\lambda)$ is a linear pencil of order $r \times r$,

$$D_{22}(\lambda) = A_1 - \lambda B_1.$$

If A_1 is singular, we repeat the same process on $D_{22}(\lambda)$ until we get

$$A_m - \lambda B_m$$

with a nonsingular A_m.

The third algorithm is a finite modification of the AB algorithm. We make a QR decomposition,

$$\theta[A,B]T = [L_0 | 0],$$

with the permutation chosen in the same way as if we made a normalized decomposition of A alone. This can be effected e.g. by first decomposing A, and then applying rotations to annihilate the B elements row by row.

Let Q_0 as usual be the last n columns of T, and denote,

$$Q_0 = \begin{bmatrix} -\bar{B}_1 \\ \bar{A}_1 \end{bmatrix},$$

where $\bar{A}_1$ will by construction have $n-r$ zero columns.

Continue now by decomposing,

$$\tilde{A}_1 - \lambda\tilde{B}_1 = T_1(\bar{A}_1 - \lambda\bar{B}_1)\theta_1,$$

where T_1, θ_1 are the orthogonal matrix and permutation of a normalized decomposition of $\bar{B}_1$.

Now the transformed pencil has the form

$$\tilde{A}_1 - \lambda\tilde{B}_1 = \left[0 \left| \frac{*}{A_1}\right.\right] - \lambda\left[\begin{array}{c|c} I_{n-r} & * \\ \hline 0 & B_1 \end{array}\right].$$

We can continue and repeat the same process with the pencil $A_1 - \lambda B_1$, until finally we have A_m nonsingular.

It might be noted that the second and third algorithm can be used also to separate large and small eigenvalues, not just zero and non-zero. A tolerance has to be set when assigning ranks in that case.

It is further possible to develop a series expansion of the eigen-value of a perturbed matrix pencil, see [4].

4. Computing Jordan vector chains of a regular linear pencil.

Let $A - \lambda B$ be a regular pencil with B nonsingular, and let λ be an eigenvalue. A Jordan chain is a sequence of vectors $u_0, u_1, \ldots, u_s$ satisfying

$$(4.1) \qquad \begin{cases} (A-\lambda B)u_0 = 0 \\ (A-\lambda B)u_1 = Bu_0 \\ \cdots\cdots \\ (A-\lambda B)u_s = Bu_{s-1} \end{cases}$$

$Bu_s \notin$ image of $(A-\lambda B)$.

We will describe two algorithms to compute a Jordan chain.

The first algorithm builds up a sequence of larger and larger matrices, and finds bases of their nullspaces. The sequence of matrices is defined by,

$$\mathbb{A}_0 = A - \lambda B, \quad \mathbb{A}_k = \begin{bmatrix} \mathbb{A}_{k-1} & 0 \\ -B & A-\lambda B \end{bmatrix}, \quad k=1,2,..,s$$

and the bases of the null spaces are

$$Q_0, Q_1, \ldots \quad , Q_k = \text{kernel}(\mathbb{A}_k).$$

We can prove that:

i) Any solution of the first $k+1$ equations of (4.1) belongs to the subspace Q_k and vice versa.

ii) If we partition Q_k into blocks of n rows each,

$$Q_k = \begin{bmatrix} X_0^k \\ X_1^k \\ \vdots \\ X_k^k \end{bmatrix},$$

the following implications hold

$$\begin{array}{l} X_0^0 \supseteq X_0^1 \supseteq \ldots \supseteq X_0^s \\ \qquad X_1^1 \supseteq \qquad \supseteq X_1^s \\ \qquad\qquad X_{s-1}^{s-1} \supseteq X_{s-1}^s . \end{array}$$

From these subspaces we can construct the Jordan chains in the following way. Those of length one are obtained from,

$$T_0 = X_0^0 \setminus X_0^1,$$

where $\setminus$ denotes complement.

To get the chains of length $k+1$, the following steps have to be carried out:

1. For $k=1,2,\ldots,s$, find $T_k = X_0^k \setminus X_0^{k+1}$, solve the system $X_0^k Y_k = T_k$, and postmultiply $\hat{Q}_k = Q_k Y_k$.

2. For $k=1,2,\ldots,s-1$, find $\hat{T}_k = \hat{X}_k^k \setminus \hat{X}_k^{k+1}$, $\hat{T}_s = \text{image } \hat{X}_s^s$, solve the system $\hat{X}_k^k Z_k = \hat{T}_k$, and postmultiply $P_k = \hat{Q}_k Z_k$.

Now the columns of P_k give the vectors $u_{0i}^k, u_{1i}^k, \ldots, u_{ki}^k$ which form a Jordan chain of length $k+1$.

<u>The second algorithm</u> proceeds in two steps. For simplicity assume that $\lambda=0$. The first step transforms the pencil into block triangular form

$$U(A-\lambda B)V = \left[\begin{array}{c|c|c|c} -\lambda I_{n-r_1} & \multicolumn{3}{c}{M_1-\lambda N_1} \\ \hline & -\lambda I_{r_1-r_2} & \multicolumn{2}{c}{M_2-\lambda N_2} \\ \hline & & \ddots & \\ & & -\lambda I_{r_{m-1}-r_m} & M_m-\lambda N_m \\ \hline & & & \bar{A}_m-\lambda \bar{B}_m \end{array}\right]$$

The second step obtains the Jordan chain from the columns of

$$V = [V_{n-r_1} \quad V_{r_1-r_2} \quad V_{r_{m-1}-r_m} \mid V_{r_m}]$$

in a way that is a generalization of the algorithm for a matrix eigenvalue problem [1], [5].

To find the matrix V, we use the algorithm for separating zero eigenvalues described in section 3 of this report.

5. <u>Finding the regular kernel of a singular pencil</u>.

Now we consider the general case when $A-\lambda B$ is a rectangular $m \times n$ matrix pencil. We can use a variant of the AB algorithm to extract a regular kernel. Precisely as in section 3 we construct a sequence of pencils of successively lower order, until we obtain a regular one. First we consider the case with full column rank, then the general case when rows and columns both are linearly dependent.

1. Let $D(\lambda) = A-\lambda B$ be a full column rank $m \times n$ pencil, the rank of

$A-\lambda B$ being n. In order to isolate the regular kernel we obtain the finite sequence of pencils $\{A_k - \lambda B_k\}$, $k=1,2,..,p$ applying AB-algorithm. The typical step is

$$\mathbb{A}_k=[A_{k-1}\ B_{k-1}],\quad A_0=A,\ B_0=B,\ (k=1,2,..,p),$$

$$Q_k = \text{nullspace}\ \mathbb{A}_k = \begin{bmatrix} -B_k \\ A_k \end{bmatrix};$$

the size of Q_k being $2(2n-r_k)x[2(2n-r_{k-1})-r_k]$, where r_k is the rank of $\mathbb{A}_k$, $r_0=n$. Here the following inequalities hold

$$d_1 \equiv \dim Q_1 = 2n-r_1,\quad r_1 \geq n$$

$$d_2 \equiv \dim Q_2 = 2d_1-r_2,\quad r_2 \geq d_1$$

$$\cdots\cdots\cdots\cdots\cdots\cdots\cdots\cdots\cdots$$

$$d_p \equiv \dim Q_p = 2d_{p-1}-r_p,\quad r_p \geq d_{p-1},$$

so that $d_1 \geq d_2 \geq \ldots \geq d_p$.

The process terminates at step p if one of the following two conditions holds $d_p=0$ or $d_p=d_{p-1}$; if $d_p=0$, $A-\lambda B$ has no regular kernel; if $d_p=d_{p-1}$ the pencil $A_p-\lambda B_p$ is the kernel of the original pencil $A-\lambda B$.

2. Let $D(\lambda) = A-\lambda B$ be a totally singular $m x n$ $(m \geq n)$ pencil. The following algorithm is proposed.

a) For $\mathbb{A}_1 = [A\ B]$ we obtain the normalized decomposition

$$\theta_1 \mathbb{A}_1 T_1 = S_1 = [L_0\ \ 0].$$

Let r_1 be the rank of $\mathbb{A}_1$,

$$T_1 = [P_0^{(1)} Q_0^{(1)}],\quad \text{where}\quad P_0^{(1)} \in C^{2nxr_1},\ Q_0^{(1)} \in C^{2nx(2n-r_1)}.$$

If $r_1 \neq n$, then to continue the process we select from $P_0^{(1)}$ and $Q_0^{(1)}$ the matrix whose size is smaller. So if $r_1 \geq n$, then we select $Q_1 = Q_0^{(1)}$, otherwise $Q_1 = P_0^{(1)}$. We have

$$Q_1 = \begin{bmatrix} A_1 \\ B_1 \end{bmatrix}, \quad \text{if} \quad Q_1 = P_0^{(1)} \quad \text{and} \quad Q_1 = \begin{bmatrix} -B_1 \\ A_1 \end{bmatrix}, \quad \text{if} \quad Q_1 = Q_0^{(1)}.$$

Evidently, Q_1 is a 2nxn matrix.

b) Let the size of $A_p - \lambda B_p$ be the same as $A_{p-1} - \lambda B_{p-1}$ at some step p. In this case we must see whether or not $A_p - \lambda B_p$ is a singular pencil. To answer the question we check up whether or not $A_p - \lambda B_p$ is a singular matrix for any fixed λ for example, $\lambda = \pi$. If $A_p - \lambda B_p$ is a regular pencil, then $A_p - \lambda B_p$ is also the regular kernel of the pencil $A-\lambda B$.

References

1. KUBLANOVSKAYA, V.N. On a method of solving the complete eigenvalue problem for a degenerate matrix. Zh. Vychisl. Mat. Mat. Fiz. 6 (1966), 611-620; Transl. in USSR Comput. Math. Math. Phys. 6, 4 (1968), 1-14.
2. KUBLANOVSKAYA, V.N. The AB-algorithm and its properties. (In Russian), Trans. of the Steklov Inst. of Math., Leningrad 102, 42-60 (1980).
3. KUBLANOVSKAYA, V.N. AB-algorithm and its modifications for the spectral problems of linear pencils of matrices. LOMI preprints E-10-81, Leningrad (1981).
4. KUBLANOVSKAYA, V.N. On algorithms for the solution of spectral problems of linear matrix pencils. LOMI preprints E-1-82, Leningrad (1982).
5. KÅGSTRÖM, BO and RUHE, AXEL. An algorithm for numerical computation of the Jordan normal form of a complex matrix. ACM, TOMS, 6, 398-419 (1980).
5. MOLER, C.B. and STEWART, G.W. An algorithm for the generalized matrix eigenvalue problem, SIAM J. Num. Anal. 10, 241-256 (1973).

On computing the Kronecker canonical form of regular (A-λB)-pencils

Bo Kågström
Institute of Information Processing
University of Umeå
S-901 87 Umeå, Sweden

1. Introduction

In this paper we consider the problem of computing a canonical decomposition of a regular pencil $A-\lambda B$ where A and B are complex matrices. By definition (see e.g. [2]) a pencil $A-\lambda B$ is _regular_ if and only if A and B are square $(\in C^{n\times n})$ and $\det(A-\lambda B) \not\equiv 0$ i.e. a nonzero determinant for all except a finite number of $\lambda \in C$. The finite eigenvalues correspond to the generalized eigenvalue problem $Ax = \lambda Bx$. Transformations preserving the eigenvalues of $A-\lambda B$ and their multiplicities are called _strictly equivalent transformations_ [2]:

$$(1.1) \qquad P(A-\lambda B)Q = \tilde{A}-\lambda\tilde{B}$$

where P and $Q \in C^{n\times n}$ are constant and nonsingular. Correspondingly $A-\lambda B$ and $\tilde{A}-\lambda\tilde{B}$ in (1.1) are _strictly equivalent pencils_. For regular pencils we have the following canonical decomposition under strictly equivalent transformations (see section 3 for a derivation):

$$(1.2) \qquad P(A-\lambda B)Q = \left[\begin{array}{c|c} J & O \\ \hline O & I \end{array}\right] - \lambda \left[\begin{array}{c|c} I & O \\ \hline O & N \end{array}\right]$$

where J corresponds to the _finite eigenvalues_ of $A-\lambda B$ (including zero eigenvalues) and the nilpotent N corresponds to the _infinite eigenvalues_ of $A-\lambda B$ (corresponds to zero-eigenvalues of B). The "most diagonal" form of (1.2) is the _Weierstrass Kronecker canonical form_ [2] (W-KCF) where J and N are direct sums of Jordan blocks corresponding to the finite and infinite eigenvalues respectively. Assume that $A-\lambda B$, except for zero and infinite eigenvalues of multiplicities t_0 and t_∞ respectively, has p distinct nonzero and finite eigenvalues $\lambda_1, \lambda_2, \dots, \lambda_p$, each having algebraic multiplicity $t_k \geq 1$. We introduce the following notation for the structure of an eigenvalue (a generalization from $A-\lambda I$, see [7], [8])

$$(1.3) \qquad \alpha_k = (n_1^{(k)}, n_2^{(k)}, \dots, n_{h_k}^{(k)}), \quad k=0,1,\dots,p,\infty$$

where $n_1^{(k)}$ is the geometric multiplicity i.e. the number of eigenvectors, $n_j^{(k)}$ for $j \geq 2$ is the number of principal vectors of grade j and h_k is the maximal height of the principal chains i.e. the maximal order of a Jordan block corresponding to λ_k. In the context of implicit systems of differential equations,

$$(1.4) \qquad B\mathring{x} = Ax$$

h_∞ is called the nilpotency index of N and is often denoted by m ($N^{m-1} \neq 0$, $N^m \equiv 0$). The nilpotency index plays an important role when defining admissible initial conditions in order to obtain an unique solution of (1.4). The eq. (1.4) is also called a differential/algebraic system or a descriptor system [11].

We collect the structure indices α_k (1.3) from different eigenvalues to the multiindex

$$(1.5) \qquad \alpha := (\alpha_0, \alpha_1, \alpha_2, \dots, \alpha_p, \alpha_\infty)$$

and we denote the set of pencils $A-\lambda B$ with structure α (1.5) by E_α. Recently the problem of computing α (1.5) of a given regular pencil $A-\lambda B$ and at the same time reducing the pencil to a simpler but strictly equivalent form has been studied in papers by Kublanovskaya [5,6], Van Dooren [16,17], and Wilkinson [18,19]. We will study two of the proposed methods more closely. The first one originates from the paper by Wilkinson [18] (implicitly also in Gantmacher [2]) and is at the same time a derivation of the W-KCF (1.2). In order to be a practical algorithm the method requires the knowledge of the Jordan normal form of certain matrices. Secondly we study the AB-algorithm due to Kublanovskaya [5] and make a formulation in terms of singular value decompositions, which computes the structures α_0 and α_∞ of the zero and infinite eigenvalues respectively of a regular pencil. The behaviour of the algorithms in finite precision arithmetic will be illustrated by a numerical example. All computations have been carried out on a CDC Cyber 170/730 computer at the Umeå University Computing Centre, UMDAC. The relative precision (machep) is 48 bits i.e. machep $\approx 0.36_{10}-14$. However first we start by discussing what we can expect to compute in the presence of rounding errors.

2. Perturbation theory

A straight forward generalization of the $(A-\lambda I)$-problem (see Kågström, Ruhe [7]) gives that for each α, the set E_α consitutes a manifold in the space of regular pencils $A-\lambda B$ where A and $B \in C^{n\times n}$. When working in finite precision arithmetic it is impossible to decide whether a given pencil $A-\lambda B$ belongs to a certain E_α. It is more appropriate to require

$$A-\lambda B \in E_\alpha(\varepsilon)$$

where

$$(2.1) \qquad E_\alpha(\varepsilon) = \{\text{regular } X-\lambda Y:\ \|X-C\|/\|X\| + \|Y-D\|/\|Y\| \leq \varepsilon \text{ and } C-\lambda D \in E_\alpha\}$$

Suppose that we have computed a W-KCF (1.2) with structure α (1.5) corresponding to a regular pencil, say $C-\lambda D \in E_\alpha$ such that

$$(2.2) \qquad P(A-\lambda B)Q = \left[\begin{array}{c|c} J+E_{11} & E_{12} \\ \hline E_{21} & I+E_{22} \end{array}\right] - \lambda \left[\begin{array}{c|c} I+F_{11} & F_{12} \\ \hline F_{21} & N+F_{22} \end{array}\right]$$

or equivalently

$$(2.3) \qquad P(A-\lambda B)Q = P(C-\lambda D)Q + (E-\lambda F)$$

where

$$E = \begin{bmatrix} E_{11} & E_{12} \\ E_{21} & E_{22} \end{bmatrix} \quad \text{and} \quad F = \begin{bmatrix} F_{11} & F_{12} \\ F_{21} & F_{22} \end{bmatrix} .$$

The following upper bounds are easily verified:

(2.4a) $$||A-C||_F \leq ||P^{-1}||_2 \cdot ||Q^{-1}||_2 \cdot ||E||_F$$

(2.4b) $$||B-D||_F \leq ||P^{-1}||_2 \cdot ||Q^{-1}||_2 \cdot ||F||_F$$

The equation (2.4a-b) and the definition of the set $E_\alpha(\varepsilon)$ (2.1) give the following theorem.

<u>Theorem 2.1</u>

If the equations (2.2) and (2.3) hold then

(2.5a) $$A-\lambda B \in E_\alpha(\varepsilon)$$

where

(2.5b) $$\varepsilon = \left(\frac{||E||_F}{||A||_F} + \frac{||F||_F}{||B||_F}\right) ||P^{-1}||_2 \cdot ||Q^{-1}||_2 .$$

Assuming the knowledge of reliable estimates of $||E||_F$ and $||F||_F$ from our algorithms, this theorem give us a good way of validating and assessing the computed structure α and the associated canonical form. In practice, for the algorithms studied here, the dominating contributions to the perturbations E and F originate from deleted singular values when computing nullities of certain matrices. In this way the vector of indices α_k in (1.3) is determined. The sensitivity of a computed structure α_k is extremely dependent on the gap(β/δ) between the singular values we interpret as zero(δ) and nonzero(β) respectively. We get a <u>well-conditioned structure</u> α_k if the quotient β/δ is large enough and the ideal case is when δ is close to the machine precision and β is of order 1. If there is no appreciable gap $\delta \approx \beta$, the problem of determining the structure of $A-\lambda B$ is ill-conditioned in the respect that $A-\lambda B$ as well belongs to $E_{\alpha'}(\varepsilon')$ (different E, F, P and Q in (2.5b)) for another structure α' and an ε' of the same size as ε. The choice of structure is then to some extent arbitrary and this pathological behaviour is inherent in the pencil $A-\lambda B$. In the algorithms we discuss, the computed structure α_k is controlled by a tolerance and we seek $n_j^{(k+1)}$ i.e. the number of generalized vectors of grade $j+1$ such that $\beta/\delta \geq 1000$. See [7,8] for a discussion of the practical calculation of the Jordan structure of $A-\lambda I$.

The sensitivity of the problem of computing the structure α can also be explained in terms of perturbation analysis of the generalized eigenvalue problem $Ax = \lambda Bx$ ([13-14],[15]). For example,

if $K(B) = ||B||_2 \cdot ||B^{-1}||_2$ is large we can get large changes in some of the eigenvalues of $B^{-1}A$, warning us from working with $B^{-1}A-\lambda I$ when computing the structure of $A-\lambda B$. More difficult is the situation when A and B have almost intersecting nullspaces, since these problems can give rize to ill-conditioned eigenvalues that can affect otherwise well-conditioned eigenvalues. In theory a regular pencil $A-\lambda B$ has nonintersecting nullspaces i.e. $N(A) \cap N(B) = \{0\}$ but in the presence of rounding errors this is not always obvious and the pencil might behave as a singular one (see examples in Wilkinson [19] and Van Dooren [16]). We mention that $N(A) \cap N(B) = \{0\}$ does not imply that $A-\lambda B$ is regular which is obvious from the following example:

$$A = \begin{bmatrix} 1 & 1 & 1 \\ 0 & 0 & 1 \\ 0 & 0 & 1 \end{bmatrix} \qquad B = \begin{bmatrix} 1 & 2 & 1 \\ 0 & 0 & 2 \\ 0 & 0 & 1 \end{bmatrix}$$

Singular pencils will not be discussed further in this paper.

3. Computing W-KCF by using JNF

Is it possible to compute the W-KCF (1.2) by existing reliable software? The answer is yes. We will give a constructive proof including reliable error estimates and illustrate the derived algorithm in the MATLAB-environment [10], extended with the JNF-function ([1],[8]) for computing the Jordan normal form of a general matrix. The presentation of the algorithm in exact arithmetic is based on the paper by Wilkinson [18]. We consider the general case where both A and B may be singular. Consequently the cases with ill conditioned A

and/or B with respect to inversion can be handled with this method.

Since the method we discuss here includes the computation of Jordan normal forms in finite precision arithmetic, it inherits all the numerical difficulties of that problem (see [3], [4], [7-9]).

3.1. Algorithm in exact arithmetic

Since $A-\lambda B$ is regular there exists a c, such that $\det(A-cB) \neq 0$ and we consider the shifted pencil

$$A_1 - (\lambda-c)B \tag{3.1}$$

where $A_1 = A-cB$ and A_1 is nonsingular. By making the strictly equivalent transformation

$$A_1^{-1}(A_1 - (\lambda-c)B)I \tag{3.2}$$

we get the pencil $I - (\lambda-c)A_1^{-1}B$ where $A_1^{-1}B$ is singular with the same zero-structure as B and we compute the JNF of $A_1^{-1}B$:

$$(A_1^{-1}B)S = S\left[\begin{array}{c|c} J_1 & 0 \\ \hline 0 & J_2(0) \end{array}\right] \tag{3.3}$$

Here $J_2(0)$ is the nilpotent part with the structure α_∞ (1.3), corresponding to the infinite eigenvalues of $A-\lambda B$. It is now easily verified that the strictly equivalent pencil $I - (\lambda-c)A_1^{-1}B$ can be written in the canonical form:

$$I - (\lambda-c)A_1^{-1}B = S\left[\begin{array}{c|c} J_1 & 0 \\ \hline 0 & I+cJ_2(0) \end{array}\right]\left(\left[\begin{array}{c|c} U_1 & 0 \\ \hline 0 & I \end{array}\right] - \lambda\left[\begin{array}{c|c} I & 0 \\ \hline 0 & N_2(0) \end{array}\right]\right)S^{-1} \tag{3.4}$$

where

$$U_1 = J_1^{-1} + cI \tag{3.4a}$$

$$N_2(0) = (I + cJ_2(0))^{-1}J_2(0). \tag{3.4b}$$

Here the upper triangular U_1 corresponds to the finite eigenvalues of $A-\lambda B(=\lambda_i^{-1}(J_1)+c)$ and has the same structure as J_1; $N_2(0)$ is nilpotent with the structure of $J_2(0)$. By knowing the Jordan normal forms,

$$U_1X_1 = X_1D_1 \tag{3.5}$$

$$N_2(0)X_2 = X_2D_2(0) \tag{3.6}$$

we can get an explicit expression of the W-KCF (1.2). Since the structure indices α_k [see eq's (1.3) and (1.5)] of U_1 and $N_2(0)$ are settled beforehand by the JNF of $A_1^{-1}B$ (3.3), the computation of their JNF's can be considerably simplified. Notice that by knowing c in eq. (3.4b), X_2 and $D_2(0)$ can explicitly be computed. The general case is obvious by illustrating with one Jordan block; in this case of order 5:

$$J_2(0) = \begin{bmatrix} 0 & r_1 & 0 & 0 & 0 \\ 0 & 0 & r_2 & 0 & 0 \\ 0 & 0 & 0 & r_3 & 0 \\ 0 & 0 & 0 & 0 & r_4 \\ 0 & 0 & 0 & 0 & 0 \end{bmatrix} \tag{3.7}$$

and

$$(I+cJ_2(0))^{-1}J_2(0) = \begin{bmatrix} 0 & r_1 & -cr_1r_2 & c^2r_1r_2r_3 & -c^3r_1r_2r_3r_4 \\ 0 & 0 & r_2 & -cr_2r_3 & c^2r_2r_3r_4 \\ 0 & 0 & 0 & r_3 & -cr_3r_4 \\ 0 & 0 & 0 & 0 & r_4 \\ 0 & 0 & 0 & 0 & 0 \end{bmatrix} \tag{3.8}$$

By eliminating the elements above the superdiagonal, column by column, from the right to the left we get $D_2(0) = J_2(0)$ and

$$X_2 = \begin{bmatrix} 1 & 0 & 0 & 0 & 0 \\ 0 & 1 & cr_2 & c^2r_2r_3 & c^3r_2r_3r_4 \\ 0 & 0 & 1 & 2cr_3 & 3c^2r_3r_4 \\ 0 & 0 & 0 & 1 & 3cr_4 \\ 0 & 0 & 0 & 0 & 1 \end{bmatrix}, \tag{3.9}$$

which corresponds to the execution of step 7 of our JNF-algorithm [7]. To get X_1 and D_1 we have to perform steps 6 and 7.

Finally we have that

$$A-\lambda B = A_1(I - (\lambda-c)A_1^{-1}B) =$$

$$= A_1 S \left[\begin{array}{c|c} J_1X_1 & 0 \\ \hline 0 & (I+cJ_2(0))X_2 \end{array}\right] \left(\left[\begin{array}{c|c} D_1 & 0 \\ \hline 0 & I \end{array}\right] - \right.$$

$$\left. - \lambda \left[\begin{array}{c|c} I & 0 \\ \hline 0 & D_2(0) \end{array}\right] \right) \left[\begin{array}{c|c} X_1^{-1} & 0 \\ \hline 0 & X_2^{-1} \end{array}\right] S^{-1}$$

or

$$(3.10) \qquad P(A-\lambda B)Q = \left[\begin{array}{c|c} D_1 & 0 \\ \hline 0 & I \end{array}\right] - \lambda \left[\begin{array}{c|c} I & 0 \\ \hline 0 & D_2(0) \end{array}\right]$$

where

$$(3.11) \qquad Q = S \left[\begin{array}{c|c} X_1 & 0 \\ \hline 0 & X_2 \end{array}\right]$$

and

$$(3.12) \qquad P = \left[\begin{array}{c|c} (J_1X_1)^{-1} & 0 \\ \hline 0 & [(I+cJ_2(0))X_2]^{-1} \end{array}\right] S^{-1}(A-cB)^{-1},$$

which is the W-KCF (1.2) of $A-\lambda B$.

From D_1 we get the structures α_k (1.3) of the finite eigenvalues λ_k, $k=0,1,\ldots,p$ and $D_2(0)$ give us the structure α_∞ of the infinite eigenvalue of $A-\lambda B$. The columns of Q corresponding to the finite eigenvalues constitute the right generalized eigen- and principal vectors satisfying

$$(3.13a) \qquad (A-\lambda_k B)q_i = 0 \qquad \text{(eigenvectors)}$$

$$(3.13b) \qquad (A-\lambda_k B)q_{j+1} = Bq_j \qquad \text{(principal vectors).}$$

For $\lambda_0=0$ we have that $Aq_i = 0$ and $Aq_{j+1} = Bq_j$. In the same way the rows of P corresponding to the infinite eigenvalues constitute the left generalized eigen- and principal vectors of λ_∞ satisfying $p_i^H B = 0$ and $p_{j+1}^H B = p_j^H A$ respectively.

3.2. Sources of error

We continue by investigating the sensitivity of the algorithm in the presence of rounding errors, where we only take account of the effects from major operations like matrix-inversions and Jordan normal form computations.

When choosing the shift in (3.1) we want one that makes $A-cB$ as wellconditioned as possible with respect to inversion; but without too much overhead computations. So far we have not studied this sub-problem thoroughly.

In the following we assume that we have a shift that makes A_1 (1.3) wellconditioned and the computed $A_1^{-1}B$ fulfills

$$\widetilde{A_1^{-1}B} = A_1^{-1}B + F \tag{3.14}$$

where, $||F||_2 \leq f(n)\,||A_1^{-1}||_2||B||_2\cdot\text{machep}$, and f is a low-degree polynomial in n, the order of the matrices:

The next step is to compute the JNF of $\widetilde{A_1^{-1}B}$ and we get

$$(\widetilde{A_1^{-1}B} + E_1)S = S \left[\begin{array}{c|c} J_1 & 0 \\ \hline 0 & J_2(0) \end{array}\right] \tag{3.15}$$

and partition the perturbation matrix E_1 accordingly

$$E_1 = \left[\begin{array}{c|c} E_{11}^{(1)} & E_{12}^{(1)} \\ \hline E_{21}^{(1)} & E_{22}^{(1)} \end{array}\right] . \tag{3.16}$$

Here $||E_{12}^{(1)}||$ and $||E_{21}^{(1)}||$ are usually (assuming a proper clus-tering) much smaller than $||E_{11}^{(1)}||$ and $||E_{22}^{(1)}||$ since, except for errors made when transforming $\widetilde{A_1^{-1}B}$ to block diagonal form, the diagonal blocks also accumulate deleted singular values from the deflation processes, when computing the Jordan structure α of $A_1^{-1}B$. To summarize, the size of $||E_1||$ and $K(S) = ||S||\,||S^{-1}||$ are decided by

- the conditioning of the problem of clustering the eigenvalues of $A_1^{-1}B$ in blocks, which is determined by the separation between the corresponding subspaces (see Stewart [13]). A good measure of this separation is the spectral projector P_k on the invariant subspace corresponding to the k-th diagonal block [4]. A necessary condition (but not sufficient for a proper clustering) is that all spectral projectors P_k have a moderate norm.

- how well the clustered eigenvalues, corresponding to diagonal blocks of $A_1^{-1}B$, fulfill the conditions for a numerical multiple eigenvalue, giving the structures α_k associated with the blocks (see [7], [9]). For well-conditioned matrices, with respect to the definition of a numerical multiple eigenvalue,

$$(3.17)\qquad ||E_1||_F = \mathcal{O}\Big(\sum_{k=0}^{p} t_k\,||P_k||\,\text{machep}\Big)$$

where t_k is the size of the k-th diagonal block.

Now introduce $E_S = -S^{-1}E_1S$ and partion accordingly

$$(3.18)\qquad E_S = \left[\begin{array}{c|c} E_{11} & E_{12} \\ \hline E_{21} & E_{22} \end{array}\right]$$

and we have that

$$(3.19)\qquad I - (\lambda-c)\widetilde{A_1^{-1}B} = S\left(I - (\lambda-c)\left[\begin{array}{c|c} J_1+E_{11} & E_{12} \\ \hline E_{21} & J_2(0)+E_{22} \end{array}\right]\right)S^{-1}.$$

Rewrite the right hand side of (3.19) according to (3.4), and do the necessary computations of U_1 (3.4a) and $N_2(0)$ (3.4b). Introduce the following notations

(3.20a) $G_1 = J_1^{-1} + F_{11}; \quad ||F_{11}||_2 \leq f(n-t_\infty)\,||G_1||_2$ machep

(3.20b) $G_2(0) = [I + cJ_2(0)]^{-1} + F_{22}; \quad ||F_{22}||_2 \leq f(t_\infty)\,||G_2^{(0)}||_2$ machep

where f is a low-degree polynomial in the orders of J_1 and $J_2(0)$ respectively. We now obtain

(3.21) $$S\left[\begin{array}{c|c} J_1 & 0 \\ \hline 0 & I+cJ_2(0) \end{array}\right]\left(\left[\begin{array}{c|c} G_1+cI+cG_1E_{11} & cGE_{12} \\ \hline cG_2(0)E_{21} & I+cG_2(0)E_{22} \end{array}\right] - \right.$$

$$\left. - \lambda \left[\begin{array}{c|c} I+G_1E_{11} & G_1E_{12} \\ \hline G_2(0)E_{21} & G_2(0)J_2(0)+G_2(0)E_{22} \end{array}\right]\right) \quad S^{-1}.$$

Compute the JNF of $\tilde{U}_1 = G_1+cI$ and $\tilde{N}_2(0) = G_2(0)J_2(0)$:

(3.22) $[\tilde{U}_1 + E_1^{(D)}]\; X_1 = X_1D_1$

(3.23) $[\tilde{N}_2(0) + E_2^{(D)}]\; X_2 = X_2D_2(0)$

where $E_i^{(D)}$ mostly corresponds to deleted singular values.

In section 3.1 we pointed out that, since D_1 and $D_2(0)$ have the same Jordan structures as J_1 and $J_2(0)$, these JNF computations could be considerably simplified. However in the numerical example presented later we make two full Jordan decompositions, but since we already have forced J_1 and $J_2(0)$ to be exact Jordan matrices, the sizes of $||E_1^{(D)}||$ and $||E_2^{(D)}||$ will be as in (3.17).

Now, by introducing the perturbation matrices ΔA_{ij} and ΔB_{ij},

$$\Delta A_{11} = X_1^{-1}(cG_1E_{11})X_1 - X_1^{-1}E_1^{(D)}X_1$$

$$\Delta A_{12} = X_1^{-1}(cG_1E_{12})X_2 \equiv \frac{1}{c}\,\Delta B_{12}$$

$$\Delta A_{21} = X_2^{-1}(cG_2(0)E_{21})X_1 \equiv \frac{1}{c}\,\Delta B_{21}$$

$$\Delta A_{22} = X_2^{-1}(cG_2(0)E_{22})X_2$$

$$\Delta B_{11} = X_1^{-1}(G_1E_{11})X_1$$

$$\Delta B_{22} = X_2^{-1}(G_2(0)E_{22})X_2 - X_2^{-1}E_2^{(0)}X_2$$

we can finally rewrite (3.21) as

$$(3.24)\qquad S\left[\begin{array}{c|c} J_1X_1 & 0 \\ \hline 0 & (I+cJ_2(0))X_2 \end{array}\right]\left(\left[\begin{array}{c|c} D_1+\Delta A_{11} & \Delta A_{12} \\ \hline \Delta A_{21} & I+\Delta A_{22} \end{array}\right] - \lambda\left[\begin{array}{c|c} I+\Delta B_{11} & \Delta B_{12} \\ \hline \Delta B_{21} & D_2(0)+\Delta B_{22} \end{array}\right]\right)\left[\begin{array}{c|c} X_1^{-1} & 0 \\ \hline 0 & X_2^{-1} \end{array}\right] S^{-1}.$$

Altogether our computed W-KCF

$$(3.25)\qquad \left[\begin{array}{c|c} D_1 & 0 \\ \hline 0 & I \end{array}\right] - \lambda\left[\begin{array}{c|c} I & 0 \\ \hline 0 & D_2(0) \end{array}\right] \equiv P(C-\lambda D)Q,$$

where P and Q are defined in eq's (3.10-12), is an exact W-KCF of a perturbed pencil

$$(3.26)\qquad (A+\Delta A) - \lambda(B+\Delta A).$$

By deleting second order terms $\mathcal{O}(\text{const}\|F_{ii}E_{ij}\|)$ in ΔA_{ij} and ΔB_{ij} we get

$$(3.27a)\qquad \Delta A = (A-cB)(-cE_1 - S\left[\begin{array}{c|c} J_1E_1^{(D)} & 0 \\ \hline 0 & 0 \end{array}\right] S^{-1})$$

$$(3.27b)\qquad \Delta B = (A-cB)(-E_1 - S\left[\begin{array}{c|c} 0 & 0 \\ \hline 0 & [I+cJ_2(0)]E_2^{(D)} \end{array}\right] S^{-1})$$

where E_1 and S are defined in equations (3.15-17). By estimating ΔA and ΔB,

$$(3.28a)\qquad \|\Delta A\| \le (|c|\cdot\|E_1\| + K(S)\cdot\|J_1\|\cdot\|E_1^{(D)}\|)\|A-cB\|$$

(3.28b) $$||\Delta B|| \le (||E_1|| + K(S)\cdot||I + cJ_2(0)||\cdot||E_2^{(D)}||)\,||A-cB||$$

and we have proved the following theorem.

Theorem 3.1

By applying the algorithm described in section 3.1 to a pencil $A-\lambda B$ in finite precision arithmetic we compute a nearby W-KCF (3.25) corresponding to a $C-\lambda D \in E_\alpha$ such that $A-\lambda B \in E_\alpha(\varepsilon)$, where

$$(3.29)\qquad \varepsilon = \left[||E_1||\left(\frac{c}{||A||} + \frac{1}{||B||}\right) + K(S)\left\{\frac{||E_1^{(D)}||\cdot||J_1||}{||A||} + \frac{||E_2^{(D)}||\,||I+cJ_2(0)||}{||B||}\right\}\right]||A-cB||.$$

For explicit expression of E_1, $E_i^{(D)}$ in terms of singular values see eq's (2.16-19) in [9].

Estimates of the critical terms $K(S) = ||S||\,||S^{-1}||$, $||E_1||$, $||E_1^{(D)}||$ and $||E_2^{(D)}||$ are given to the user from our JNF-routine, that makes it easy to validate and assess the computed W-KCF (for more details see [7-9] and the following example).

3.3. A numerical example

Here we illustrate the behaviour of the algorithm on a regular pencil $A-\lambda B$ with only 0 and ∞ as eigenvalues. The pencil is constructed like $P^{-1}(JA-\lambda JB)Q^{-1}$ where

$$JA = \left[\begin{array}{ccccc|ccc|c} 1 & 0 & 0 & 0 & 0 & 0 & 0 & 0 & 0 \\ 0 & 1 & 0 & 0 & 0 & 0 & 0 & 0 & 0 \\ 0 & 0 & 1 & 0 & 0 & 0 & 0 & 0 & 0 \\ 0 & 0 & 0 & 1 & 0 & 0 & 0 & 0 & 0 \\ 0 & 0 & 0 & 0 & 1 & 0 & 0 & 0 & 0 \\ \hline 0 & 0 & 0 & 0 & 0 & 0 & 1 & 0 & 0 \\ 0 & 0 & 0 & 0 & 0 & 0 & 0 & 1 & 0 \\ 0 & 0 & 0 & 0 & 0 & 0 & 0 & 0 & 0 \\ \hline 0 & 0 & 0 & 0 & 0 & 0 & 0 & 0 & 0 \end{array}\right]$$

$$JB = \left[\begin{array}{ccc|cc|cccc} 0 & 1 & 0 & 0 & 0 & 0 & 0 & 0 & 0 \\ 0 & 0 & 1 & 0 & 0 & 0 & 0 & 0 & 0 \\ 0 & 0 & 0 & 0 & 0 & 0 & 0 & 0 & 0 \\ \hline 0 & 0 & 0 & 0 & 1 & 0 & 0 & 0 & 0 \\ 0 & 0 & 0 & 0 & 0 & 0 & 0 & 0 & 0 \\ \hline 0 & 0 & 0 & 0 & 0 & 1 & 0 & 0 & 0 \\ 0 & 0 & 0 & 0 & 0 & 0 & 1 & 0 & 0 \\ 0 & 0 & 0 & 0 & 0 & 0 & 0 & 1 & 0 \\ 0 & 0 & 0 & 0 & 0 & 0 & 0 & 0 & 1 \end{array}\right]$$

and P and Q are uniformly distributed random matrices with $K(Q) = ||Q||_2 \cdot ||Q^{-1}||_2 = 139.9$ and $K(P) = ||P||_2 ||P^{-1}||_2 = 88.4$, generated in MATLAB [10].

We have used $c = -4$ as shift in (3.1), giving $K(A_1) = ||A_1||_2 \cdot ||A_1^{-1}||_2 = 2.8_{10}+4$ and the computed eigenvalues of $A_1^{-1}B$ before clustering are:

Table 3.1

-.00007718237845	-.00013374359627*I
-.00007718237846	.00013374359627*I
.00015436476320	-.00000000000000*I
-.00000000000010	.00000006685024*I
-.00000000000062	-.00000006685024*I
.25000304495032	.00000000000000*I
.24999597752417	.00000697695751*I
.24999597752417	-.00000697695751*I
.25000000000256	.00000000000000*I
$Re(\lambda_i)$	$Im(\lambda_i)$

Computed eigenvalues of $A_1^{-1}B$ from MATLAB (COMQR2-routine).

We report the explicitly computed D_1 and $D_2(0)$ together with $\tilde{P}A\tilde{Q}$ and $\tilde{P}B\tilde{Q}$, computed a posteriori, where $\tilde{P}$ and $\tilde{Q}$ are the computed transformation matrices (see eq's (3.10-12) and (3.24-27)) and $K(\tilde{P}) = 5.8_{10}+3$, $K(\tilde{Q}) = 78.9$.

Table 3.2

.99999999996911	+	.00000000000005*I			
1.00000000004442	+	.00000000000108*I			
.99999999998599	-	.00000000000112*I			
1.00000000001060	+	.00000000000009*I			
_.99999999999991_	+_	_.00000000000006*I_			
-.00000000000414	-	.00000000000020*I	-.00000000000142	+	.0*I
.00000000000438	-	.00000000000005*I	-.00000000000142	+	.0*I
-.00000000000136	-	.00000000000005*I	-.00000000000142	+	.0*I
.00000000000030	-	.00000000000011*I	-.00000000000142	+	.0*I

Diagonals of $\tilde{P}A\tilde{Q}$ and computed D_1.

.00000000000727	-	.00000000000011*I	
-.00000000248563	-	.00000000020385*I	
.00000000000055	+	.00000000000001*I	
-.00000000000391	-	.00000000000062*I	
_.00000000000005_	+_	_.00000000000016*I_	
1.59322235649433	+	.00000000000057*I	1.593222356484759
1.75196470873101	+	.00000000000018*I	1.751964708726888
-.00000000000002	+	.00000000000029*I	0.000000000000000

Super diagonals of $\tilde{P}A\tilde{Q}$ and computed D_1.

.00000000000987	-	.00000000000010*I	.00000000000134	+	.0*I
-.00000000001848	-	.00000000000037*I	.00000000000134	+	.0*I
.00000000000235	+	.00000000000023*I	.00000000000134	+	.0*I
-.00000000000071	+	.00000000000008*I	.00000000000134	+	.0*I
_.00000000000130_	+_	_.00000000000004*I_	.00000000000134	+	.0*I
.99999999999974	-	.00000000000007*I			
.99999999999832	+	.00000000000005*I			
1.00000000000209	-	.00000000000002*I			
1.00000000000023	-	.00000000000002*I			

Diagonals of $\tilde{P}B\tilde{Q}$ and computed $D_2(0)$.

.18278995995502 -	.00000000000009*I	.182789959956579
44.77384883699619 +	.00000000003669*I	44.773848836796788
-.00000000000003 -	.00000000000001*I	.000000000000000
1.82057787921332 +	.00000000000056*I	1.820577879213651
.00000000000003 -	.00000000000002*I	
-.00000000000247 +	.00000000000003*I	
.00000000000051 -	.00000000000002*I	
.00000000000002 +	.00000000000007*I	

Super diagonals of $\tilde{P}B\tilde{Q}$ and computed $D_2(0)$.

The distance δ (2.1) from $A-\lambda B$ to the nearby pencil $C-\lambda D$ with the computed W-KCF as the exact one is

$$\frac{\|A-\tilde{A}\|_F}{\|A\|_F} + \frac{\|B-\tilde{B}\|_F}{\|B\|_F} = 2.8_{10}{-11} + 1.0_{10}{-11} = 3.8_{10}{-11}$$

where

$$\tilde{A} = \tilde{P}\left[\begin{array}{c|c} D_1 & 0 \\ \hline 0 & I \end{array}\right]\tilde{Q} \quad \text{and} \quad \tilde{B} = \tilde{P}\left[\begin{array}{c|c} I & 0 \\ \hline 0 & D_2(0) \end{array}\right]\tilde{Q}.$$

Our algorithm dependent ε (3.29) is $1.8_{10}{-10}$, ($K(S) = 100.4$; $\|E_1\|_F = 1.8_{10}{-11}$; $\|E_1^{(D)}\|_F = 0.9_{10}{-13}$; $\|E_2^{(D)}\|_F = 1.7_{10}{-12}$), which is a reliable estimate of δ. Whichever algorithm we choose for computing the W-KCF of $A-\lambda B$, we can never expect δ to be less than $\max\{K(\tilde{Q}),K(\tilde{P})\}\cdot$machep, which, in this case is $2.1_{10}{-11}$.

4. The AB-algorithm in terms of SVD

Recently V.N.Kublanovskaya [5-6] presented the AB-algorithm, a basic tool for handling spectral problems of linear matrix pencils. The AB-algorithm computes two sequences of matrices $\{A_k\}$ and $\{B_k\}$ satisfying

(4.1) $$A_k B_{k+1} = B_k A_{k+1} \qquad k=0,1,\ldots,\ A_0=A,\ B_0=B$$

where A_{k+1} and B_{k+1} are blocks (one of them upper triangular) of the nullspace of the augmented matrix $C_k = [A_k \mid B_k]$ in the following way:

$$(4.2) \qquad \mathcal{N}(C_k) = \text{nullspace } \{C_k\} = \begin{bmatrix} -B_{k+1} \\ \hline A_{k+1} \end{bmatrix}.$$

By applying the AB-algorithm to $A-\lambda B$ we get information about the structure $\alpha_0(1.3)$ of the zero-eigenvalue in the W-KCF(1.2). A finite sequence of range-nullspace separations deflates the singularity, and different ways to perform a range-nullspace separation give rise to different algorithms. In [5], Kublanovskaya presents the AB-algorithm in terms of the QR and LR algorithms (see also the paper [6b] in this volume). Our experiences from computing the Jordan structures (see [7-9]) demonstrate the necessity of using the singular value decomposition (SVD), in order to obtain a numerically stable range-nullspace separation. This motivates a formulation of a modified AB-algorithm [5] in terms of SVD, and we show some of its properties and demonstrate its behaviour on two regular $(A-\lambda B)$-problems. As much as possible we follow the notations in [5-6].

4.1 The algorithm in theory

To start with we assume that $\det B \neq 0$ and $\det A = 0$ i.e $A-\lambda B$ has zero as an eigenvalue. First we are looking for the nullspace of the n by $2n$ matrix $C_0 = [A \mid B]$ (see eq's (4.1-4.2)) and by utilizing the SVD of C_0

$$(4.3) \qquad C_0 = U\,[\underbrace{0}_{n} \mid \underbrace{\Sigma}_{n}] \begin{bmatrix} W_1^H \\ \hline W_2^H \end{bmatrix} \begin{matrix} \}\, n \\ \}\, n \end{matrix}$$

we have that

$$\mathcal{N}(C_0) = \text{span}\{W_1\}$$

Now let,

$$(4.4) \qquad \begin{matrix} n\{ \\ n\{ \end{matrix} \underbrace{\begin{bmatrix} -B_1 \\ \hline A_1 \end{bmatrix}}_{n} = W_1,$$

and since the columns of W_1 are orthonormal A_1 and B_1 satisfy

$$(4.5a) \qquad AB_1 = BA_1$$

$$(4.5b) \qquad A_1^H A_1 + B_1^H B_1 = I$$

Since B is nonsingular by assumption, A and A_1 have the same zero-structure α_0, which implies that A and A_1 have the same number of zero-singular values ($=n_{A_1}$), which is the number of Jordan blocks corresponding to the zero-eigenvalue λ_0 of $A-\lambda B$.
However the orthonormality condition (4.5b) gives us more information about the singular values of A_1 and B_1, namely

$$(4.6) \qquad (\sigma_{A_1}^{(i)})^2 + (\sigma_{B_1}^{(i)})^2 = 1.$$

Further A_1 and B_1 have the same right singular vectors, and we summarize (see Stewart [12] for a proof):

$$(4.7a) \qquad U_{A_1}^H A_1 V_1 = \underbrace{\left[\begin{array}{c|c} 0 & 0 \\ \hline 0 & \Sigma_{A_1} \end{array}\right]}_{n_{A_1}}, \quad \Sigma_{A_1} = \operatorname{diag}\{\sigma_{A_1}^{(i)}\} \text{ nonsingular}$$

$$(4.7b) \qquad U_{B_1}^H B_1 V_1 = \overbrace{\left[\begin{array}{c|c} I & 0 \\ \hline 0 & \Sigma_{B_1} \end{array}\right]}, \quad \Sigma_{B_1} = \operatorname{diag}\{\sigma_{B_1}^{(i)}\}$$

and

$$(4.8) \qquad \Sigma_{A_1}^2 + \Sigma_{B_1}^2 = I$$

where Σ_{B_1} is nonsingular by assumption.

Unfortunately the eq's (4.7a-b) are not a strictly equivalent transformation of the pencil $A_1 - \lambda B_1$. What to do? Multiply W_1 (4.4) with V_1 (4.7a-b) from the right and we still have an orthonormal basis of $N(C_0)$:

$$(4.9) \qquad \left[\begin{array}{c} -B_1 \\ \hline A_1 \end{array}\right] V_1 = \left[\begin{array}{c|c} -B_{11} & -B_{12} \\ \hline 0 & A_{12} \end{array}\right]$$
$$\qquad\qquad\qquad\qquad \underbrace{\quad}_{n_{A_1}} \quad \underbrace{\quad}_{n-n_{A_1}}$$

Now by only utilizing the SVD of B_1 (4.7b) we have that

$$(4.10) \qquad U_{B_1}^H (A_1 - \lambda B_1) V_1 = \begin{bmatrix} 0 & M_1 \\ \hline 0 & A^{(1)} \end{bmatrix} - \lambda \begin{bmatrix} I & 0 \\ \hline 0 & B^{(1)} \end{bmatrix}$$
$$\underbrace{\hphantom{xxx}}_{n_{A_1}} \underbrace{\hphantom{xxx}}_{n-n_{A_1}} \qquad \underbrace{\hphantom{xxx}}_{n_{A_1}} \underbrace{\hphantom{xxx}}_{n-n_{A_1}}$$

where $B^{(1)}$ is diagonal and equal to Σ_{B_1} (4.7b). From eq's (4.9-10) and (4.5a) we see that

$$(4.11) \qquad P_1 (A - \lambda B) Q_1 = U_{B_1}^H (A_1 - \lambda B_1) V_1$$

where $P_1 = U_{B_1}^H B^{-1}$ and $Q_1 = B_1 V_1$ i.e we have eliminated n_{A_1} of the zero-eigenvalues of $A - \lambda B$ by a strictly equivalent transformation, and the n_{A_1} first columns of Q_1 are orthonormal right general-ized eigenvectors. The remaining eigenvalues of $A - \lambda B$ coincide with the eigenvalues of $A^{(1)} - \lambda B^{(1)}$ (4.10), which has zero-eigenvalues if $A - \lambda B$ has Jordan blocks of size ≥ 2. Continue with $A^{(1)} - \lambda B^{(1)}$ i.e. repeat the operations from eq' (4.3), and further on now with A and B replaced by $A^{(1)}$ and $B^{(1)}$ respectively and we end up with

$$(4.12) \qquad \hat{P}_2 (A^{(1)} - \lambda B^{(1)}) \hat{Q}_2 = \begin{bmatrix} 0 & M_2 \\ \hline 0 & A^{(2)} \end{bmatrix} - \lambda \begin{bmatrix} I & 0 \\ \hline 0 & B^{(2)} \end{bmatrix}$$
$$\underbrace{\hphantom{xxx}}_{n_{A_2}} \underbrace{\hphantom{xxx}}_{n-(n_{A_1}+n_{A_2})} \underbrace{\hphantom{xxx}}_{n_{A_2}}$$

Here n_{A_2} is the number of zero-singular values of A_2 (is equal to the number of unit-singular values of B_2) and $B^{(2)}$ is diagonal and equal to Σ_{B_2} (see eq. (4.7a-b)) where

$$(4.13) \qquad N(C_1) = N([A^{(1)} \mid B^{(1)}]) = \text{span} \{W_2\}$$

and

$$(4.14) \qquad \begin{matrix} n-n_{A_1} \left\{ \\ n-n_{A_1} \left\{ \right. \right. \end{matrix} \begin{bmatrix} -B_2 \\ \hline A_2 \end{bmatrix} = W_2 .$$
$$\underbrace{\hphantom{xxxx}}_{n-n_{A_1}}$$

Now let

$$(4.15) \qquad P_2 = \left[\begin{array}{c|c} I_{n_{A_1}} & 0 \\ \hline 0 & \hat{P}_2 \end{array}\right], \quad Q_2 = \left[\begin{array}{c|c} I_{n_{A_1}} & 0 \\ \hline 0 & \hat{Q}_2 \end{array}\right]$$

where

$$(4.16) \qquad \hat{P}_2 = U_{B_2}^H B^{(1)^{-1}}, \qquad \hat{Q}_2 = B_2 V_2$$

and we have that

$$(4.17) \qquad P_2 P_1 (A-\lambda B) Q_1 Q_2 = \left[\begin{array}{c|c|c} 0 & \multicolumn{2}{c}{M_1} \\ \hline 0 & 0 & M_2 \\ \cline{2-3} & 0 & A^{(2)} \end{array}\right] - \lambda \left[\begin{array}{c|c|c} I & \multicolumn{2}{c}{0} \\ \hline 0 & I & 0 \\ \cline{2-3} & 0 & B^{(2)} \end{array}\right]$$

$$\underbrace{}_{n_{A_1} \geq}\ \underbrace{}_{n_{A_2}}\ \underbrace{}_{n-(n_{A_1}+n_{A_2})}$$

The process continues until a step h where $A^{(h)}$ is nonsingular ($n_{A_{h+1}} = 0$ from eq's (4.7a-b)) and the structure α_0 of the zero-eigenvalue of $A-\lambda B$ is given by

$$(4.18) \qquad \alpha_0 = (n_{A_1}, n_{A_2}, \ldots, n_{A_h}).$$

Let $\quad t = \sum_{i=1}^{h} n_{A_i}$, then the first t columns of

$$(4.19) \qquad Q = Q_1 Q_2 \ldots Q_h$$

form a basis of the right generalized subspace (is a deflating subspace) corresponding to the zero-eigenvalue of $A-\lambda B$. The n_{A_1} eigenvectors are still orthonormal but not vectors of higher grade. Analogously the first t rows of

$$(4.20) \qquad P = P_h P_{h-1} \ldots P_2 P_1$$

form a basis of the left generalized subspace. From eq's (4.11), (4.15-16) we see that the left transformation matrix P is affected by the conditioning of B and successive $B^{(k)}$ with respect to inversion.

Note that the matrix $B^{(k)}$ is diagonal in each step, with the non-unit singular values of $B^{(k-1)}$ as diagonal elements, and that the deflated part of $A-\lambda B$ is successively transformed into standard form [see eq. (4.17)]. If we instead make use of the QR-decomposition for each range-nullspace separation we get a sequence $\{B^{(k)}\}$ of upper-triangular matrices (see Kublanovskaya [5-6]).

In this context it is interesting to compare this approach to the method presented by Van Dooren [16], which is a straight forward generalization of Kublanovskaya's algorithm for determining the Jordan structure of $A-\lambda I$ as used in [7-8]. The structure index α_0 (4.18) is obtained under unitary transformations P_i and Q_i and the relation corresponding to (4.17) looks like

$$(4.21)\qquad P_2P_1(A-\lambda B)Q_1Q_2 = \left[\begin{array}{c|c|c} 0 & \multicolumn{2}{c}{M_1} \\ \hline & 0 & M_2 \\ \cline{2-3} 0 & 0 & A^{(2)} \end{array}\right] - \lambda \left[\begin{array}{c|c|c} E_{11} & \multicolumn{2}{c}{N_1} \\ \hline & E_{22} & N_2 \\ \cline{2-3} 0 & 0 & B^{(2)} \end{array}\right]$$

where E_{ii}, N_i and $B^{(2)}$ are full matrices. A further reduction giving the right eigenvectors or the whole right eigenspace of the zero-eigenvalue, will include Gaussian-type eliminations without pivoting (see Van Dooren [16]). By using the AB-algorithm approach, we avoid these possibly unstable transformations [see eq's (4.11-12)], and the right transformation matrices Q_i are the product of unitary and blocks taken from unitary matrices:

$$(4.22)\qquad Q_i = \left[\begin{array}{c|c} I_{t_{i-1}} & 0 \\ \hline 0 & B_iV_i \end{array}\right], \quad t_{i-1} = n_{A_1} + \dots + n_{A_{i-1}}$$

where V_i is the right singular vectors of A_i and B_i [see eq's (4.7a-b)] and $-B_i$ is the upper square $(n-t_{i-1})$-block of the orthonormal basis of $N(C_{i-1})$ [see eq's (4.13-14)].

In the general case, when B may be singular, we get (4.7-8) of the form

$$(4.23a)\qquad U_{A_1}^H A_1 V_1 = \left[\begin{array}{c|c|c} 0 & 0 & 0 \\ \hline 0 & \Sigma_{A_1} & 0 \\ \hline 0 & 0 & I \end{array}\right], \quad \Sigma_{A_1} = \operatorname{diag}\{\sigma_{A_1}^{(i)}\}$$

$$\qquad\qquad\qquad\quad \underbrace{}_{n_{A_1}} \qquad\quad \underbrace{}_{n_{B_1}}$$

$$(4.23b)\qquad U_{B_1}^H B_1 V_1 = \left[\begin{array}{c|c|c} I & 0 & 0 \\ \hline 0 & \Sigma_{B_1} & 0 \\ \hline 0 & 0 & 0 \end{array}\right], \quad \Sigma_{B_1} = \operatorname{diag}\{\sigma_{B_1}^{(i)}\}$$

$$(4.23c)\qquad \Sigma_{A_1}^2 + \Sigma_{B_1}^2 = I$$

where $0 < \sigma_{A_1}^{(i)}, \sigma_{B_1}^{(i)} < 1$.

The singularity, n_{B_1} of B moves to B_1 [see eq. (4.9)] and we have that

$$(4.24)\qquad U_{B_1}^H (A_1 - \lambda B_1) V_1 = \left[\begin{array}{c|c} 0 & M_1 \\ \hline 0 & A^{(1)} \end{array}\right] - \lambda \left[\begin{array}{c|c} I & 0 \\ \hline 0 & B^{(1)} \end{array}\right] =$$

$$= \left[\begin{array}{c|c|c} 0 & M_{12} & M_{13} \\ \hline 0 & A_{22}^{(1)} & A_{23}^{(1)} \\ \hline 0 & A_{32}^{(1)} & A_{33}^{(1)} \end{array}\right] - \lambda \left[\begin{array}{c|c|c} I & 0 & 0 \\ \hline 0 & \Sigma_{B_1} & 0 \\ \hline 0 & 0 & 0 \end{array}\right].$$

$$\underbrace{}_{n_{A_1}} \qquad\qquad \underbrace{}_{n_{B_1}} \qquad\qquad \underbrace{}_{n_{A_1}} \qquad\qquad \underbrace{}_{n_{B_1}}$$

As in the case $\det B \neq 0$, the algorithm continues until $A^{(h)}$ is nonsingular. Each $B^{(k)}$ and Q_k will have n_{B_1} zero columns. Anyhow the first t columns of Q (4.19) span the right generalized subspace of λ_0. Now it is possible to deflate the infinite eigenvalues by repeating the algorithm on $C_0 = [B^{(h)} | A^{(h)}]$ i.e. the structure α_∞ of λ_∞ is found by computing the zero-structure of the pencil $B^{(h)} - \mu A^{(h)}$, $\mu = \lambda^{-1}$. If desirable we can start by deflating the infinite eigenvalues.

4.2. Two numerical examples

We have implemented the described algorithm in MATLAB[10]. The most crucial step is the range-nullspace separations, controlled by repeated singular value decompositions [see eq's (4.7) and (4.23)]. How should we compute the nullity of A_1? In theory

$$(4.25)\qquad (\sigma_{A_1}^{(i)})^2 + (\sigma_{B_1}^{(i)})^2 = 1$$

i.e. if e.g. $\sigma_{B_1}^{(i)} = 1$ then $\sigma_{A_1}^{(i)} = 0$. However in the presence of rounding errors (4.25) is also satisfied to working accuracy for some $\sigma_{A_1}^{(i)} = \eta^{1/2}$, where η is the relative floating point accuracy. Shall we interpret such a nonsingular value as nonzero or not? Our experiences from computing the Jordan structure of a matrix say that, if we are concerned with stability and a well-conditioned transformation matrix, in the first place, it should be interpreted as zero. By leaving a singular value of order $\eta^{1/2}$, we get a super diagonal element of a Jordan block of the same size i.e. a weak coupling in the corresponding principal chain, resulting in an ill-conditioned transformation matrix. In the pencil case when deleting a singular value of order $\eta^{1/2}$, we compute a structure α of a nearby pencil such that $A-\lambda B \in E_\alpha(\mathcal{O}(\eta^{1/2}))$ [see eq. (2.5)]. In the numerical examples presented here, the criterion for determining the nullity of A_1 is the number of unit singular values of B_1. The criterion only requests one SVD, that of B_1, since A_1 and B_1 have the same right singular vectors. We do not claim that this is an optimal criterion, but it works well for the examples studied so far. Since only the SVD of B_1 is necessary in the subsequent computations [see eq. (4.10)], the computational cost is close to optimal.

Now let us report the following two examples:

Example 4.1. $\det A = 0$, $\det B \neq 0$. The pencil $A-\lambda B$ is constructed like the example in section 3.3 i.e. $A-\lambda B = P^{-1}(JA - \lambda JB)Q^{-1}$. The only difference is that the first five diagonal elements of JB are replaced by 3.0. First we applied the algorithm on $C_0 = [A \mid B]$. The computed structure of the zero-eigenvalue is $\alpha_0 = (2,1,1)$ which agrees with the block structure of JA (see section 3.3).

Introduce $\tilde{A}-\lambda\tilde{B}$ for the transformed pencil corresponding to the right hand side of eq. (4.17) where

(4.26a)

$$\tilde{A} = \left[\begin{matrix} .0000 & .0000 & .1349 & .2171 \\ -.0000 & -.0000 & .8836 & .0013 \\ .0000 & -.0000 & -.0000 & .8529 \\ -.0000 & -.0000 & .0000 & -.0000 \\ .0000 & -.0000 & -.0000 & .0000 \\ -.0000 & -.0000 & -.0000 & .0000 \\ -.0000 & .0000 & .0000 & -.0000 \\ -.0000 & .0000 & -.0000 & -.0000 \\ -.0000 & .0000 & .0000 & -.0000 \end{matrix} \; \underbrace{\begin{matrix} .0482 & -.0418 & .0006 & -.221 & -.1410 \\ -.0202 & -.1017 & -.1801 & .058 & -.1612 \\ .0399 & -.1914 & .0811 & .033 & -.0351 \\ .0099 & .0927 & -.0351 & -.113 & -.1235 \\ .1101 & .0161 & -.0319 & -.102 & .3215 \\ -.0084 & .2687 & -.0298 & -.018 & -.0325 \\ .0121 & .0441 & .3209 & .034 & .0416 \\ .0242 & .0428 & -.0497 & .384 & .0770 \\ -.0864 & .0232 & -.0108 & -.029 & .4257 \end{matrix}}_{A^{(3)}} \right]$$

and

(4.26b)

$$\tilde{B} = \left[\begin{matrix} 1.0000 & .0000 & -.0000 & .0000 \\ -.0000 & 1.0000 & .0000 & -.0000 \\ -.0000 & -.0000 & 1.0000 & -.0000 \\ -.0000 & -.0000 & .0000 & 1.0000 \\ .0000 & .0000 & -.0000 & -.0000 \\ -.0000 & .0000 & .0000 & .0000 \\ -.0000 & .0000 & -.0000 & -.0000 \\ .0000 & .0000 & -.0000 & -.0000 \\ -.0000 & -.0000 & .0000 & -.0000 \end{matrix} \; \underbrace{\begin{matrix} -.0000 & .0000 & .0000 & -.0000 & .0000 \\ .0000 & -.0000 & -.0000 & -.0000 & .0000 \\ -.0000 & .0000 & -.0000 & .0000 & .0000 \\ .0000 & .0000 & .0000 & .0000 & -.0000 \\ .9897 & -.0000 & .0000 & .0000 & .0000 \\ .0000 & .9564 & -.0000 & .0000 & .0000 \\ -.0000 & .0000 & .9441 & .0000 & -.0000 \\ .0000 & .0000 & -.0000 & .9091 & .0000 \\ .0000 & .0000 & -.0000 & .0000 & .8315 \end{matrix}}_{B^{(3)}} \right]$$

Other computed characteristics are:

$$K(\tilde{Q}) = 4.854_{10}{+3}; \qquad K(\tilde{P}) = 15.2;$$

$$\|\tilde{A} - \tilde{P}A\tilde{Q}\|_F = 0.7_{10}{-12}; \qquad \|\tilde{B} - \tilde{P}B\tilde{Q}\|_F = 1.2_{10}{-11}.$$

From these facts, it is possible to conclude that $A-\lambda B \in E_{\alpha_0}(\delta)$, where $\alpha_0 = (2,1,1)$ and $\delta = 4.0_{10}{-12}$ [see eq. (2.1) and theorem 2.1], i.e. the computed structure of $\lambda_0 = 0$ is welldefined.

The structure of a nonzero but finite eigenvalue λ_k can be computed in the same way by studying the shifted pencil

$$(4.26) \qquad A - (\lambda-\lambda_k)B$$

which has zero as an eigenvalue with the structure α_k of λ_k. Since $\lambda_1 = 1/3$ is an eigenvalue of $A-\lambda B$ we continue to apply the algorithm on $C_0 = [A^{(3)} - 0.333\ldots B^{(3)} \,|\, B^{(3)}]$ and the computed structure is $\alpha_1 = (2,2,1)$, which agree with the block structure of JB (see section 3.3). The corresponding computed characteristics for $\tilde{A}^{(3)} - (\lambda-\lambda_1)\tilde{B}^{(3)}$ are:

$$K(\tilde{P}) = 1.3; \qquad K(\tilde{Q}) = 1.2;$$

$$||\tilde{A}^{(3)} - \tilde{P}A^{(3)}\tilde{Q}||_F = 3.2_{10}{-13}; \qquad ||\tilde{B}^{(3)} - \tilde{P}B^{(3)}\tilde{Q}||_F = 2.9_{10}{-14}.$$

<u>Example 4.2.</u> $\det A = 0$, $\det B = 0$, $N(A) \cap N(B) = \{0\}$. Here we study the pencil in section 3.3 with λ_0 of multiplicity 4 and λ_∞ of multiplicity 5. As in example 4.1 we first computed the structure $\alpha_0 = (2,1,1)$, giving the transformed pencil $\tilde{A}-\lambda\tilde{B}$ where

(4.27a)

$$\tilde{A} = \left[\begin{array}{rr|rr|rrrrr}
.0000 & -.0000 & .6557 & -.1707 & .2772 & .0175 & .0145 & -.0000 & .0000 \\
-.0000 & .0000 & .6075 & .1341 & -.1183 & -.0461 & .0060 & .0000 & .0000 \\
\hline
.0000 & -.0000 & .0000 & -.8529 & -.1209 & -.1619 & .0425 & -.0000 & -.0000 \\
-.0000 & .0000 & -.0000 & .0000 & -.0203 & .1928 & .0932 & -.6253 & -.2701 \\
\hline
.0000 & -.0000 & -.0000 & .0000 & -.1143 & -.6144 & -.3660 & -.2325 & -.2074 \\
-.0000 & .0000 & .0000 & -.0000 & .0624 & .0498 & .0454 & -.7268 & .0885 \\
.0000 & -.0000 & -.0000 & .0000 & -.0371 & -.0852 & -.0571 & -.1632 & -.9360 \\
.0000 & -.0000 & .0000 & -.0000 & .4655 & -.3146 & .4398 & -.0000 & -.0000 \\
.0000 & -.0000 & -.0000 & -.0000 & -.4046 & -.1825 & .6167 & -.0000 & .0000
\end{array}\right]$$

(columns 5–9 braced underneath: $A^{(3)}$)

and

(4.27b)

$$
\tilde{B} = \left[\begin{array}{cccc}
1.0000 & -.0000 & .0000 & -.0000 \\
.0000 & 1.0000 & -.0000 & .0000 \\
.0000 & .0000 & 1.0000 & -.0000 \\
-.0000 & .0000 & -.0000 & 1.0000 \\
-.0000 & .0000 & .0000 & .0000 \\
-.0000 & .0000 & .0000 & .0000 \\
-.0000 & -.0000 & -.0000 & .0000 \\
.0000 & .0000 & .0000 & -.0000 \\
.0000 & .0000 & .0000 & .0000
\end{array}
\underbrace{\begin{array}{ccccc}
.0000 & .0000 & -.0000 & .0000 & .0000 \\
.0000 & .0000 & .0000 & -.0000 & -.0000 \\
-.0000 & -.0000 & -.0000 & .0000 & -.0000 \\
.0000 & -.0000 & -.0000 & -.0000 & .0000 \\
.7752 & .0000 & -.0000 & -.0000 & .0000 \\
.0000 & .6652 & .0000 & -.0000 & .0000 \\
.0000 & -.0000 & .5266 & -.0000 & .0000 \\
.0000 & .0000 & -.0000 & .0000 & -.0000 \\
-.0000 & .0000 & .0000 & .0000 & .0000
\end{array}}_{B^{(3)}}\right]
$$

The two zero columns of $\tilde{B}$ correspond to the original singularity of B and is the number of nilpotent blocks in the W-KCF (1.2). In order to get the full structure α_∞ of λ_∞ we applied the algorithm on $C_0 = [B^{(3)} | A^{(3)}]$ i.e. we compute the zero-structure of $B^{(3)} - \mu A^{(3)}$.

Introduce the resulting transformed pencil $\tilde{B}^{(3)} - \mu\tilde{A}^{(3)}$ where

$$
\tilde{B}^{(3)} = \left[\begin{array}{ccccc}
.0000 & -.0000 & -.3035 & .6031 & .0428 \\
-.0000 & .0000 & .4202 & .2216 & .0197 \\
.0000 & -.0000 & -.0000 & -.0000 & .0642 \\
-.0000 & -.0000 & .0000 & .0000 & .6796 \\
-.0000 & -.0000 & -.0000 & .0000 & -.0000
\end{array}\right]
$$

and

$$
\tilde{A}^{(3)} = \left[\begin{array}{ccccc}
1.0000 & -.0000 & -.0000 & .0000 & .0000 \\
.0000 & 1.0000 & .0000 & .0000 & .0000 \\
.0000 & .0000 & 1.0000 & -.0000 & .0000 \\
-.0000 & -.0000 & .0000 & 1.0000 & -.0000 \\
-.0000 & .0000 & .0000 & .0000 & .7308
\end{array}\right],
$$

and from $\tilde{B}^{(3)}$ we see that $\alpha_\infty = (2,2,1)$.

By replacing $A^{(3)}$ in (4.27a) and $B^{(3)}$ in (4.27b) by $\tilde{A}^{(3)}$ and $\tilde{B}^{(3)}$ respectively, we obtain the final transformed pencil.

Acknowledgements

The author is grateful to Axel Ruhe and Charles Van Loan for their constructive comments.

Financial support has been received from the Swedish Institute of Applied Mathematics (ITM).

References

1. EDSTEDT, P., WESTIN, M., Interactive computation of the Jordan normal form (JNF) in MATLAB, Report UMNAD-1-81, Inst. of Information Processing, Umeå University, Sweden (1981) (in Swedish).
2. GANTMACHER, F.R., The Theory of Matrices, Vol. I and II (Transl.), Chelsea, New York (1959).
3. GOLUB, G.H., WILKINSON, J.H., Ill-conditioned eigensystems and the computation of the Jordan canonical form, SIAM Review, Vol. 18, No. 4 (1976), 578-619.
4. KAHAN, W., Conserving confluence curbs ill-condition, Tech. Report 6, Dep. of Comput. Sci., University of California, Berkeley, Calif., (1972), 1-54.
5. KUBLANOVSKAYA, V.N., The AB-algorithm and its modifications for the spectral problem of linear pencils of matrices, LOMI-preprint E-10-81, USSR Academy of Sciences, Leningrad (1981), 1-25.
6a. KUBLANOVSKAYA, V.N., On algorithms for the solution of spectral problems of linear matrix pencils, LOMI-preprint E-1-82, USSR Academy of Sciences, Leningrad (1982), 1-43.
6b. KUBLANOVSKAYA, V.N., An approach to solving the spectral problem of $A-\lambda B$. This volume.
7. KÅGSTRÖM, B., RUHE, A., An algorithm for numerical computation of the Jordan normal form of a complex matrix, ACM Transactions on Mathematical Software, Vol. 6, No. 3, (1980), 398-419.
8. KÅGSTRÖM, B., RUHE, A., ALGORITHM 560, JNF, An algorithm for numerical computation of the Jordan normal form of a complex matrix [F2], ACM Transactions on Mathematical

Software, Vol. 6, No. 3 (1980), 437-443.

9. KÅGSTRÖM, B., How to compute the Jordan normal form - the choice between similarity transformations and methods using the chain relations, Report UMINF.91-81, Inst. of Information Processing, Umeå University, Sweden (1981), 1-48.

10. MOLER, C., MATLAB - An interactive matrix laboratory, Dept. of Computer Science, University of New Mexico, Albuquerque, New Mexico.

11. SINCOVEC, R.F., DEMBART, B., EPTON, M.A., ERISMAN, A.M., MANKE, J.W. and YIP, E.L., Solvability of large scale descriptor systems, Final Report DOE Contract ET-78-C-01-2876, Boeing Computer Services Co., Seattle, USA (1979).

12. STEWART, G.W., On the perturbation of pseudo-inverses, projections and linear least squares problems, SIAM Review 19 (1977), 634-662.

13. STEWART, G.W., Error and perturbation bounds for subspaces associated with certain eigenvalue problems, SIAM Review, Vol. 15 (1973), 752-764.

14. STEWART, G.W., Perturbation theory for the generalized eigenvalue problem, Recent Advances in Numerical Analysis, Ed. C. de Boor, G. Golub, Academic Press, New York (1978), 193-206.

15. SUN, JI-GUANG, Perturbation theorems for the generalized eigenvalue problem, to appear in Lin. Alg. Appl. (1982).

16. VAN DOOREN, P., The computation of Kronecker's canonical form of a singular pencil, Lin. Alg. Appl., Vol. 27 (1979), 103-141.

17. VAN DOOREN, P., The generalized eigenstructure problem in linear system theory, IEEE Trans. Aut. Contr., Vol. AC-26, No. 1 (1981), 111-129.

18. WILKINSON, J.H., Linear differential equations and Kronecker's canonical form, Recent Advances in Numerical Analysis, Ed. C. de Boor, G. Golub, Academic Press, New York (1978), 231-265.

19. WILKINSON, J.H., Kronecker's canonical form and the QZ algorithm, Lin. Alg. Appl., Vol. 28 (1979), 285-303.

Reducing subspaces : definitions, properties and algorithms.

Paul Van Dooren
Philips Research Laboratory
2, Av. Van Becelaere, Box 8
1170 Brussels, Belgium

Abstract

In this paper we introduce the new concept of reducing subspaces of a singular pencil, which extends the notion of deflating subspaces to the singular case. We briefly discuss uniqueness of such subspaces and we give an algorithm for computing them. The algorithm also gives the Kronecker canonical form of the singular pencil.

1. Introduction

The last few years, the numerical literature has started to show some interest in the computation of the Kronecker canonical form of singular pencils because of its relevance in several applications [1][6][10][12][13] and stable algorithms have been developed recently [2][6][9][10][15][16]. Since this problem is (often) ill conditioned an appropriate reformulation of conditioning has to be made in order to give a meaning to the computed results (see e.g. [10]).

Other concepts that were not extended to the singular case are those of eigenvectors and invariant subspaces, or more generally, of deflating subspaces. In Section 2, we introduce the concept of reducing subspaces and show it generalizes the notion of deflating subspaces to the singular case. Such subspaces reduce the eigenstructure problem of a singular pencil $\lambda B-A$ to the eigenstructure problem of two smaller pencils. Under eigenstructure we understand here all the invariants of the $m\times n$ pencil $\lambda B-A$ under equivalence transformations (i.e. invertible column and row transformations). The eigenstructure of $\lambda B-A$ is retrieved in its Kronecker canonical form [4]:

$$M(\lambda B-A)N = \text{diag}\{L_{\ell_1},\ldots,L_{\ell_s},\ L^T_{r_1},\ldots,L^T_{r_t},\ I-\lambda N,\ \lambda I-J\} \tag{1.1}$$

where i) L_k is the $(k+1)\times k$ bidiagonal pencil

$$L_k = \underbrace{\begin{bmatrix} \lambda & & \\ -1 & \ddots & \\ & \ddots & \lambda \\ & & -1 \end{bmatrix}}_{k} \Bigg\} k+1 \tag{1.2}$$

and ii) N is nilpotent and both N and J are in Jordan canonical form.
The 'eigenstructure' of the pencil $\lambda B-A$ is then given by :

i) the finite elementary divisors $(\lambda-\alpha_i)^{d_j}$ of $\lambda B-A$, reflected by the Jordan blocks of size d_j at α_i in J.

ii) the infinite elementary divisors $(\mu)^{d_j}$ of $\lambda B-A$, reflected by the Jordan blocks of size d_j at 0 in N.

iii) the left and right minimal indices $\{\ell_1,\ldots,\ell_s\}$ and $\{r_1,\ldots,r_t\}$ of $\lambda B-A$, reflected by the blocks L_{ℓ_i} and $L^T_{r_j}$, respectively.

For the computation of the eigenstructure of a singular pencil, algorithms have been developed previously [9][12][15][16] . They compute a quasi triangular form :

$$(1.3)\quad M(\lambda B-A)N = \begin{bmatrix} \lambda B_r-A_r & * & * & * \\ 0 & \lambda B_i-A_i & * & * \\ 0 & 0 & \lambda B_f-A_f & * \\ 0 & 0 & 0 & \lambda B_\ell-A_\ell \end{bmatrix}$$

where λB_r-A_r and $\lambda B_\ell-A_\ell$ are singular pencils containing the right and left minimal indices of $\lambda B-A$, respectively, and λB_f-A_f and λB_i-A_i are regular pencils containing the finite and infinite elementary divisors of $\lambda B-A$, respectively. It is shown in [9] that this form can be obtained under unitary transformations M and N, thus guaranteeing the numerical stability of the method. In Section 3 we show how to combine this algorithm with a recent algorithm derived for the computation of deflating subspaces of a regular pencil, in order to obtain an algorithm for computing reducing subspaces of a singular pencil.

The following notation and conventions will be used throughout the paper. We denote a block diagonal matrix by diag $\{A_{11},\ldots,A_{kk}\}$. We use A^* for the conjugate transpose of a matrix A and A^T for the transpose of a matrix A. A complex (real) square matrix A is called unitary (orthogonal) when $A^*A=I$ ($A^TA=I$). When no explicit distinction is made between the complex and the real case, we use the term unitary and the notation A^* for the real case as well. Script is used for vectorspaces. $\mathcal{H}_n$ will denote the spaces $\mathbb{C}^n$ or $\mathbb{R}^n$, depending on the context of the problem. $A\mathcal{X}$ is the image of $\mathcal{X}$ under A ; $\mathcal{Im}$ A and $\mathcal{Ker}$ A are the image and kernel of A, respectively. $\mathcal{X}+\mathcal{Y}$ and $\mathcal{X}\oplus\mathcal{Y}$ are the sum and the direct sum, respectively of the spaces $\mathcal{X}$ and $\mathcal{Y}$. Two vectorspaces of special interest are $\mathcal{N}_r(\lambda B-A)$ and $\mathcal{N}_\ell(\lambda B-A)$, the right and left null spaces, respectively, of a pencil $\lambda B-A$. These are vectorspaces over the field of rational functions in λ and are of dimension

$$(1.4\ a;b)\qquad \dim.\mathcal{N}_r(\lambda B-A)=n-r;\quad \dim.\mathcal{N}_\ell(\lambda B-A)=m-r$$

respectively, when $\lambda B-A$ is a m×n pencil of <u>normal rank</u> r [3]. These dimensions are also called the right and left nullity of $\lambda B-A$, respectively, and the pencil $\lambda B-A$ is said to be right and left invertible, respectively, when the corresponding nullity is zero. When the columns of a matrix X form a basis for the space $\mathcal{X}$, this is denoted by $\mathcal{X} = \langle X \rangle$. The space spanned by the null vector only is denoted by $\{0\}$. By $\Lambda(B,A)$ we denote the <u>spectrum</u> of the pencil $\lambda B-A$, i.e. the collection of generalized eigenvalues, multiplicities counted. By $E(B,A)$ we denote the complete <u>eigenstructure</u> of the pencil $\lambda B-A$, i.e. all the structural elements as described in (1.1).

2. Deflating and reducing subspaces.

Let $\mathcal{X}$ and $\mathcal{Y}$ be subspaces of $\mathcal{H}_n$ and $\mathcal{H}_m$, respectively, such that

$$\mathcal{Y} = B\mathcal{X} + A\mathcal{X} \tag{2.1}$$

Let ℓ and k be their respective dimensions and construct the unitary matrices Q and Z, partitioned as :

$$Z = [\underbrace{Z_1}_{\ell} \mid Z_2] \quad ; \quad Q = [\underbrace{Q_1}_{k} \mid Q_2] \tag{2.2}$$

such that

$$\mathcal{X} = \langle Z_1 \rangle \quad ; \quad \mathcal{Y} = \langle Q_1 \rangle \tag{2.3}$$

Then it follows from (2.1) that $Q_2^* A Z_1 = Q_2^* B Z_1 = 0$ and thus

$$Q^*(\lambda B-A)Z \triangleq \lambda\hat{B}-\hat{A} \triangleq \lambda \left[\begin{array}{c|c} \hat{B}_{11} & \hat{B}_{12} \\ \hline 0 & \hat{B}_{22} \end{array}\right] - \left[\begin{array}{c|c} \hat{A}_{11} & \hat{A}_{12} \\ \hline 0 & \hat{A}_{22} \end{array}\right] \tag{2.4}$$

In this new coordinate system $\mathcal{X}$ and $\mathcal{Y}$ are now represented by

$$\mathcal{X} = \left\langle \begin{bmatrix} I_\ell \\ 0 \end{bmatrix} \right\rangle \; ; \quad \mathcal{Y} = \left\langle \begin{bmatrix} I_k \\ 0 \end{bmatrix} \right\rangle \tag{2.5 a;b}$$

The map $\lambda B-A$ restricted to the spaces $\mathcal{X}$ and $\mathcal{Y}$ and its spectrum are also denoted by

$$\lambda\hat{B}_{11}-\hat{A}_{11} = (\lambda B-A)\big|_{\mathcal{X},\mathcal{Y}} \quad ; \quad \Lambda(\hat{B}_{11},\hat{A}_{11}) = \Lambda(B,A)\big|_{\mathcal{X},\mathcal{Y}} \tag{2.6 a;b}$$

In the regular case, i.e. when m and n are equal to the normal rank, the dimensions of $\mathcal{X}$ and $\mathcal{Y}$ satisfy the inequality [8] :

$$\dim.\mathcal{Y} \geq \dim.\mathcal{X} \tag{2.7}$$

and it is only in the case of equality that such spaces become of interest. They are called <u>deflating subspaces</u> and possess the following property (see [8] for a proof)

<u>Theorem 2.1.</u>

Let $\mathcal{X},\mathcal{Y}$ be a pair of deflating subspaces and perfom the corresponding transformation (2.4), then the diagonal pencils $\lambda\hat{B}_{ii}-\hat{A}_{ii}$, i=1,2 are regular and $\Lambda(\hat{B}_{11},\hat{A}_{11}) \cup \Lambda(\hat{B}_{22},\hat{A}_{22}) = \Lambda(B,A)$. □

This theorem justifies the terminology "deflating subspaces", since the problem of computing $\Lambda(B,A)$ is now deflated to two eigenvalue problems of smaller dimension. The following results are important for the characterization of some specific pairs of deflating subspaces.

<u>Lemma 2.1.</u> [8]

The equation in M and L :

$$M(\lambda B-A) + (\lambda\hat{B}-\hat{A})L = \lambda\tilde{B}-\tilde{A} \tag{2.8}$$

where $\lambda B-A$ and $\lambda\hat{B}-\hat{A}$ are <u>regular</u> pencils and $\lambda\tilde{B}-\tilde{A}$ is arbitrary, has a <u>unique</u> solution when $\Lambda(B,A) \cap \Lambda(\hat{B},\hat{A})=\emptyset$.

□

<u>Lemma 2.2.</u>

Let the pencils $\lambda B-A$ and $\lambda\hat{B}-\hat{A}$ be conformably partitioned and upper block triangular:

$$\lambda B-A = \left[\begin{array}{c|c} \lambda B_{11}-A_{11} & \lambda B_{12}-A_{12} \\ \hline 0 & \lambda B_{22}-A_{22} \end{array}\right]; \quad \lambda\hat{B}-\hat{A} = \left[\begin{array}{c|c} \lambda\hat{B}_{11}-\hat{A}_{11} & \lambda\hat{B}_{12}-\hat{A}_{12} \\ \hline 0 & \lambda\hat{B}_{22}-\hat{A}_{22} \end{array}\right] \tag{2.9 a;b}$$

and let $\lambda B-A$ and $\lambda\hat{B}-\hat{A}$ be equivalent, i.e. there exist invertible matrices M and N such that :

$$M(\lambda B-A)N = \lambda\hat{B}-\hat{A} \tag{2.10}$$

Then M and N are also upper block triangular if $\Lambda(B_{11},A_{11})$ and $\Lambda(\hat{B}_{22},\hat{A}_{22})$ have no common points.

Proof :

Using $L=N^{-1}$, we rewrite (2.10) with a conformable partitioning of M and L :

$$\left[\begin{array}{c|c} M_{11} & M_{12} \\ \hline M_{21} & M_{22} \end{array}\right] \cdot \left[\begin{array}{c|c} \lambda B_{11}-A_{11} & \lambda B_{12}-A_{12} \\ \hline 0 & \lambda B_{22}-A_{22} \end{array}\right]$$

(2.11)

$$= \left[\begin{array}{c|c} \lambda \hat{B}_{11}-\hat{A}_{11} & \lambda \hat{B}_{12}-\hat{A}_{12} \\ \hline 0 & \lambda \hat{B}_{22}-\hat{A}_{22} \end{array}\right] \cdot \left[\begin{array}{c|c} L_{11} & L_{12} \\ \hline L_{21} & L_{22} \end{array}\right]$$

This yields the equation

(2.12) $$M_{21}(\lambda B_{11}-A_{11}) - (\lambda \hat{B}_{22}-\hat{A}_{22})L_{21} = 0 .$$

Because of Lemma 2.1 and $\Lambda(B_{11},A_{11}) \cap \Lambda(\hat{B}_{22},\hat{A}_{22})=\emptyset$, (2.12) has a unique solution for M_{21} and L_{21} which is clearly $M_{21}=L_{21}=0$. This completes the proof.

□

This lemma leads directly to the following theorem.

Theorem 2.2.

Let Λ_1 be a subset of $\Lambda(B,A)$ disjoint from the remaining eigenvalues $\Lambda_2=\Lambda(B,A)\setminus\Lambda_1$. Then there exists a unique pair of deflating subspaces such that $\Lambda(B,A)\big|_{X,Y} = \Lambda_1$.

Proof :

Let ℓ_1 be the number of generalized eigenvalues in Λ_1. It is known by construction (see [11]) that there always exist one pair X_1,Y_1 of dimension ℓ_1 satisfying $\Lambda(B,A)\big|_{X_1,Y_1} = \Lambda_1$. Its uniqueness follows from Lemma 2.2. Indeed, without loss of generality we may assume that (see (2.3)-(2.5) for the appropriate coordinate system)

(2.13) $$X_1 = Y_1 = \left< \begin{bmatrix} I_{\ell_1} \\ 0 \end{bmatrix} \right>$$

and thus that $\lambda B-A$ has the block triangular form (2.9a). If there is a second pair X_2, Y_2 satisfying $\Lambda(B,A)\big|_{X_2,Y_2} = \Lambda_1$, then there exist updating transformations Q and Z as in (2.3) (2.4) and such that

(2.14 a;b) $$X_2 = \langle Z_1 \rangle \quad ; \quad Y_2 = \langle Q_1 \rangle$$

$$(2.14c) \qquad Q \cdot \left[\begin{array}{c|c} \lambda B_{11}-A_{11} & \lambda B_{12}-A_{12} \\ \hline 0 & \lambda B_{22}-A_{22} \end{array}\right] \cdot Z = \left[\begin{array}{c|c} \lambda \hat{B}_{11}-\hat{A}_{11} & \lambda \hat{B}_{12}-\hat{A}_{12} \\ \hline 0 & \lambda \hat{B}_{22}-\hat{A}_{22} \end{array}\right]$$

Since $\Lambda_1 \cap \Lambda_2 = \emptyset$ we are in the situation of Lemma 2.2 and thus Q and Z are both upper block triangular. Therefore

$$(2.15) \qquad \langle Z_1 \rangle = \langle Q_1 \rangle = \left\langle \begin{bmatrix} I_{\ell_1} \\ \\ 0 \end{bmatrix} \right\rangle$$

which establishes the unicity of the deflating subspaces.

□

Theorem 2.2. can also be retrieved from the work of Stewart [8] but Lemma's 2.1 and 2.2 are also useful for the extension of the above results to the singular case. Let $\lambda B-A$ be a m×n singular pencil with normal rank r. We will show that for any pair X,Y as in (2.1), the following inequality is always satisfied :

$$(2.16) \qquad \dim.Y \geq \dim.X - \dim.N_r$$

In the case of equality the pair X,Y plays a role comparable to that of deflating subspaces in the regular case. Such spaces are given the name of reducing subspaces of the pencil $\lambda B-A$. Notice that this concept reduces to that of deflating subspaces in the regular case since then $N_r = \{0\}$. We first prove the following extension of Lemma 2.2 :

Lemma 2.3.

The equation in M and L :

$$(2.17) \qquad M(\lambda B-A) + (\lambda D-C)L = \lambda F-E$$

where $\lambda B-A$ and $\lambda D-C$ are left and right invertible respectively, has a solution when $\Lambda(B,A) \cap \Lambda(D,C) = \emptyset$.

Proof :

First transform $\lambda B-A$ and $\lambda D-C$ to their Kronecker canonical form via the equivalence transformations :

$$(2.18\ a;b) \qquad M_1(\lambda B-A)N_1 = \lambda B_c-A_c \quad ; \quad M_2(\lambda D-C)N_2 = \lambda D_c-C_c$$

which reduces the equation (2.17) to the equivalent equation

$$(2.19)\qquad M_c(\lambda B_c - A_c) + (\lambda D_c - C_c)L_c = \lambda F_c - E_c$$

with $M_c = M_2 M M_1^{-1}$, $L_c = N_2^{-1} L N_1$ and $\lambda F_c - E_c = M_2(\lambda F - E)N_1$.
When partitioning M_c, L_c and $\lambda F_c - E_c$ conformably with the blocks on the diagonal forms $\lambda B_c - A_c$ and $\lambda D_c - E_c$, equation (2.19) reduces to the set of (independent) equations :

$$(2.20)\qquad [M_c]_{ij}[\lambda B_c - A_c]_j + [\lambda D_c - C_c]_i [L_c]_{ij} = [\lambda F_c - E_c]_{ij}$$

When the canonical blocks $[\lambda B_c - A_c]_j$ and $[\lambda D_c - C_c]_i$ are regular, we are in the situation of Lemma 2.1 and (2.20) has a unique solution since by assumption these blocks have disjoint spectrum. When one or both blocks are singular we now show how to reduce the problem to a regular one. Because of the assumptions of left and right invertibility, the only singular blocks that may occur are :

$$(2.21\ a;b)\qquad [\lambda B_c - A_c]_j = L_k \quad ; \quad [\lambda D_c - C_c]_i = L_\ell^T$$

By deleting the first or last row in L_k the truncated block $[\overline{\lambda B_c - A_c}]_j$ is regular with k eigenvalues at ∞ or zero, respectively. This corresponds to taking the first or last column in $[M_c]_{ij}$ equal to zero and solving for the truncated matrix $[\overline{M_c}]_{ij}$. A dual technique can be used for the second term in (2.20) such that this equation is replaced by :

$$(2.22)\qquad [\overline{M_c}]_{ij}[\overline{\lambda B_c - A_c}]_j + [\overline{\lambda D_c - C_c}]_i [\overline{L_c}]_{ij} = [\lambda F_c - E_c]_{ij}$$

where the upperbar indicates that the matrix has been truncated if needed. As indicated above, the truncation(s) may always be performed such that the blocks $[\overline{\lambda B_c - A_c}]_j$ and $[\overline{\lambda D_c - C_c}]_i$ have disjoint spectrum. We thus satisfy Lemma 2.1 and the solution $[\overline{M_c}]_{ij}$, $[\overline{L_c}]_{ij}$ of (2.22) yields also a solution to (2.20) by merely adding a zero column and row to reconstruct $[M_c]_{ij}$ and $[L_c]_{ij}$, respectively. Putting all these solutions together, we thus constructed (nonunique) matrices M_c and L_c satisfying (2.19). □

This now leads to the following generalization of Theorem 2.1.

<u>Theorem 2.3.</u>

Let $\mathcal{X},\mathcal{Y}$ be a pair of reducing subspaces and perform the coordinate transformation (2.3)(2.4), then the diagonal pencils have zero left and right nullity, respectively and $\Lambda(\hat{B}_{11},\hat{A}_{11}) \cup \Lambda(\hat{B}_{22},\hat{A}_{22}) = \Lambda(B,A)$.

Proof:

Let r_i be the normal rank of the pencils $\lambda\hat{B}_{ii}-\hat{A}_{ii}$, $i=1,2$. First we prove the inequality (2.16) and show that equality also implies $r_1=\ell$ and $r_2=n-k$. Clearly

(2.23 a;b) $$r_1 \leq \ell \quad ; \quad r_2 \leq n-k$$

and, because of the structure of (2.4), the following holds :

(2.24 a) $$k-r_1 = \dim.\ \mathcal{N}_r(\lambda\hat{B}_{11}-\hat{A}_{11}) \leq \dim.\ \mathcal{N}_r(\lambda B-A) = n-r$$

(2.24 b) $$m-\ell-r_2 = \dim.\mathcal{N}_\ell(\lambda\hat{B}_{22}-\hat{A}_{22}) \leq \dim.\mathcal{N}_\ell(\lambda B-A) = m-r$$

Combining these inequalities we find

(2.25 a) $$k-\ell \leq k-r_1 \leq n-r$$

(2.25 b) $$m-\ell-n+k \leq m-\ell-r_2 \leq m-r$$

From this it easily follows that (2.16) holds since $k-\ell = \dim.\mathcal{X}-\dim.\mathcal{Y}$ and $n-r = \dim.\mathcal{N}_r$. Moreover equality implies the middle terms in (2.25) to be equal to their upper and lower bounds, which then gives $r_1=\ell$ and $r_2=n-k$. In order to prove the second part of the theorem, we show the existence of conformable transformations M and N such that :

(2.26) $$\left[\begin{array}{c|c} M_{11} & M_{12} \\ \hline 0 & M_{22} \end{array}\right] \cdot \left[\begin{array}{c|c} \lambda\hat{B}_{11}-\hat{A}_{11} & \lambda\hat{B}_{12}-\hat{A}_{12} \\ \hline 0 & \lambda\hat{B}_{22}-\hat{A}_{22} \end{array}\right] \cdot \left[\begin{array}{c|c} N_{11} & N_{12} \\ \hline 0 & N_{22} \end{array}\right] =$$

$$\left[\begin{array}{cc|cc} \lambda\tilde{B}_r-\tilde{A}_r & 0 & 0 & 0 \\ 0 & \lambda\tilde{B}_{11}-\tilde{A}_{11} & \lambda\tilde{B}_{12}-\tilde{A}_{12} & 0 \\ \hline 0 & 0 & \lambda\tilde{B}_{22}-\tilde{A}_{22} & 0 \\ 0 & 0 & 0 & \lambda\tilde{B}_\ell-\tilde{A}_\ell \end{array}\right]$$

where $\lambda\tilde{B}_{ii}-\tilde{A}_{ii}$, $i=1,2$ are regular and $\lambda\tilde{B}_r-\tilde{A}_r$ and $\lambda\tilde{B}_\ell-\tilde{A}_\ell$ contain the right and left minimal indices of $\lambda B-A$. For this we first choose M_{ii}, N_{ii}, $i=1,2$ such that the pencils $\lambda\hat{B}_{ii}-\hat{A}_{ii}$ are transformed to their Kronecker canonical form

(2.27 a) $$M_{11}(\lambda\hat{B}_{11}-\hat{A}_{11})N_{11} = \mathrm{diag}\{\lambda\tilde{B}_r-\tilde{A}_r,\ \lambda\tilde{B}_{11}-\tilde{A}_{11}\}$$

(2.27 b) $$M_{22}(\lambda\hat{B}_{22}-\hat{A}_{22})N_{22} = \mathrm{diag}\{\lambda\tilde{B}_{22}-\tilde{A}_{22},\lambda\tilde{B}_{\ell}-\tilde{A}_{\ell}\}$$

The remaining three zero blocks in (2.26) are then obtained by an appropriate choice of M_{12} and N_{12}. This is possible by virtue of the previous lemma since the spectrum of the pairs $\lambda\tilde{B}_r-\tilde{A}_r$ and $\lambda\tilde{B}_{22}-\tilde{A}_{22}$, $\lambda\tilde{B}_{11}-\tilde{A}_{11}$ and $\lambda\tilde{B}_{\ell}-\tilde{A}_{\ell}$, $\lambda\tilde{B}_r-\tilde{A}_r$ and $\lambda\tilde{B}_{\ell}-\tilde{A}_{\ell}$, are mutually disjoint. From the form (2.26) we now easily see that the central pencil

(2.28) $$\left[\begin{array}{c|c} \lambda\tilde{B}_{11}-\tilde{A}_{11} & \lambda\tilde{B}_{12}-\tilde{A}_{12} \\ \hline 0 & \lambda\tilde{B}_{22}-\tilde{A}_{22} \end{array}\right]$$

is the regular part of $\lambda B-A$, and the second part of this theorem then immediately follows from Theorem 2.1.

□

Notice also that there is a one to one correspondence between pairs of reducing subspaces of the pencil $\lambda B-A$ and pairs of deflating subspaces of its regular part, as shown in the proof of the above theorem. This remark leads directly to a generalization of Theorem 2.2. to the singular case.

Theorem 2.4.

Let Λ_1 be a subset of $\Lambda(B,A)$ disjoint from the remaining spectrum $\Lambda_2=\Lambda(B,A)\setminus\Lambda_1$. Then there exists a unique pair of reducing subspaces such that $\Lambda(B,A)\big|_{X,Y} = \Lambda_1$.

Proof :

This follows immediately from Theorem 2.2. and the observation that to every pair of reducing subspaces there corresponds a pair of deflating subspaces of the regular part of $\lambda B-A$.

□

When a set of reducing subspaces of a pencil $\lambda B-A$ performs a separation in the spectrum $\Lambda(B,A)=\Lambda_1 \cup \Lambda_2$, $\Lambda_1 \cap \Lambda_2 = \emptyset$, then M_{21} and N_{21} in (2.26) may be chosen such that $\lambda\tilde{B}_{12}-\tilde{A}_{12}$ is eliminated as well since $\Lambda(\tilde{B}_{11},\tilde{A}_{11})$ and $\Lambda(\tilde{B}_{22},\tilde{A}_{22})$ are disjoint. This proves thus that the reduction (2.4) obtained by this pair of subspaces has the property

(2.29) $$E(B_{11},A_{11}) \cup E(B_{22},A_{22}) = E(B,A)$$

We thus proved the following result.

Corollary 2.1.

When a set of reducing subspaces performs a separation in the spectrum of the pencil $\lambda B-A$, then $E(B,A)\big|_{X,Y}$ is a subset of $E(B,A)$

□

As shown in Theorem 2.3, the right and left null space structures are always separated by a pair of reducing subspaces. The minimal and maximal pairs of reducing subspaces are easily seen to be those separating $\lambda\tilde{B}_r-\tilde{A}_r$ and $\lambda\tilde{B}_\ell-\tilde{A}_\ell$, respectively, from the rest of the pencil. We also have that any pair of reducing subspaces $\mathcal{X},\mathcal{Y}$ satisfies

(2.30 a) $$\{0\} \subset \mathcal{X}_{min} \subset \mathcal{X} \subset \mathcal{X}_{max} \subset \mathcal{H}_n$$

(2.30 b) $$\{0\} \subset \mathcal{Y}_{min} \subset \mathcal{Y} \subset \mathcal{Y}_{max} \subset \mathcal{H}_m$$

as easily follows from the proof of Theorem 2.3.
The computation of deflating subspaces with specified spectrum Λ_1 has been described in [11] and a stable algorithm, based on an updating of the QZ decomposition, was given there. This can be used to compute reducing subspaces with specified spectrum, as soon as one has an algorithm to compute the pairs $\mathcal{X}_{min}$, $\mathcal{Y}_{min}$ and $\mathcal{X}_{max}$, $\mathcal{Y}_{max}$, or in other words, as soon as one has an algorithm to extract the regular part $\lambda B_{reg}-A_{reg}$ of the pencil $\lambda B-A$. Since to each pair of deflating subspaces $\mathcal{X}_{reg}$, $\mathcal{Y}_{reg}$ of this regular part there corresponds a pair of reducing subspaces $\mathcal{X},\mathcal{Y}$ with the property

(2.31) $$\Lambda(B,A)\big|_{\mathcal{X},\mathcal{Y}} = \Lambda(B_{reg},A_{reg})\big|_{\mathcal{X}_{reg},\mathcal{Y}_{reg}}$$

this indeed solves the problem of computing reducing subspaces with specified spectrum.

3. Algorithms.

In this section we show how ideas of previous algorithms [9][11] can be combined to yield an algorithm for computing pairs of reducing subspaces. We first show that the constructions of the pairs $\mathcal{X}_{min}$, $\mathcal{Y}_{min}$ and $\mathcal{X}_{max}$, $\mathcal{Y}_{max}$ can be solved recursively by building a chain of decompositions of the type (2.4) but where only the last decomposition of this chain corresponds to a pair of reducing subspaces. At each stage of the recursion, information about the structure of the pencils $\lambda\hat{B}_{11}-\hat{A}_{11}$ and $\lambda\hat{B}_{22}-\hat{A}_{22}$ is recovered. The results relie on the following theorem, implicitly proved in [9].

Theorem 3.1.

Let $\mathcal{X} = \mathrm{Ker}\, B$, $\mathcal{Y} = A\mathcal{X}$, then the corresponding decomposition (2.4) has the property that $E(\hat{B}_{22},\hat{A}_{22})$ and $E(B,A)$ are equal except for the infinite elementary divisors and right minimal indices of $E(\hat{B}_{22},\hat{A}_{22})$ which are those of $E(B,A)$ reduced by 1. □

Such a decomposition is easily obtained by the following construction. Choose Z such that $\langle Z_1\rangle = \mathit{Ker}\ B$. We then have (with $s_1 = \dim.\mathit{Ker}\ B$) :

$$(3.1)\qquad (\lambda B-A)Z = \left[\begin{array}{c|c} \underbrace{-A_1}_{s_1} & \lambda B_2 - A_2 \end{array}\right]$$

where B_2 has full column rank. Choose then Q such that $\langle Q_1\rangle = \mathit{Im}\ A_1 = A\mathcal{X}$. We then have (wirh $r_1 = \dim.\mathit{Im}\ A_1$) :

$$(3.2)\qquad Q^*(\lambda B-A)Z = \left[\begin{array}{c|c} -\hat{A}_{11} & \lambda B_{12}-A_{12} \\ \hline \underbrace{0}_{s_1} & \lambda B_{22}-A_{22} \end{array}\right]\begin{array}{l}\}r_1 \\ \\ \end{array}$$

where $\hat{A}_{11}$ has full row rank and $\begin{bmatrix} B_{12} \\ B_{22}\end{bmatrix}$ has full column rank.

This reduction step (3.1)(3.2) can be performed on the bottom pencil $\lambda B_{22}-A_{22}$ and can be repeated recursively like this, until one obtains, say at step ℓ, a pencil $\lambda B_{\ell+1,\ell+1} - A_{\ell+1,\ell+1}$ where $\mathit{Ker}\ B_{\ell+1,\ell+1} = \{0\}$. No further reduction can thus be obtained and the complete decomposition then looks like (see [9] for more details) :

$$(3.3\text{ a})\qquad \hat{Q}^*(\lambda B-A)\hat{Z} = \left[\begin{array}{c|c} \lambda B_{ri}-A_{ri} & * \\ \hline 0 & \lambda B_{f\ell}-A_{f\ell} \end{array}\right] =$$

$$(3.3\text{ b})\qquad \left[\begin{array}{cccc|c} -\hat{A}_{11} & * & \cdots & * & \\ & -\hat{A}_{22} & \cdots & * & \\ & & \ddots & \vdots & * \\ & 0 & & -\hat{A}_{\ell\ell} & \\ \hline \underbrace{}_{s_1} & \underbrace{}_{s_2} & 0\ \cdots & \underbrace{}_{s_\ell} & \lambda B_{\ell+1,\ell+1} - A_{\ell+1,\ell+1} \end{array}\right]\begin{array}{l}\}r_1 \\ \}r_2 \\ \vdots \\ \}r_\ell \\ \\ \end{array}$$

Because of the construction, all $\hat{A}_{ii}$ have full row rank and $B_{\ell+1,\ell+1}$ has full column rank. This ensures the following inequalities to hold (when defining $s_{\ell+1}=0$) :

$$(3.4)\qquad s_1 \geq r_1 \geq s_2 \geq r_2 \geq \cdots \geq s_\ell \geq r_\ell \geq s_{\ell+1}$$

The dimensions $\{s_i\}$ and $\{r_i\}$ produced by the algorithm can be shown to yield the following information about the eigenstructure of the top pencil : $\lambda B_{ri}-A_{ri}$ has

$$\begin{cases} e_i = (s_i-r_i) \text{ right minimal indices equal to } i-1 \\ d_i = (r_i-s_{i+1}) \text{ infinite elementary divisors of degree } i \end{cases} \qquad (i=1,\ldots,\ell)$$

and has no other structural elements. Furthermore, $\lambda B_{ri}-A_{ri}$ and $\lambda B_{f\ell}-A_{f\ell}$ are easily seen to be right invertible and left invertible, respectively, because of the rank properties of the $\hat{A}_{ii}$ and of $B_{\ell+1,\ell+1}$, respectively. We are then in the situation of Theorem 2.3. and have thus constructed a pair of reducing subspaces. Finally it follows from the full column rank of $B_{\ell+1,\ell+1}$ that the bottom pencil $\lambda B_{f\ell}-A_{f\ell}$ has no infinite elementary divisors. The constructed pair of reducing subspaces is thus the unique pair X_i, Y_i whose spectrum contains all the infinite eigenvalues of $\lambda B-A$:

$$(3.5) \qquad (\lambda B-A)\Big|_{X_i,Y_i} = \{\infty,\ldots,\infty\}$$

The spaces are spanned by the first $\hat{s}$ columns of $\hat{Z}$ and $\hat{r}$ columns of $\hat{Q}$, respectively, where

$$(3.6\ a;b) \qquad \hat{s} = \sum_{i=1}^{\ell} s_i \quad ; \quad \hat{r} = \sum_{i=1}^{\ell} r_i$$

It is well known that the minimal indices of the pencils

$$(3.7\ a;b;c;d) \qquad \lambda B-A \ ; \ \mu A-B \ ; \ \mu'(A-\alpha B)-B \ ; \ \lambda'(B-1/\alpha A)-A$$

are all the same and that their elementary divisors are related via the transformations $\mu=1/\lambda$; $\mu'=1/(\lambda-\alpha)$; $1/\lambda' = 1/\lambda-1/\alpha$.
Therefore the infinite elementary divisors of $\mu'(A-\alpha B)-B$ and $\lambda'(B-1/\alpha A)-A$ are those of $\lambda B-A$ at $\lambda=\alpha$. Notice that the cases (3.7b) and (3.7a) are special cases of the latter two for $\alpha=0$ and $\alpha=\infty$, respectively. We can thus obtain the unique pair of reducing subspaces X_α,Y_α whose spectrum contains all the eigenvalues α of $\lambda B-A$:

$$(3.8) \qquad (\lambda B-A)\Big|_{X_\alpha,Y_\alpha} = \{\alpha,\ldots,\alpha\}$$

by using the above method on the pencils $\mu'(A-\alpha B)-B$ or $\lambda'(B-1/\alpha A)-A$. The pencil (3.7c) is chosen when $|\alpha| \leq \|B\| \,/\, \|A\|$ and the pencil (3.7d) is chosen otherwise (here $\|\cdot\|$ stands for any norm invariant under unitary transformations). This choice ensures the backward stability of the method. Indeed, the algorithm implementing the decomposition (3.3) on a pencil $\lambda D-C$ can be shown to yield exactly

$$(3.9) \qquad \bar{Q}^*(\lambda\bar{D}-\bar{C})\bar{Z} = \left[\begin{array}{c|c} \lambda D_{11}-C_{11} & \lambda D_{12}-C_{12} \\ \hline 0 & \lambda D_{22}-C_{22} \end{array}\right]$$

for a slightly perturbed pencil $\lambda\bar{D}-\bar{C}$ satisfying :

$$(3.10) \qquad \begin{aligned} \|\bar{C}-C\| &\leq \Pi.\varepsilon.\|C\| \\ \|\bar{D}-D\| &\leq \Pi.\varepsilon.\|D\| \end{aligned}$$

Here ε is the machine accuracy of the computer and Π is some polynomial expression in the dimensions m and n of the pencil $\lambda D-C$. Moreover, the matrices $\bar{Q}$ and $\bar{Z}$ are $\Pi.\varepsilon$-close to being unitary. With the choice of pencil (3.7c) or (3.7d) proposed above, this yields that the computed spaces $\bar{X}_\alpha$, $\bar{Y}_\alpha$, spanned by the nearly orthonormal columns of $\bar{Q}_1$ and $\bar{Z}_1$, are the exact spaces with spectrum at α of a slightly perturbed pencil $\lambda\bar{B}-\bar{A}$, satisfying

$$(3.11) \qquad \begin{aligned} \|\bar{A}-A\| &\leq 3\Pi.\varepsilon\ \|A\| \\ \|\bar{B}-B\| &\leq 3\Pi.\varepsilon\ \|B\| \end{aligned}$$

In going from (3.10) to (3.11) it is important that α or $1/\alpha$, respectively, can be appropriately bounded, which explains the appropriate choice of pencil (3.7c) or (3.7d) (see [9]). This thus yields a stable algorithm for the computation of a pair of reducing subspaces whose spectrum consists of <u>one</u> point α only.

When one wants to compute pairs of reducing subspaces corresponding to more of less points or, more generally, to compute <u>all</u> reducing subspaces with specified spectrum, one may proceed as follows. First one extracts the regular part of the pencil $\lambda B-A$ via the above algorithm and its dual form. The 'dual' algorithm consists of inverting the role of columns and rows in the above method. This is obtained by using the above method on the 'pertransposed' (i.e. transposed over the second diagonal) and then 'pertransposing' the obtained result (see [9]). This then yields a decomposition of the type :

$$(3.12) \qquad \tilde{Q}^*(\lambda B-A)\tilde{Z} = \left[\begin{array}{c|c} \lambda B_{rf}-A_{rf} & * \\ \hline 0 & \lambda B_{i\ell}-A_{i\ell} \end{array}\right]$$

where now $\lambda B_{rf}-A_{rf}$ contains all the <u>finite</u> elementary divisors and the <u>right</u> minimal indices of $\lambda B-A$, and where $\lambda B_{i\ell}-A_{i\ell}$ contains all the <u>infinite</u> elementary divisors and the <u>left</u> minimal indices of $\lambda B-A$. The constructed pair of reducing subspaces $\mathcal{X}_f,\mathcal{Y}_f$ is thus the unique pair whose spectrum contains all the finite eigenvalues $\lambda B-A$:

$$(3.13)\qquad (\lambda B-A)\Big|_{\mathcal{X}_f,\mathcal{Y}_f} = \{\alpha_1,\dots,\alpha_k\}$$

Using this dual decomposition on the diagonal blocks $\lambda B_{ri}-A_{ri}$ and $\lambda B_{f\ell}-A_{f\ell}$ of (3.3a) one then separates the right minimal indices and infinite elementary divisors of $\lambda B_{ri}-A_{ri}$ in two diagonal blocks λB_r-A_r and λB_i-A_i, and the finite elementary divisors and left minimal indices of $\lambda B_{f\ell}-A_{f\ell}$ in two diagonal blocks λB_f-A_f and $\lambda B_\ell-A_\ell$:

$$(3.14)\qquad Q^*(\lambda B-A)Z=\begin{bmatrix} \lambda B_r-A_r & * & * & * \\ 0 & \lambda B_i-A_i & * & * \\ 0 & 0 & \lambda B_f-A_f & * \\ 0 & 0 & 0 & \lambda B_\ell-A_\ell \end{bmatrix}\begin{matrix} \}m_1 \\ \}m_2 \\ \}m_3 \\ \}m_4 \end{matrix}$$
$$\qquad\qquad\qquad\qquad\quad \underbrace{}_{n_1}\quad \underbrace{}_{n_2}\quad \underbrace{}_{n_3}\quad \underbrace{}_{n_4}$$

This decomposition yields the regular part of $\lambda B-A$ (see [9]) :

$$(3.15)\qquad \lambda B_{reg}-A_{reg} = \underbrace{\left[\begin{array}{c|c} \lambda B_i-A_i & * \\ \hline 0 & \lambda B_f-A_f \end{array}\right]}_{d=n_2+n_3}\Bigg\} d=m_2+m_3$$

and the normal rank r of $\lambda B-A$ is given by $r=m_1+d+n_4$. The reducing subspaces $\mathcal{X}_{min}$, $\mathcal{Y}_{min}$, $\mathcal{X}_{max}$ and $\mathcal{Y}_{max}$ are spanned by the first n_1 columns of Z, m_1 columns of Q, $(n_1+n_2+n_3)$ columns of Z and $(m_1+m_2+m_3)$ columns of Q, respectively.
As discussed in the previous section, there corresponds a pair of reducing subspaces for $\lambda B-A$ to each pair of deflating subspaces of its regular part $\lambda B_{reg}-A_{reg}$, and conversely.The problem is thus reduced now to the computation of deflating subspaces of a regular pencil, which is essentially solved in [11] . These subspaces are obtained by an efficient update of the QZ decomposition [5] in order to obtain any requested ordering of eigenvalues along the diagonal of the decomposition. The method is also adapted to cope with the specific problem of real pencils (see [11] for more details). The numerical stability of this QZ update is proved in [11], which together with the above mentioned method thus yield a stable method for computing pairs of reducing subspaces of an arbitrary singular pencil.

4. Concluding remarks.

In the previous section we have presented a method to compute pairs of reducing subspaces with prescribed spectrum, as introduced in Section 2. The method consists of two steps : first, the extraction of the regular part of the pencil $\lambda B-A$, and, second, the computation of a pair of deflating subspaces of this regular part. The latter part can be performed in $O(d^3)$ operations (where d is the dimension of the regular part) using the QZ algorithm [5] and the update in [11] for obtaining the correct spectrum. The method described here for the extraction of the regular part, though, may require a number of operations which is not cubic in the dimensions m and/or n of $\lambda B-A$ but quartic (see [9]) since up to $O(\min\{m,n\})$ rank determinations of full matrices may be required. When efficiently exploiting the computations of previous rank determinations at each step, to overall amount of operations may be reduced to $O(mn^2)$. This is e.g. done in [2] for a specific class of pencils often occurring in linear system theory, but the idea can be extrapolated to the general case. Similar ideas may be found in the work of Kublanovskaya on dense and sparse pencils [15][16] .

Another link with linear system theory is of a more theoretic nature. All the geometric concepts introduced by Wonham [14] can be shown to be special cases of the concept of reducing subspaces introduced here. Reducing subspaces also enter the picture naturally when trying to extend some results of factorization to the singular case (see e.g. [10]). These remarks thus tend to indicate that the concept of reducing subspaces, as defined here, is an appropriate extension of the concept of deflating subspaces, since it occurs in several practical problems.

A last remark ought to be made about the possible ill-posedness of the spaces we are trying to compute. It is indeed shown via some simple examples in [9] that singular pencils may have an ill-posed eigenstructure and that one must be careful when interpreting the computed results. Yet when one fixes the normal rank of a pencil $\lambda B-A$ to the minimal possible one within ε perturbations of A and B, then the problem of computing reducing subspaces becomes well-posed (in a 'restricted' sense, of course, [10]). This is comparable to the problem of computing a generalized inverse A^* of a m×n matrix A which becomes well-posed when fixing its ε-rank. Moreover, there is hope to derive perturbation bounds for reducing subspaces in the style of Stewart's work on deflating subspaces [7][8] since there is a strong parallelism between both concepts.

References

[1] BOLEY D., Computing the controllability/observability decomposition of a linear time invariant dynamic system, a numerical approach, Ph. D. Thesis, Stanford University, 1981.

[2] EMAMI-NAEINI A., VAN DOOREN P., Computation of zeros of linear multivariable systems, to appear Automatica, 1982.

[3] FORNEY, G. D. Jr., Minimal bases of rational vector spaces with applications to multivariable linear systems, SIAM J. Contr., Vol. 13, pp. 493-520, 1975.

[4] GANTMACHER F. R., Theory of matrices I & II, Chelsea, New York, 1959.

[5] MOLER C., STEWART G., An algorithm for the generalized matrix eigenvalue problem, SIAM J. Num. Anal., Vol. 10, pp. 241-256, 1973.

[6] PAIGE C., Properties of numerical algorithms related to computing controllability, IEEE Trans. Aut. Contr., Vol. AC-26, pp. 130-138.

[7] STEWART G., Error and perturbation bounds for subspaces associated with certain eigenvalue problems, SIAM Rev., Vol. 15, pp. 727-764, 1973.

[8] STEWART G., On the sensitivity of the eigenvalue problem $Ax=\lambda Bx$, SIAM Num. Anal. Vol. 9, pp. 669-686, 1972.

[9] VAN DOOREN P., The computation of Kronecker's canonical form of a singular pencil, Lin. Alg. & Appl., Vol. 27, pp. 103-141, 1979.

[10] VAN DOOREN P., The generalized eigenstructure problem in linear system theory, IEEE Trans. Aut. Contr., Vol. AC-26, pp. 111-129, 1981.

[11] VAN DOOREN P., A generalized eigenvalue approach for solving Riccati equations, SIAM Sci. St. Comp., Vol. 2, pp. 121-135, 1981.

[12] WILKINSON J., Linear differential equations and Kronecker's canonical form, Recent Advances in Numerical Analysis, Ed. C. de Boor, G. Golub, Academic Press, New York, 1978.

[13] WILKINSON J., Kronecker's canonical form and the QZ algorithm, Lin. Alg. & Appl., Vol. 28, pp. 285-303, 1979.

[14] WONHAM W., Linear multivariable theory. A geometric approach, (2nd Ed.) Springer, New York, 1979.

[15] KUBLANOVSKAYA V., AB algorithm and its modifications for the spectral problem of linear pencils of matrices, LOMI-preprint E-10-81, USSR Academy of Sciences, 1981.

[16] KUBLANOVSKAYA V., On an algorithm for the solution of spectral problems of linear matrix pencils, LOMI-preprint E-1-82, USSR Academy of Sciences, 1982.

SECTION A.2

OF

GENERAL (A-λB)-PENCILS

ASPECTS FROM DIFFERENTIAL EQUATIONS

DIFFERENTIAL/ALGEBRAIC SYSTEMS AND MATRIX PENCILS*

C. W. Gear
Department of Computer Science, University of Illinois at Urbana-Champaign

L. R. Petzold
Sandia National Laboratories, Livermore

Abstract

In this paper we study the numerical solution of the differential/algebraic systems $F(t, y, y') = 0$. Many of these systems can be solved conveniently and economically using a range of ODE methods. Others can be solved only by a small subset of ODE methods, and still others present insurmountable difficulty for all current ODE methods. We examine the first two groups of problems and indicate which methods we believe to be best for them. Then we explore the properties of the third group which cause the methods to fail.

The important factor which determines the solvability of systems of linear problems is a quantity called the global nilpotency. This differs from the usual nilpotency for matrix pencils when the problem is time dependent, so that techniques based on matrix transformations are unlikely to be successful.

1. INTRODUCTION

We are interested in initial value problems for the differential/algebraic equation (DAE)

$$(1.1) \qquad F(t, y, y') = 0 ,$$

where F, y, and y′ are s-dimensional vectors. F will be assumed to be suitably differentiable. Many of these problems can be solved conveniently and economically using numerical ODE methods. Other problems cause serious difficulties for these methods. Our purpose in this paper is first to examine those classes of problems that are solvable by ODE methods, and to indicate which methods are most advantageous for this purpose. Secondly, we want to describe the problems which are not solvable by ODE methods, and the properties of these problems which cause the methods to fail.

The idea of using ODE methods for solving DAE systems directly was introduced in [3], and is best illustrated by considering the simplest possible algorithm,

*Supported in part by the U.S. Department of Energy, Grant DEAC0276ER02383 and by the U.S. Department of Energy Office of Basic Energy Sciences.

based on the backward Euler method. In this method the derivative $y'(t_{n+1})$ at time t_{n+1} is approximated by a backward difference of $y(t)$, and the resulting system of nonlinear equations is solved for y_{n+1},

$$(1.2) \qquad F(t_{n+1}, y_{n+1}, (y_{n+1} - y_n)/(t_{n+1} - t_n)) = 0 .$$

In this way the solution is advanced from time t_n to time t_{n+1}. Higher order techniques such as backward differentiation formulas (BDF), Runge-Kutta methods, and extrapolation methods are generalizations of this simple idea.

One of the main advantages in using ODE methods directly for solving DAE systems is that these methods preserve the sparsity of the system. For example, one set of DAE systems which is particularly simple to solve consists of systems which are really ODEs in disguise. If, in (1.1), $\partial F/\partial y'$ is nonsingular, then the system can, in principle, be inverted to obtain an explicit system of ODEs

$$(1.3) \qquad y' = f(t, y) .$$

However, if $\partial F/\partial y'$ is a sparse matrix, its inverse may not be sparse. Thus it is preferable to solve the system directly in its original form. Similarly, it is possible to reduce more complex DAE systems to a standard form which, though not as simple as (1.3), may be handled via well known techniques. This approach also tends to destroy the natural sparsity of the system.

The most challenging difficulties for solving DAE systems occur when $\partial F/\partial y'$ is singular. These are the systems with which we are concerned here. In some sense the simplest, or at least the best understood, class of DAE systems is that which is linear with constant-coefficients. These systems,

$$(1.4) \qquad Ay'(t) + By(t) = g(t) ,$$

can be completely understood via the Kronecker canonical form of the matrix pencil $(B + \lambda A)$. The important characteristic of equation (1.4) that determines the behavior of the system and numerical methods is the nilpotency of the matrix pencil $B + \lambda A$. Numerical methods such as (1.2) can be used to solve linear and nonlinear systems of nilpotency no greater than one with no great difficulty. Algorithms based on these methods experience problems when the nilpotency is greater than one. With some care techniques based on higher order methods such as extrapolation can be constructed for solving systems of the form (1.4), even if the nilpotency exceeds one. We consider these issues in Section 2.

One might hope that the study of (1.4) could be used as a guide for understanding more complicated DAE systems. In general this fails to be true. The structure of the local constant-coefficient system may not describe the behavior of solutions to the DAE, for nonlinear or even linear, non-constant-coefficient systems whose index is greater than one. Numerical methods which work for (1.4) break down when the matrices are time-dependent and the nilpotency is greater than one. In fact, we are not aware of any numerical methods (based on ODE techniques or otherwise) for solving general linear DAE systems, let alone nonlinear systems. In Section 3 we examine time-dependent problems and show where the difficulties with conventional methods arise.

We will not examine the general nonlinear problem (1.1) here because we do not yet know how to solve the nonconstant coefficient linear problem. We do know that the nonconstant-coefficient case is not a simple extension of the constant-coefficient case when the nilpotency exceeds one, so there is no guarantee that methods found to solve the nonconstant-coefficient case will extend to the nonlinear case.

2. CONSTANT-COEFFICIENT PROBLEMS

The existence and solution of linear constant-coefficient systems (1.4) is easily understood by transforming the system to Kronecker canonical form (KCF). For details see [11]. We give only an overview. The matrix pencil $B + \lambda A$ will be written as (A,B). The main idea is that there exist nonsingular matrices P and Q which reduce (A,B) to canonical form. When P and Q are applied to (1.4), we obtain

$$PAQQ^{-1}y' + PBQQ^{-1}y = Pg(t) \quad (2.1)$$

where (PAQ,PBQ) is the canonical form. When $B + \lambda A$ is singular for all values of λ, no solutions exist, or infinitely many solutions exist. It is not even reasonable to try to solve these systems numerically. Fortunately, numerical ODE methods reject these problems almost automatically because they have to solve a linear system involving the matrix $A + h\beta B$ (where h is the stepsize and β is a scalar which depends on the method and recent stepsize history), and this is singular for all values of h. When $\det(A + B/\lambda)$ is not identically zero, the system is called "solvable" because solutions to the differential/algebraic equation exist and two solutions which share the same initial values must be identical. In the following we will deal only with solvable systems.

For solvable systems the KCF of (2.1) can be written as

$$y_1'(t) + Cy_1(t) = g_1(t) \;, \quad (2.2a)$$

$$Ey_2'(t) + y_2(t) = g_2(t) \;, \quad (2.2b)$$

where

$$Q^{-1}y(t) = \begin{bmatrix} y_1(t) \\ y_2(t) \end{bmatrix} \;, \quad Pg(t) = \begin{bmatrix} g_1(t) \\ g_2(t) \end{bmatrix} \;,$$

and E has the property that there exists an integer m such that $E^m = 0$, $E^{m-1} \neq 0$. The value of m is defined to be the nilpotency of the system. The matrix E is composed of Jordan blocks of the form

$$(2.3) \qquad \begin{bmatrix} 0 & & & \\ 1 & \ddots & & \\ & \ddots & \ddots & \\ & & 1 & 0 \end{bmatrix},$$

and m is the size of the largest of these blocks. Note that the nilpotency does not exceed the number of infinite eigenvalues of $B + \lambda A$, but is less if E contains more than one Jordan block.

The behavior of numerical methods for solving standard ODE systems (2.2a) is well understood and will not be discussed here. Since the systems (2.2a) and (2.2b) are completely uncoupled and the methods we are interested in are linear, it suffices for understanding (1.4) to study the action of numerical methods on subsystems of the form (2.2b), where E is a single block of form (2.3). When E is a matrix of the form (2.3) and size n, the system is referred to as a canonical (m = n) subsystem.

Systems whose nilpotency does not exceed one are the most easily understood, and they seem to occur far more frequently in solving practical problems than the other (> 1) subsystems. When the nilpotency does not exceed one, the matrix E in (2.2b) is identically zero. Thus the system reduces to a system of standard form ODEs plus some variables which are completely determined by simple linear relations.

What kinds of methods are most useful for solving these problems? Since, for the DAE, the values of the algebraic components are completely determined at all times (there are no arbitrary initial conditions for these variables), it is desirable for the numerical solution to share this property.

Most automatic codes for solving DAE systems [7] are designed to handle nonlinear systems of nilpotency ≤ 1. These codes cannot handle systems of higher nilpotency, and it would be desirable in such codes to detect higher nilpotency problems and stop. However, detection of these systems in practice seems to be a fairly difficult problem at present.

Systems of nilpotency greater than one have several properties which are not shared by the lower nilpotency systems. We can understand many of the properties of (1.4) and of numerical methods by studying the simplest nilpotency 3 problem,

$$(2.4) \qquad \begin{aligned} z_1 &= g(t) \\ z_1' - z_2 &= 0 \\ z_2' - z_3 &= 0 \,. \end{aligned}$$

The solution to this problem is $z_1 = g(t)$, $z_2 = g'(t)$, $z_3 = g''(t)$. If initial values are specified for the z_i, the solution has a jump discontinuity unless these initial values are compatible with the solution. If the driving term g(t) is not twice differentiable everywhere, the solution will not exist everywhere. For example, if g(t) has a simple jump discontinuity at some point, z_2 is a dirac delta function, and z_3 is the derivative of a dirac delta.

What happens when a numerical method is applied to one of these problems? It is surprising that some of the numerical ODE methods work so well on these problems which are so unlike ODEs. We can best explain how the methods work by example. When the backward Euler method is used to solve the nilpotency = 3 problem (2.4), we find that the solution at time t_n is given in terms of the solution at time t_{n-1} by

$$\begin{aligned} z_{1,n} &= g_n \\ z_{2,n} &= (z_{1,n} - z_{1,n-1})/h \\ z_{3,n} &= (z_{2,n} - z_{2,n-1})/h \end{aligned} \tag{2.5}$$

where $h = t_n - t_{n-1}$. The values of z_1 will be correct at all steps (if roundoff error is ignored), although the initial value $z_{1,0}$ may be incorrect. If the initial values (which need not be specified for the original problem but must be specified for the numerical procedure) are inconsistent, the values of $z_{2,1}$ and $z_{3,1}$ are incorrect. In fact, as $h \to 0$ they diverge. However, after two steps we obtain an O(h) correct value of $z_{2,2}$ because it is obtained by the divided difference of g(t). Finally, after the third step we obtain a good approximation to z_3 which is given by the second divided difference of g(t). After the third step all the components will be O(h) accurate.

The behavior of a general BDF method is very similar to that of backward Euler for fixed stepsize as shown in the following theorem, proved in [11].

Theorem 2.1. If the k-step constant-stepsize BDF method is applied to the constant-coefficient linear problem (1.4) with $k < 7$, the solution is $O(h^k)$ accurate globally after a maximum of $(m-1)k + 1$ steps.

Unfortunately, these results for BDF break down when the stepsize is not constant, as shown in the next theorem, proved in [5].

Theorem 2.2. If the k-step BDF method is applied to (1.4) with $k < 7$ and the ratio of adjacent steps is bounded, then the global error is $O(h_{max}^q)$, where $q = \min(k, k-m+2)$.

Although, in principle, a problem of nilpotency no greater than seven could be solved by the six-step BDF method with variable stepsize, the hypothesis in Theorem 2.2 that the ratio of adjacent steps is bounded is not a reasonable model in practice. When a code is attempting to take the next step, all previous stepsizes are now fixed, and the next step must be chosen to achieve the desired error. In this model the error of a BDF formula used for numerical differentiation is O(h), where h is the current stepsize. Consequently, if the nilpotency exceeds 2, the error of one step does not converge as that stepsize goes to zero, and diverges if the nilpotency exceeds 3.

The above results suggest that variable-stepsize BDF is not a suitable method for solving constant-coefficient DAEs with arbitrary nilpotency. There is an elegant way of handling these problems based on extrapolation of the backward Euler method, which we will now describe. All ODE methods, however, break down when the coefficients of the system are not constant (this is discussed in the next section).

Extrapolation is a technique for improving the order of a numerical solution by trying to eliminate the error term. The reason for using the backward Euler method as the basis for extrapolation in our particular situation is that the global error has an asymptotic expansion in the stepsize h. That is,

$$(2.6) \qquad y(H, h) = y(H) + \sum_{i=1}^{m} \tau_i(H)h^i + O(h^{m+1})$$

where y(H, h) is the numerical solution at time H which is computed with stepsize h, and y(H) is the solution to the DAE at time H. It is easy to see that this expansion exists for linear constant-coefficient DAEs by noting that there is such an expansion for each of the canonical subsystems once the initial errors have disappeared.

In the algorithm a sequence of approximations $\{y(H, h_j)\}$, $j = 1, 2, \ldots$, to the solution at time H are formed using the backward Euler method with stepsize h_j. The stepsizes h_j are related to the basic stepsize H by an integer sequence $\{n_j\}$ such that $h_j = \frac{H}{n_j}$, $j = 1, 2, \ldots$. $\{n_j\}$ is an increasing sequence of integers; for example, $\{n_j\} = \{1, 2, 4, 8, \ldots\}$. The idea underlying extrapolation is that by taking linear combinations of the $y(H, h_j)$, $j = 1, 2, \ldots$, we can cancel out terms in the error expansion (2.6) and obtain a more accurate approximation. The simplest way to do this is to use the Aitken interpolation process to define an extrapolation tableau whose first column consists of $\{y(H, h_j)\}$. Each succeeding column contains approximations to y which are one order of accuracy higher than the column to the left of it.

$$(2.7) \qquad \begin{array}{lllll} y(H, h_1) = T_{11} & \searrow & & & \\ y(H, h_2) = T_{21} & \longrightarrow & T_{22} & \searrow & \\ y(H, h_3) = T_{31} & \longrightarrow & T_{32} & \longrightarrow & T_{33} \end{array}$$

The columns are related to one another by a recursion relation which is defined by the interpolation algorithm

$$(2.8) \qquad T_{i,k} = T_{i,k-1} + \frac{(T_{i,k-1} - T_{i-1,k-1})}{(\frac{n_i}{n_{i-k+1}}) - 1} .$$

Normally, when extrapolation is used for solving ODE systems, the diagonal elements $T_{i,i}$ are the best approximations to the solution. Error estimates are generated by comparing each diagonal element to the diagonal element immediately above it, or to the subdiagonal element to the left of it. Steps are accepted or rejected, and new stepsizes and orders are selected based on comparisons such as these [2].

What happens when extrapolation is used for solving a linear constant-coefficient DAE of nilpotency > 1? Consider, for example, the nilpotency 3 problem (2.4) and the sequence $\{n_j\} = \{1, 2, 4, 8, \ldots\}$. Now since z_3 is incorrect until after the second step has been taken, it follows that $y(H, h_1) = T_{11}$ is not a good approximation to the solution. All of the other elements in the first column, however, satisfy the expansion (2.6). Since the diagonal elements are formed from linear combinations involving T_{11}, these elements too will be in error. It is clear that in this example, in contrast to the situation for ODEs, the best approximation occurs on the subdiagonal. The diagonal should be ignored (once we have determined that it contains large errors). Note, however, that if the sequence $\{n_j\} = \{2, 4, 8, \ldots,\}$ had been used instead, then all of the elements in the first column would have satisfied (2.6), and we could have proceeded as usual. Thus, if an upper bound for the nilpotency of the system is known in advance, we can use extrapolation with the usual stepsize selection strategies, provided the sequence $\{n_j\}$ starts out with a sufficiently large integer n_0. If we do not know the maximum nilpotency of the system, then extrapolation may be used as explained above, but the strategies must be modified so that subdiagonal approximations are sometimes accepted instead of the diagonal approximations. In practice the use of this technique is complicated somewhat by the possibility of discontinuities in the function g, and also by the fact that, for higher nilpotency systems, the matrices needed for solving for the solution of the backward Euler formula are severely ill-conditioned. This technique is the best approach that we know of for solving linear constant-coefficient DAE systems.

3. NONCONSTANT-COEFFICIENT PROBLEMS

In this section we study the nonconstant-coefficient linear problem,

$$(3.1) \qquad A(t)y'(t) + B(t)y(t) = g(t) ,$$

and examine the reasons why these systems have proven to be so difficult to solve.

When the coefficients are not constant as in (3.1), there are several possible ways to define the nilpotency of the system. We can clearly define the local nilpotency, $\ell(t)$ = nilpotency($A(t)$, $B(t)$), whenever the pencil $(A(t), B(t))$ is nonsingular. (The notation $(A(t), B(t))$ refers to the time-dependent matrix pencil $B(t) + \lambda A(t)$.) We can also define the global nilpotency, when it exists, in terms of possible reductions of the DAE to a canonical form. By making a change of variables $y = H(t)z$ and scaling the system by $G(t)$, where $G(t)$ and $H(t)$ are nonsingular, we obtain from (3.1)

$$G(t)A(t)H(t)z' + (G(t)B(t)H(t) + G(t)A(t)H'(t))z = G(t)g(t) \tag{3.2}$$

Now, if there exist $G(t)$ and $H(t)$ so that

$$G(t)A(t)H(t) = \begin{bmatrix} I_1 & 0 \\ 0 & E \end{bmatrix}, \tag{3.3}$$

$$G(t)B(t)H(t) + G(t)A(t)H'(t) = \begin{bmatrix} C(t) & 0 \\ 0 & I_2 \end{bmatrix},$$

and the nilpotency of E is m, we will say that the system has global nilpotency of m. Note that the global nilpotency is the local nilpotency in this semi-canonical form.

Clearly, it is the global nilpotency that determines the behavior of the solution. If the global nilpotency is a constant m, we know that n_1 independent initial values can be chosen, where n_1 is the dimension of the "differential" part of the system, and that the driving term can be subject to differentiation $m-1$ times. (Changes in the nilpotency or the structure of the system are called turning points. Problems with turning points are of importance in electrical network analysis. See Sastry, et al. [10] for a discussion in that context, and Campbell [1] for a discussion of types of turning points.)

The local nilpotency in some sense governs the behavior of the numerical method. For example, if the matrix pencil is singular, then numerical ODE methods cannot solve the problem because they will be faced with the solution of singular linear equations. In understanding why numerical ODE methods break down, it is natural to ask how the local nilpotency and global nilpotency are related. The next theorem answers this question.

Theorem 3.1. If the local nilpotency is not greater than one, then it is not changed by a smooth transformation. If the local nilpotency is greater than one, then almost all smooth nonconstant transformations of variables in (3.1) will yield a system whose local nilpotency is two. On a lower dimension manifold the nilpotency may be greater than two, or the pencil may be singular. When a transformation to the semicanonical form (3.3) is used, this shows the relationship between the local and global nilpotencies.

Proof. Suppose we make the smooth transformation $y = Sz$ in (3.1) to get $ASz' + (BS + AS')z = g(t)$. Suppose P and Q transform (A,B) to canonical form.

(In what follows all matrices will be taken to depend on t except I_i and E.) The local nilpotency of the new system is given by

$$\ell(t) = \text{nilpotency } (AS,\ BS + AS')$$

Since the local nilpotency of a pencil is unchanged by pre- and post-multiplication by nonsingular matrices, multiply by P and Q to get

$$(3.4)\qquad \ell(t) = \text{nilpotency } (PAQ, PBQ + PAQQ^{-1}S'S^{-1}Q)$$

$$= \text{nilpotency } \left(\begin{bmatrix} I_1 & 0 \\ 0 & E \end{bmatrix},\ \begin{bmatrix} C & 0 \\ 0 & I_2 \end{bmatrix} + PAQ(Q^{-1}S'S^{-1}Q) \right).$$

$$= \text{nilpotency } \left(\begin{bmatrix} I_1 & 0 \\ 0 & E \end{bmatrix},\ \begin{bmatrix} C & 0 \\ 0 & I_2 \end{bmatrix} + \begin{bmatrix} I_1 & 0 \\ 0 & E \end{bmatrix} D \right),$$

Now $D = Q^{-1}S'S^{-1}Q$ is essentially an arbitrary matrix. That is, if we view the original system as fixed and the transformation S open to choice, it can be chosen to give D any value we wish for a particular value of t by solving $S' = QDQ^{-1}S$ as an initial value problem. Partition D into $D_{i,j}$, $i,j = 1, 2$. From (3.4):

$$(3.5)\qquad \ell(t) = \text{nilpotency } \left(\begin{bmatrix} I_1 & 0 \\ 0 & E \end{bmatrix},\ \begin{bmatrix} C + D_{11} & D_{12} \\ ED_{21} & I_2 + ED_{22} \end{bmatrix} \right)$$

If nilpotency $(E, I_2) \leq 1$, then $E \equiv 0$ or is null, and (3.5) shows that $\ell(t)$ is given by

$$(3.6)\qquad \ell(t) = \text{nilpotency } \left(\begin{bmatrix} I_1 & 0 \\ 0 & E \end{bmatrix},\ \begin{bmatrix} D_{11} & D_{12} \\ 0 & I_2 \end{bmatrix} \right) .$$

Lemma 3.1 below shows that $\ell(t) =$ nilpotency (E, I_2), proving the first part of the theorem.

For the second part of the theorem, we will consider two cases: nilpotency $(E, I_2) = 2$ and nilpotency $(E, I_2) > 2$.

If nilpotency $(E, I_2) = 2$, we must examine the right hand side of (3.5) more carefully to observe that almost always it is possible to do row (P) and column (Q) operations to reduce it to the form in (3.6). E consists of a set of diagonal blocks. The nilpotency 1 blocks are 0. For these, (3.5) is already in the form

(3.6). The 2 × 2 blocks are

$$\begin{bmatrix} 0 & 0 \\ 1 & 0 \end{bmatrix},$$

which lead to rows in (3.5) of the form

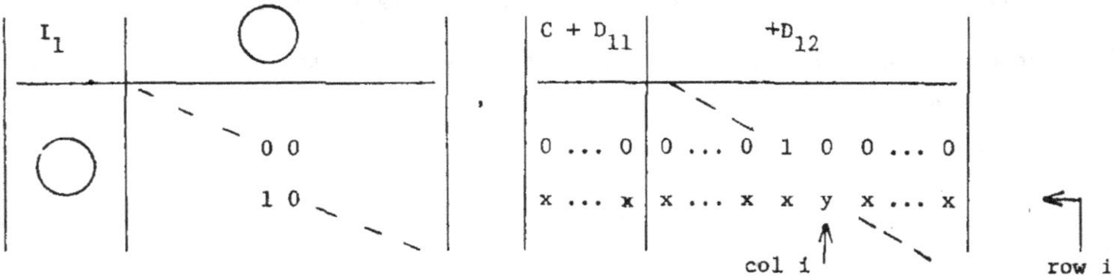

where x is a generic arbitrary element and y is an arbitrary element which happens to occupy the position shown. If $y \neq 0$ we can subtract multiples of the i-th column from each of the other columns to cancel the "x" terms and divide the i-th column by y. Since this is a nonsingular "Q" transformation, it does not change the local nilpotency so that

$$\ell(t) = \text{nilpotency}\left(\begin{bmatrix} I_1 & 0 \\ 0 & E_2 \end{bmatrix}, \begin{bmatrix} \tilde{C} & \tilde{D}_{12} \\ 0 & I \end{bmatrix}\right),$$

and the result follows as before.

If $y = 0$ the nilpotency can increase, as can be verified by example. It cannot decrease, as shown in Campbell [1]. If the column containing y is zero, the pencil is singular.

If nilpotency $(E, I_2) \geq 3$, then m rows of the pencil have the form

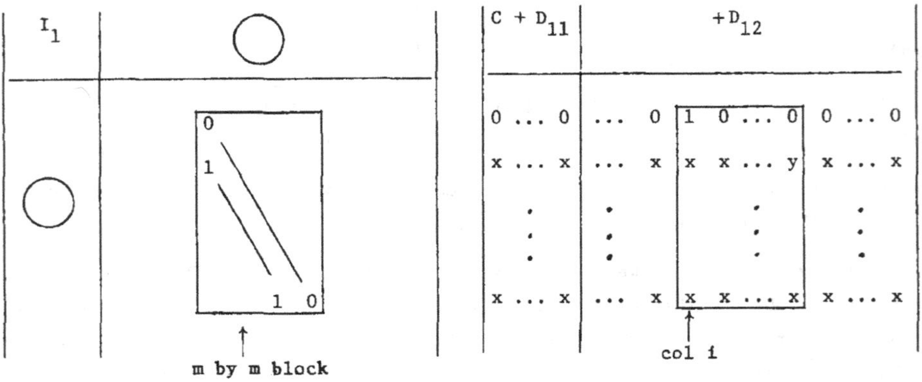

Now if y is nonzero we can use column operations to reduce the second row of "x's" to zeroes, except for a 1 in place of y. This does not affect the left hand matrix because column i+m-1 in that matrix is identically zero. Then, moving column i+m-1 immediately to the right of column i, we obtain

$$\left[\begin{array}{c|c} I_1 & \\ \hline & \boxed{\begin{array}{ccc} 0 & 0 & \\ 1 & 0 & \\ & & I \\ & & \end{array}} \end{array}\right] , \quad \left[\begin{array}{c|c} C + D_{11} & \\ \hline \begin{array}{ccc} 0 & \dots & 0 \\ 0 & \dots & 0 \\ x & \dots & x \\ & \vdots & \\ x & \dots & x \end{array} & \begin{array}{c|c|c} \dots \ 0 & 1 \ \ 0 \ \dots \ 0 & 0 \ \dots \ 0 \\ \dots \ 0 & 0 \ \ 1 \ \dots \ 0 & 0 \ \dots \ 0 \\ \dots \ x & x \ \ x \ \dots \ x & x \ \dots \ x \\ & \vdots & \\ \dots \ x & x \ \ x \ \dots \ x & x \ \dots \ x \end{array} \end{array}\right]$$

Now move columns i+2 to i+m-1 inclusive and the corresponding rows to the last position in the "algebraic" part of the matrix, and observe that we have just increased the "differential" part of the system by m-2 rows. The nilpotency of the block treated has decreased to two.

If y = 0, again we may have a singular pencil, or the nilpotency may decrease (but will always be greater than one [1]), as can be seen by examples. Q.E.D.

<u>Lemma 3.1.</u>

$$\text{nilpotency} \left(\begin{bmatrix} I_1 & 0 \\ 0 & E \end{bmatrix} , \begin{bmatrix} C & D \\ 0 & I_2 \end{bmatrix} \right) = \text{nilpotency } (E, I_2)$$

<u>Proof.</u> The result follows by simple reductions to nullify D. Premultiply the pencil by

$$\begin{bmatrix} I_1 & -D \\ 0 & I_2 \end{bmatrix}$$

and postmultiply by

$$\begin{bmatrix} I_1 & DE \\ 0 & I_2 \end{bmatrix}$$

to obtain

$$\left(\begin{bmatrix} I_1 & 0 \\ 0 & E \end{bmatrix} , \begin{bmatrix} C & CDE \\ 0 & I_2 \end{bmatrix} \right) .$$

A similar transformation can be applied to reduce the upper right corner to $C(CDE)E = C^2DE^2$. This can be repeated m times to obtain $C^mDE^m = 0$.

Q.E.D.

The following example illustrates a change of local nilpotency from 2 to m by means of a time-dependent transformation. Suppose we start with the nilpotency two problem

$$\left[\begin{array}{c|cc} I_2 & & \\ \hline & 0 & 0 \\ & 1 & 0 \end{array}\right] y' + \left[\begin{array}{c|c} I_1 & \\ \hline & I_2 \end{array}\right] y = 0$$

where I_2 is an (m-2) by (m-2) identity matrix. Substitute $z = Qy$ where $D = Q'Q^{-1}$ is given by

$$D = \left[\begin{array}{cccc|cc} -1 & 1 & & & & \\ & \ddots & \ddots & & & \\ & & & 1 & & \\ & & & -1 & & \\ \hline 1 & 0 & \cdots & 0 & 0 & -1 \\ 0 & 0 & \cdots & 0 & 0 & 0 \end{array}\right]$$

The corresponding Q is obtained by solving $Q' = DQ$, $Q(0) = I$. The pencil of the transformed system can be postmultiplied by Q^{-1} to get the pencil

$$\left(\left[\begin{array}{c|cc} I_1 & & \\ \hline & 0 & 0 \\ & 1 & 0 \end{array}\right], \left[\begin{array}{cccc|cc} 0 & 1 & & & & \\ & \ddots & \ddots & & & \\ & & & 1 & & \\ & & & 0 & 0 & 1 \\ \hline 0 & 0 & \cdots & 0 & 1 & 0 \\ 1 & 0 & \cdots & 0 & 0 & 0 \end{array}\right] \right)$$

Now move rows m-1 and m to the top and then move column m-1 before column 1 to get

$$\left(\left[\begin{array}{cccc} 0 & & & \\ 1 & \ddots & & \\ & \ddots & \ddots & \\ & & 1 & 0 \end{array}\right], \left[\begin{array}{c} I \end{array}\right] \right)$$

which has nilpotency m.

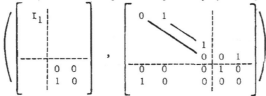

We have seen that the constant step BDF method can be used for constant coefficient problems. What happens when it is applied to nonconstant-coefficient problems? If the local nilpotency is two we may have a stability problem depending on the rate of change of the coefficients. If the local nilpotency is greater than two, we almost always have a stability problem. We want to stress that this is a stability problem and not an accuracy question, so it does not appear that higher order methods will help. Also note that the behavior of the numerical method depends on the local nilpotency while the behavior of the underlying equation depends on the global nilpotency. This indicates that if the global nilpotency exceeds two, the numerical results are almost certainly meaningless, either because of instability or because they reflect the wrong nilpotency.

We will illustrate the instability by example. A detailed discussion appears in [4].

Example 1

Consider the system

$$(3.8)\qquad \begin{bmatrix} 0 & 0 \\ 1 & \eta t \end{bmatrix} z' + \begin{bmatrix} 1 & \eta t \\ 0 & 1+\eta \end{bmatrix} z = \begin{bmatrix} g(t) \\ 0 \end{bmatrix}$$

This was obtained from

$$\begin{bmatrix} 0 & 0 \\ 1 & 0 \end{bmatrix} y' + y = \begin{bmatrix} g(t) \\ 0 \end{bmatrix}$$

by the change of variables $y = Hz$ where

$$H = \begin{bmatrix} 1 & \eta t \\ 0 & 1 \end{bmatrix}$$

so the global nilpotency is 2. If $\eta = -1$, the pencil in (3.8) is singular, otherwise it is easy to verify that its local nilpotency is also 2 using the transformation $P = I$,

$$Q = \begin{bmatrix} 1 & -\eta t/(1+\eta) \\ 0 & 1/(1+\eta) \end{bmatrix}$$

If this is solved with the backward Euler method, we get the solution

$$z_{2,n} = \frac{\eta}{1+\eta} z_{2,n-1} - \frac{g_n - g_{n-1}}{h(1+\eta)}$$
$$z_{1,n} = g_n - \eta t_n z_{2,n}$$

The correct solution is $z_2(t_n) = -g_n'$, $z_1(t_n) = g_n + \eta t_n g_n'$. The numerical solution is clearly unstable if $\eta < -1/2$. If $\eta > -1/2$ it converges to the true solution.

Example 2

Consider the system

$$(3.9)\qquad \begin{bmatrix} 0 & 0 & 0 \\ 1 & 0 & \eta t \\ 0 & 1 & 0 \end{bmatrix} y' + \begin{bmatrix} 1 & 0 & \eta t \\ 0 & 1 & 0 \\ 0 & 0 & 1 \end{bmatrix} y = \begin{bmatrix} g \\ 0 \\ 0 \end{bmatrix}$$

This can be seen to have local nilpotency 3 by the transformation $P = I$,

$$Q = \begin{bmatrix} 1 & 0 & -\eta t \\ 0 & 1 & 0 \\ 0 & 0 & 1 \end{bmatrix}$$

and global nilpotency 2 by the transformation

$$y = \begin{bmatrix} 1 & t & -t \\ 0 & 0 & 1 \\ 0 & -1/\eta & 1/\eta \end{bmatrix} z$$

If we examine the canonical form after the transformation from y to z and write $z = [u, v, w]^T$ we find we have the equations

$$u = g$$
$$v = -u'$$
$$w' = (v-w)/\eta$$

for the underlying nilpotency 2 system. Clearly the ODE component w is unstable if $\eta < 0$ and stable if $\eta > 0$. If we apply the backward Euler method to (3.9) and write $y = [p, q, r]^T$, we find that we get the recurrence relations

$$r_n = -\frac{\eta}{h}(r_{n-1} - r_{n-2}) + \frac{g_n - 2g_{n-1} + g_{n-2}}{h^2}$$
$$q_n = \eta r_n - \frac{g_n - g_{n-1}}{h}$$
$$p_n = g_n - \eta t_n r_n$$

These recurrences are unstable whenever $h < 2\eta$ or $h < -\eta$. Consequently the method cannot converge as h approaches zero.

In summary, high nilpotency problems currently pose serious difficulties. It is not clear how common they are because there have not been widely distributed codes available for DAEs until recently so scientists have been forced to eliminate the algebraic equations by differentiation. It is reported in [6] that a common formulation of the Navier Stokes equation has nilpotency 2, and the authors have been told of higher nilpotency problems in simulation, usually of mechanical systems. It is quite possible that the availability of codes for DAEs (e.g. [8]) will uncover many such problems.

REFERENCES

1. CAMPBELL, S. L., Linear time varying singular systems of differential equations, Dept. Mathematics, North Carolina State Univ., Raleigh, 1981.

2. DEUFLHARD, P., Order and stepsize control in extrapolation methods, Preprint No. 93, Univ. Heidelberg, 1980.

3. GEAR, C. W., The simultaneous numerical solution of differential-algebraic equations, IEEE Trans. Circuit Theory TC-18, (1), 1971, 89-95.

4. GEAR, C. W. and L. R. PETZOLD, ODE methods for differential/algebraic systems. In preparation.

5. GEAR, C. W., HSU, H. H. and L. PETZOLD, Differential-algebraic equations revisited, Proc. Numerical Methods for Solving Stiff Initial Value Problems, Oberwolfach, W. Germany, June 28-July 4, 1981.

6. Painter, J. F., Solving the Navier-Stokes equations with LSODI and the method of lines, Lawrence Livermore Laboratory Rpt. UCID-19262, 1981.

7. PETZOLD, L. R., Differential/algebraic equations are not ODEs, Rpt. SAND81-8668, Sandia National Laboratories, Livermore, CA, April 1981. *

8. PETZOLD, L. R., A description of DASSL: A differential/algebraic system solver, to appear, Proceedings of IMACS World Congress, Montreal, Canada, August 1982.

9. STARNER, J. W., A numerical algorithm for the solution of implicit algebraic-differential systems of equations, Tech. Rpt. 318, Dept. Mathematics and Statistics, Univ. New Mexico, May 1976.

10. SASTRY, S. S., DESOER, C. A. and P. P. VARAIYA, Jump behavior of circuits and systems, Memorandum No. UCB/ERL M80/44, Electronics Research Laboratory, University of California-Berkeley, CA, October 1980.

11. SINCOVEC, R. F., DEMBART, B., EPTON, M. A., ERISMAN, A. M., MANKE, J. W. and E. L. YIP, Solvability of large-scale descriptor systems, Final Report DOE Contract ET-78-C-01-2876, Boeing Computer Services Co., Seattle, WA.

* Published in SIAM J. Sci. Stat. Comp. 3, 367-384 (1982).

APPROXIMATION OF EIGENVALUES DEFINED BY ORDINARY DIFFERENTIAL EQUATIONS WITH THE TAU METHOD

Kam-Moon Liu and Eduardo L. Ortiz
Imperial College, University of London
London SW7

1. Introduction

The purpose of this paper is to present numerical examples on the application of the Tau method to the approximate solution of eigenvalue problems defined by ordinary differential equations.

We follow closely Ortiz' algebraic theory of the Tau method [7] and discuss its computer implementation by means of techniques proposed by Chaves and Ortiz [2] and more recently by Ortiz and Samara [14 - 15].

The Tau technique is applied to a variety of linear differential equations with constant or polynomial coefficients, of orders two to six, defined in finite intervals. We then consider cases where the coefficients are of nonpolynomial type, or the interval is semi- or double-infinite. In the last section of this paper we refer briefly to the numerical approximation of eigenvalues defined by functional differential equations and the use of a step by step technique for the Tau method discussed in[10] which seems to be particularly suitable when the differential eigenvalue problem behaves as a singularly perturbed one, or when the coefficients of the differential equation present discontinuities.

In all cases we obtain results of high accuracy which compare favourably with results reported in the recent literature.

2. The Tau method

Let $U = \{u_i(x)\}$, $i \in N = 0,1,2,\ldots$, be a polynomial basis where $u_i(x)$ is a polynomial of degree i. Let $\mathcal{D}$ be the class of linear differential operators D with polynomial coefficients and let ν be the order of D, i.e. the maximum order of differentiation in the expression of D.

We give now a brief account of Ortiz' algebraic theory of the Tau method, more details can be found in [7 - 8]. Let us call the <u>height</u> h of an operator $D \in \mathcal{D}$ the maximum difference, for all $i \in N$, between the degrees of the polynomials $Du_i(x)$ and $u_i(x)$. For reasons that will be clear later on, we call the $Du_i(x)$ <u>generating polynomials</u>. Any operator $D \in \mathcal{D}$ ([7], Th.3.2) can be uniquely associated with a sequence $Q = \{q_i(x)\}$ of <u>canonical polynomials</u> $q_i(x)$ such that

$$\forall\, i \in N - S, \qquad Dq_i(x) = u_i(x) + r_i(x),$$

where

$$r_i(x) \in \operatorname*{span}_{i \in S}\{u_i(x)\} := R_S;$$

S is a finite set of indices j such that no linear combination of elements of U is mapped by D into a polynomial of degree j. The number of elements of S is bounded by $h+\nu$, as h measures the maximum rise in the degree of a polynomial when D is applied to it, and ν fixes the maximum possible number of polynomials $Du_i(x)$ of degree less than $i+h$.

The elements of Q are linked by a simple recursive relation which follows from the general form of generating polynomials. If

$$Du_i(x) = \sum_{\kappa=0}^{i+h} \alpha_\kappa^{(i)} u_\kappa(x),$$

then formally

$$q_{i+h}(x) = \frac{1}{\alpha_{i+h}^{(i)}}\Big[u_i(x) - \sum_{\kappa=0}^{i+h-1} \alpha_\kappa^{(i)} q_\kappa(x)\Big]$$

provided that $\alpha_{i+h}^{(i)} \neq 0$. If $\alpha_{i+h}^{(i)} = 0$ and $\alpha_{i+h-1}^{(i)} \neq 0$ we have two expressions for $q_{i+h-1}(x)$ which, by elimination, lead to a previously undefined canonical polynomial or to an element of $\operatorname{Ker}(D)$. As the number of cases for which $\alpha_{i+h}^{(i)}$ may be equal to zero, for $i \in N$, is $\leq \nu$, the set S is constructed in a finite (and very small) number of steps A computer program which implements this process is described and listed in [13].

Given the differential problem

$$(2.1) \qquad Dy(x) = 0, \quad (\ell_j, y) - g_j, \quad j = 1(1)\nu,$$

for x in the compact interval $[a,b]$, where ℓ_j, $j = 1(1)\nu$, are linear functionals acting on $y(x)$ which represent the supplementary conditions imposed on y for the solution to be unique, we associate with it the Tau problem

(2.2) $$Dy_n(x) = H_n(x), \quad (\ell_j, y) = g_j, \quad j = 1(1)\nu;$$

where

$$H_n(x) = \sum_{i=t}^{n} \tau_i u_i(x), \qquad t = n+1-(s+\nu),$$

is a polynomial of degree n, which depends on $s+\nu$ free parameters τ_i. Let us assume for simplicity that $i = t(1)n \notin S$, on account of the functional relation satisfied by the $q_i(x)$, the exact polynomial solution of (2.2) is simply given by

(2.3) $$y_n(x) = \sum_{i=t}^{n} \tau_i q_i(x) \ .$$

If D is applied to $y_n(x)$ and $R_s \neq \phi$, residual terms $r_i(x) \in R_s$ will be present in the expression of $Du_i(x)$. We shall designate by ψ_i s linear conditions required to eliminate such terms. Thus $y_n(x)$, with its $s+\nu$ coefficients τ_i chosen such that the $s+\nu$ conditions

(2.4) $$\begin{cases} \psi_i = 0, & i = 1(1)s \ ; \\ (\ell_j, y_n) = g_j, & j = 1(1)\nu \end{cases}$$

are satisfied, is the required solution of (2.2). We call it a Tau approximant of order n to problem (2.1).

If (2.1) is an inhomogeneous differential equation with a right hand side $f(x) = f_0u_0(x) + \ldots + f_mu_m(x)$, $m \leq n$, a term of the form $f_0q_0(x) + \ldots + f_mq_m(x)$ should be added to the expression of $y_n(x)$ in (2.3).

$H_n(x)$ is usually chosen in such a way that it satisfies a minimality condition over $[a,b]$ or at prescribed points of it. Lanczos' choice of Chebyshev polynomials [5] for U is an attempt to satisfy the former condition in the uniform norm.

Because of the simplicity with which $y_n(x)$ can be expressed in terms of the elements of the sequence Q is that they are called canonical polynomials for D in the basis U. The canonical polynomials have other interesting properties: they do not depend on a specific interval, which makes a segmentation process extremely economic; they neither depend on the supplementary conditions, which gives great flexibility in the treatment of complex initial, boundary, or mixed conditions. If a Tau approximant of a higher order $m > n$ is required, only the canonical polynomials of orders $n+1,\dots,m$ need to be computed to produce $y_m(x)$.

It should be remarked that the construction of a Tau approximant does not involve any discretization, choice of trial functions, approximate quadratures, or large matrix inversions. The only matrix inversion required is that of the small $s+\nu$ system (2.4), which is independent of the order of the required approximation $y_n(x)$.

Chaves and Ortiz [2] developed along these lines a numerical technique for the approximate solution of eigenvalue problems defined by differential equations, and applied it to problems where the eigenvalue parameter appears linear or non-linearly in the equation. As the supplementary conditions g_j, $j = 1(1)\nu$, are now homogeneous, (2.4) leads to a $(s+\nu)\times(s+\nu)$ system of linear algebraic equations. From its secular equation

$$\text{(2.5)} \quad \det \begin{bmatrix} \psi_i, & i = 1(1)s; \\ (\ell_j, y_n), & j = 1(1)\nu \end{bmatrix} = 0,$$

or the equivalent generalized eigenvalue problem, approximate eigenvalues λ_i are determined, and with them approximate eigenfunctions $y_{n_i}(x,\lambda_i)$. Here again the order of the matrix is independent of the required order of approximation n.

It is clear that the role of $H_n(x)$ in (2.2) is to reduce the exact solution to a polynomial of degree $n+h$. In a recent paper, Ortiz and Samara [14] discuss an "operational" approach to the Tau method which is based on the construction of a vector of generating polynomials. This is done with the help of two related matrices

$$\mu = \begin{bmatrix} 0 & 1 & & & \\ & 0 & 1 & & \\ & & 0 & 1 & \\ & & & 0 & 1 \\ & & \cdots & & \end{bmatrix} \quad \text{and} \quad \eta = \begin{bmatrix} 0 & & & \\ 1 & 0 & & \\ & 2 & 0 & \\ & & 3 & 0 \\ & & \cdots & \end{bmatrix},$$

of very simple structure. The product $\eta^r \mu^s$ will transform the coefficients of a polynomial in the same way as $x^s(d^r/dx^r)$. Thus, any operator $D \in \mathcal{D}$:

$$D = \sum_{j=0}^{\nu} p_j(x) \frac{d^j}{dx^j} = \sum_{j=0}^{\nu} \sum_{k=0}^{k_j} p_{jk} x^k \frac{d^j}{dx^j}$$

can be expressed as a linear combination of matrix products of the form $\eta^j \mu^k$ with coefficients p_{jk}. Supplementary conditions can be interpreted as ν differential operators, each one representing a boundary condition, acting on y and evaluated at one or more points of $[a,b]$. Therefore, if the matrix associated to D is augmented by these vectors, a matrix P is obtained. Let $\hat{P}$ be the conjugate of P under the similarity transformation defined by U. An approximate solution of (1.2), identical to the one defined by (2.2), can be obtained considering the finite problem

$$(2.6) \qquad \hat{P}_n^t \underline{a}_n = \underline{W}_n$$

where $\hat{P}_n$ is a restriction of $\hat{P}$ to its n+1 first rows and columns and $\underline{W}$ is a vector, the elements of which are the right hand sides of the supplementary conditions, and of the equation, represented in U.

In the case of a differential eigenvalue problem defined by the differential operator

$$D = D_0 + \lambda D_1 + \lambda^2 D_2 + \dots + \lambda^r D_r, \quad \text{where } D_i \in \mathcal{D},\ i = 0(1)r,$$

equation (2.6) leads to a generalized eigenvalue problem of the form

$$\hat{P}^t \underline{a} = (A + \lambda B + \dots + \lambda^r F)\underline{a} = \underline{0}$$

which, as it is shown in [15], is equivalent to (2.5). It should be remarked that the ν vectors which represent the boundary conditions may involve the eigenvalue parameter λ. The construction of the algebraic eigenvalue problem for $\hat{P}_n$ is facilitated by the simple structure of the matrices η and μ. The matrix δ corresponding to $x^s(d^r/dx^r)$ is one with all its elements equal to zero except for those in a parallel to the main diagonal:

$$\delta = \left[\begin{array}{c|c} O_{rs} & O \\ \hline O & I^{\delta} \end{array}\right]$$

where I^{δ} is a diagonal matrix with elements $I^{\delta}_{ii} = (r+i)!/i!$, $i = 0,1,2,.$ and O_{rs} is a r x s zero matrix.

3. Numerical examples

Let us first illustrate our technique with a simple model problem:

$$\begin{cases} D\, y(x) := y''(x) + (\lambda+\lambda^2 x^2) y(x) = 0 \\ y(-1) = y(1) = 0, \quad -1 \leq x \leq 1, \end{cases}$$

(see Collatz [3]).

In this case, the matrix $M = \eta^2+\lambda I+\lambda^2\mu^2$ transforms the coefficient of any polynomial in the same way as the differential operator of (3.1). The two boundary conditions are represented by the vectors $(1,-1,1,-1,...)'$ and $(1,1,1,1,....)'$ respectively, thus

$$P = \begin{bmatrix} 1 & 1 & \lambda & 0 & \lambda^2 & & & & \\ -1 & 1 & 0 & \lambda & 0 & \lambda^2 & & & \\ 1 & 1 & 2 & 0 & \lambda & 0 & \lambda^2 & & \\ -1 & 1 & 0 & 6 & 0 & \lambda & 0 & \lambda^2 & \\ 1 & 1 & 0 & 0 & 12 & 0 & \lambda & 0 & \lambda^2 \\ \cdots & & & & & & & & \cdots \end{bmatrix};$$

U could be taken to be the Chebyshev basis for $-1 \leq x \leq 1$. Numerical results for this example are given in [2] and [15].

3.1 Constant and polynomial coefficients

In Table I we give numerical estimations of eigenvalues with our technique for second, fourth and sixth order differential equations, for n = 10 and n = 20. In this table, as in all others in this paper, we give figures rounded to N D for the approximation of the maximum order, but for all other orders of approximation retain only those figures which agree exactly with those of the former.

3.2 Nonpolynomial coefficients

We consider now two Sturm-Liouville problems where the coefficient of y(x) is a cosine function. In a first stage a Tau approximant of

TABLE I: Tau estimations of eigenvalues defined by differential problems of orders two, four and six, with constant or polynomial coefficients for n = 10, 20.

Differential problem	Approximate eigenvalues n = 10	Approximate eigenvalues n = 20
$(4 + x^2 - x^4)y''(x) +$ $+ \lambda(1 + 5x^2)y(x) = 0$ $y(-1) = y(1) = 0, \ -1 \leq x \leq 1$ $\lambda_1 = 6$, Collatz [3]	6.000000 000 16.404 39.2 67. 10	6.000000 000 16.404125 433 39.168728 541 66.802206 362 105.055214 329
$y^{IV}(x) + \lambda y''(x) + 4y(x) = 0$ $y(0) = y(\pi) = y''(0) = y''(\pi) = 0$ $0 \leq x \leq \pi$ $\lambda_1 = 5$	4.99999 5.000 10. 1	4.999999 997 4.999999 999 9.444444 444 16.250000 007 25.160002 884
$y^{VI}(x) + 49y''(x) +$ $+ \lambda(14y^{IV}(x) + 36y(x)) = 0$ $y(0) = y(\pi) = y''(0) = y''(\pi) =$ $y^{IV}(0) = y^{IV}(\pi) = 0$ $0 \leq x \leq \pi \ ; \ \lambda_1 = 1$	1.000000 1.00	0.999999 988 0.999999 989 1.000000 000 1.348066 318 1.917831 899

TABLE II: Tau approximate eigenvalues for a problem with nonpolynomial coefficients

Degree of the Tau approximant to the nonpoly-nomial coeff.	first	second	third	fourth	fifth	CP time (in secs.) required for all eigenval.
m = 4	0.4	2.0	4.	8.	12.	1.508
6	0.4900	2.0593	4.654	8.277	12.932	1.644
8	0.49004	2.0593	4.6544	8.27745	12.9322	1.849
10	0.49004	2.05935	4.65445	8.27745	12.93221	2.020

cos wx is found by using standard software for the Tau method [13]. We consider for cos wx the initial value problem

$$y''(x) + w^2 y(x) = 0, \quad y(0) = 1, \quad y'(0) = 0, \quad 0 \le x \le \pi.$$

We find Tau approximants of order m (and also degree m, as h = 0) for several values of m. Table II displays approximate eigenvalues of the differential eigenvalue problem $y''(x) + \lambda(2 + \cos x)y(x) = 0$, $y(0) = y(\pi) = 0$, obtained with n = 20, and m = 4(2)10 for cos x. In the last column we record the CP time in seconds required by the CDC CDC 7600 Imperial College computer system to evaluate all first five approximate eigenvalues. Bounds for the first eigenvalue, using Temple's theorem, were given by Collatz [3]; they are: $0.489891 \le \lambda_1 \le 0.490066$. Our estimations are within these theoretical bounds.

In the case of Mathieu equation $y''(x) + (\lambda - 2\cos 2x)y(x) = 0$, $y(0) = y(\pi) = 0$, $0 \le x \le \pi$ (odd periodic Mathieu functions) we obtain the results given in Table III.

TABLE III: Tau approximate eigenvalues for Mathieu's equation

Degree of the Tau approximant to the nonpolynomial coeff.	Approximate eigenvalues					CP time (in secs.) required for all eigenvalues
	first	second	third	fourth	fifth	
m = 4	0	3.	9.	16.	25.	1.526
6	-0.1	3.91	9.04	16.03	25.02	1.657
8	-0.1102	3.917	9.048	16.033	25.020	1.823
10	-0.11025	3.91703	9.04774	16.03299	25.02080	1.993

The next table reproduces results obtained in the same computer system by using the NAG Subroutine D02KAF [6] with a tolerance parameter TOL = 10^{-8}. Unlike ours, the NAG subroutine requires an initial guess. We have supplied good initial guesses; in the case of the first eigenvalue the guess is inside the theoretical error bound mentioned above, the NAG Subroutine output is outside of it. CP times in the same computer system as before are about 10 times larger than for our Tau technique.

TABLE IV: Same problem as in Table II solved now with the NAG subroutine DO2KAF, based on Prüfer's substitution and a shooting technique

Eigen-value	Initial guess	Approximate eigenvalue
first	0.49	0.556
second	2.06	2.089
third	4.65	4.664
fourth	8.28	8.280
fifth	12.93	12.933

The Mathieu equation $y''(x) + (\lambda - 2q \cos 2x)y(x) = 0$, $y'(0) = y'(\pi) = 0$, for $0 \leq x \leq \pi$ defines the even Mathieu functions. This equation has received considerable attention; recently Canosa [1] proposed an interesting method which enabled him to estimate the eigenvalues of this equation for a large range of values of q. In Table V we report the approximate eigenvalues, for q = 10, given by Ince [1], and Canosa's estimations by using his method with 500 steps. We also report our results obtained with the Tau technique using approximations of orders n = 19,18, with Tau approximations of orders 16 and 14 respectively for cos 2x.

TABLE V: Approximate eigenvalues of Mathieu's equation for q = 10

Eigen-value	Ince	Canosa 500 steps	Tau approximations	
			n=19, m=16	n=18, m=14
first	-13.9370	-13.9369	-13.9376	-13.9368
second	-2.39914	-2.39907	-2.398	-2.399
third	7.71727	7.71737	7.72	7.72
fourth	15.5028	15.5028	15.505	15.503
fifth	21.1046	21.1046	21.10	21.11

We will return to this problem in the next section, where a comparison with a Tau method segmented in steps will be made.

3.3 Infinite intervals

Let us consider the following semi-infinite problem

$$y''(x) + (\lambda - x)y(x) = 0, \quad y(0) = y(\infty) = 0, \quad 0 \leq x < \infty,$$

The general solution of this problem is known to be the Airy function $y(x) = k\,\mathrm{Ai}(x-\lambda)$; the first boundary condition fixes the eigenvalues as the solutions of the transcendental equation $\mathrm{Ai}(-\lambda) = 0$.

We approximate the eigenvalues of this problem by solving a series of Tau problems on the finite intervals $[0,d]$, where $d = 7(2)17$. Table VI shows results obtained for $n = 20$ and numerical estimates of the zeros of $\mathrm{Ai}(-\lambda) = 0$. It is clear that there exist a compromise between the amplitude d of the interval and the efficiency with which the eigenvalues are retrieved for a constant value of the approximating order n. A too small interval ($d < 9$ in our case) would not pick up sufficient information on the differential problem to give an accurate answer; on the other hand, a too large one (relative to the fixed n) would put a too heavy stress on the quality of our perturbation term ($d > 15$ here).

TABLE VI: Tau estimation of the eigenvalues of a differential problem defined over a semi-infinite interval

Eigen-values	Approximate eigenvalues						$\mathrm{Ai}(-\lambda)=0$ to 3D
	d = 7	d = 9	d = 11	d = 13	d = 15	d = 17	
first	2.338	2.338	2.338	2.338	2.338	2.338	2.338
second	4.088	4.088	4.088	4.088	4.088	4.087	4.088
third	5.538	5.521	5.521	5.521	5.521	5.523	5.521
fourth	6.982	6.789	6.786	6.787	6.787	6.780	6.787
fifth	8.725	7.982	7.946	7.946	7.944	7.985	7.944

Our second problem will be the quantum mechanics harmonic oscillator defined over the double-infinite interval $-\infty < x < \infty$ by

$$(3.2) \qquad y''(x) + (\lambda - c^2x^2)y(x) = 0, \quad y(-\infty) = y(\infty) = 0,$$

which we take with $c = 1/2$. The eigenvalues of this problem are given by $\lambda_k = (2k+1)c$, $k \in N$, and the normalized eigenfunctions corresponding to them are given by $y_k(x) = \exp(-cx^2/2)H_k(c^{\frac{1}{2}}x)$, where $H_n(x) = (-1)^h\exp(x^2)d^n(\exp(-x^2))/dx^n$ are Hermite's polynomials.

We solved a series of Tau eigenvalue problems over the finite intervals $[-d,d]$, for $d = 4(1)9$. For $n = 20$, $d \simeq 6$ seems to be the best interval length for all five approximate eigenvalues. Table VII displays the numerical results obtained for the first five approximate eigenvalues in each of the intervals $[-d,d]$, for $d = 6(1)9$.

TABLE VII: Approximate eigenvalues of problem (3.2), estimated on finite intervals of length 2d, for d = 6(1)9

n	Approximate eigenvalues			
	d = 6	d = 7	d = 8	d = 9
20	0.500000 1.50000 2.500 3.500 4.50	0.5000 1.500 2.50 3.50 4.5	0.500 1.50 2.5 3.5 4.	0.500 1.50 2.5 3.
30	0.500000 1.50000 2.5000 3.500 4.50	0.500000 1.500000 2.500000 3.50000 4.5000	0.500000 1.500000 2.50000 3.50000 4.5000	0.5000000 1.500000 2.5000 3.5000 4.50

4. Final remarks

Ortiz [11] has given examples on the use of his recursive formulation of the Tau method for the numerical approximation of the solution of certain types of functional differential equations. The technique discussed here can also be used in the estimation of eigenvalues for that kind of problems. Let us consider the functional differential equation

$$y''(x) + \lambda y(x/2) = 0, \quad y(-1) = y(1) = 0, \quad -1 \leq x \leq 1.$$

With a power series expansion Collatz [3] identified the eigenvalues $\lambda_1 = 2.09$, and $\lambda_2 = 13.1$. A Tau approximant of order 10 gives $\lambda_1 = 2.090663$ and $\lambda_2 = 13.054850$, estimations that are confirmed by Tau approximations of higher orders, more details are given in [12].

We have made experiments on the use of the step by step technique of [10] for the Tau method, and have been able to solve accurately problems where the coefficients of the equation present discontinuities, the solution converges very slowly to zero over an infinite interval, and in the treatment of singular perturbation problems.

At this point we wish to return briefly to Mathieu's equation ($q = 10$). In Table V ($n = 18$, $m = 14$) we recovered, with a single Tau approximant and reasonable accuracy, the values obtained by Canosa by using his technique over 500 steps. If we use a four step Tau approximant we recover exactly the values given by Canosa, with only $m = 6$, $n = 10$.

Mathieu's equation is associated with the singular perturbation problem [1]

$$\varepsilon y''(x) + (\Lambda - 2 \cos 2x) y(x) = 0, \text{ where } \varepsilon \equiv 1/q \to 0, \text{ and } \Lambda = \lambda/q$$

Canosa approximated the eigenvalues of this problem up to very large values of q and found good agreement with the asymptotic formulae of Goldstein and Ince (see [1]). For $q = 2500$ he uses 10 000 steps of his technique. With 5 Tau steps we recover his results with a relative error of 1.7×10^{-4}. Results in this direction will be published in a further note.

REFERENCES

1 CANOSA, J. Numerical solution of Mathieu's equation, J.Comp.Phys., 7, 255-272, 1971

2 CHAVES, T. and ORTIZ, E.L. On the numerical solution of two point boundary value problems for linear differential equations, Z.angew. Math.Mech. 48, 415-418, 1968

3 COLLATZ, L. Eigenwertaufgaben mit Technischen Anwendungen, Akademische Verlagsgesellschaft Geest & Portig K.-G., Leipzig 1963

4 GOULD, S.H. Variational methods for eigenvalue problems, London, Oxford University Press, 1966

5 LANCZOS, C. Trigonometric interpolation of empirical and analytical functions, J.Math.Phys., 17, 123-199, 1938

6 NAG, Mark III, Oxford, 1981

7 ORTIZ, E.L. The Tau method, SIAM J. Numer.Anal., 6, 480-492, 1969

8 ——— Canonical polynomials in the Lanczos Tau method, in Studies in Numerical Analysis, B.P.K. Scaife, Ed., pp.73-93, Academic Press, New York, 1974

9 ——— Sur quelques applications nouvelles de la méthode Tau, Seminaire Lions, IRIA, pp.247-257, Paris, 1975

10 ——— Step by step Tau method, Comp. & Math. with Appli., 1, 381-392, 1975

11 ——— On the numerical solution of nonlinear and functional differential equations with the Tau method, in Numerical Treatment of differential equations in Applications, LNinM, Eds., R. Ansorge and W. Tornig, pp.127-139, Springer-Verlag, Berlin, 1978

12 ——— Functional differential equations and the Tau method, in preparation

13 ORTIZ, E.L., PURSER, W.F.C. and RODRIGUEZ CAÑIZARES, F.J. Automation of the Tau method, Imperial College Res. Rep., 1-56, 1972

14 ORTIZ, E.L. and SAMARA, H. An operational approach to the Tau method for the numerical solution of non-linear differential equations, Computing, 27, 15-25, 1981

15 ——— An operational approach to the Tau method for the numerical solution of eigenvalue problems, Imperial College, Res. Rep. 02-06-81, 1981

SECTION A.3

OF

GENERAL $(A-\lambda B)$-PENCILS

ALGORITHMS FOR LARGE SPARSE $(A-\lambda I)$-PROBLEMS

The Two-sided Arnoldi Algorithm
for Nonsymmetric Eigenvalue Problems

by

Axel Ruhe
Institute of Information Processing
University of Umeå
S-90187 Umeå, Sweden

Abstract

Algorithms for computing a few eigenvalues of a large nonsymmetric matrix are described. An algorithm which computes both left and right eigenvector approximations, by applying the Arnoldi algorithm both to the matrix and its transpose is described. Numerical tests are reported.

1. Introduction and Summary

We are interested in developing algorithms for large sparse nonsymmetric matrix eigenvalue problems.

(1.1) $Cx = x\lambda.$

(We follow the notational convention of Parlett of letting symmetric letters stand for symmetric matrices and vice versa.) Typical applications are large initial value problems, econometric models, and finite element approximations to systems with damping, such as acoustic systems.

In the symmetric case, Spectral Transformation has proved to be a fruitful approach, and we believe that to be true also now. Let us assume that we seek all λ belonging to a region Ω of the complex plane. We then select a set of shift points $\mu_1, \mu_2, \ldots\ldots \in \bar{\Omega}$ and seek the eigenvalues of

(1.2) $B = (C - \mu I)^{-1}$

Denoting them by ν_k, we see that

$$\nu_k = \frac{1}{\lambda_k - \mu}, \tag{1.3}$$

and eigenvalues of λ_k close to μ are mapped onto ν_k of a large absolute value.

For each shift μ, we now factorize the matrix $C-\mu I$, and then apply an iterative algorithm to $(C-\mu I)^{-1}$, in order to get the eigenvalues of large absolute values. This gives us approximations to the original eigenvalues λ_k close to μ, and if we cover Ω with μ:s in an appropriate way, we will get approximations to all eigenvalues in Ω.

It is clear from this outline that the nonsymmetric problem is much more difficult than the symmetric one. Then we had an interval, and could make steps with μ from one end to the other. Moreover, we could use the inertia of the shifted matrix to indicate whether all eigenvalues had been found. For a detailed account, see [4].

In the present contribution, we will discuss possible alternatives for the choice of an iterative algorithm to be used as inner iteration of this scheme. That is, we seek the eigenvalues of large absolute value. We assume that the shift has been chosen so that they are rather well separated from the rest of the spectrum, and will converge in few iterations. Typically we should use say 50 iterations for a problem of $n = 1000$. In the computations, the heaviest work is the matrix vector multiplication, involving the factorized matrix. When developing an algorithm, we should try to reduce the number of iterations as much as possible, but can afford to use an elaborate iterative algorithm involving, e.g. reorthogonalizations.

In the nonsymmetric case, we must compute both a right and a left eigenvector approximation, in order to be able to assess the accuracy of our result. Kahan, Parlett and Jiang [5] have studied how that can be done.

We cite ([5] Theorem 2'):

Let B and two unit vectors p^H and q, with $p^H q \neq 0$ be given. For any γ define residual vectors

$$r = Bq - q\gamma, \quad u^H = p^H B - \gamma p^H.$$

Then (γ, p^H, q) are an eigentriple of B-E, and the norm of the perturbation E can be bounded by

$$(1.4) \quad ||E|| = \max \{||r||, ||u^H||\}.$$

Furthermore, we can (for small perturbations) estimate the distance from γ to an eigenvalue λ of B by

$$|\lambda - \gamma| \leq (p^H q)^{-1} ||E|| + O(||E||^2)$$

If no such p^H, q exist, the situation is considerably more complicated and we postpone the discussion of that.

Two algorithms come to mind as obvious alternatives. One is the Arnoldi algorithm [1], which yields an orthonormal basis in which the matrix is represented in Hessenberg form. It has been studied by Saad [8]. The other is the nonsymmetric Lanczos [6], which yields a pair of biorthogonal bases and a tridiagonal representation. In a recent contribution, Parlett and Taylor [7] have shown how to eliminate the breakdown that often precludes a successful termination of that algorithm. We will describe these algorithms in more detail in Section 2 of this report, and then formulate the two-sided Arnoldi algorithm, which is developed to yield both left and right eigenvectors. In Section 3, we continue by discussing how the residuals should be adjusted to give a consistent approximation to an eigentriple. We conclude with a small numerical example in Section 4. The algorithm is yet in a development stage and the example just indicates how it can behave in a case we believe to be typical.

2. The Algorithms

Let us now formulate the three algorithms we consider, and start with the Arnoldi algorithm.

<u>Algorithm A</u> (Arnoldi)

1. Start with v_1 unit length vector.
2. For $j = 1, 2, \ldots$ do
   (1) $r := Bv_j$ (Operate)
   (2) For $i = 1, 2, \ldots, j$ do
   $r := r - v_i \times (h_{ij} := v_i^H r)$ (Orthogonalize)

(3) $v_{j+1} := r/(h_{j+1j} := ||r||_2)$ (Normalize)

We see that after j steps it holds,

(2.1) $$BV_j - V_j H_{jj} = h_{j+1j} v_{j+1} e_j^T ,$$

and that we can use eigenvectors of the Hessenberg matrix H_{jj} to get approximative eigenvectors of B. If

$$H_{jj} s = s\mu, \quad y = V_j s$$

the residual is

(2.2) $$r = By - y\mu = BV_j s - V_j H_{jj} s = h_{j+1j} s_j v_{j+1}$$

$$||r|| = |h_{j+1j} s_j|$$

a bound analogous to what can be obtained for the symmetric Lanczos algorithm.

v_j makes up an orthonormal basis of the Krylov space,

$$K_j(B) = \{v_1, Bv_1, \ldots\ldots, B^{j-1} v_1\},$$

and the residual of the approximate eigenvector is orthogonal to that space.

The amount of work needed is, assuming that B has an average r_B filled elements in each row, n (r_B+2j+2) multiplicative operations in step j, and in all

(2.3) $$nj\{r_B+j+3\}$$

to perform all the steps to get V_j, H_{jj}.

It should be noted that reorthogonalization is not necessary in order to keep V_j orthogonal; this is in contrast to the symmetric Lanczos algorithm. Step 2.2 of algorithm A corresponds to a Modified Gram Schmidt orthogonalization of A,

$$A = [v_1 | BV_j] = V_{j+1} [e_1 | H_{j+1j}] = QR$$

where the last factor is upper triangular. The analysis of Björck ([2] formula (6.8)) bounds the loss of orthogonality due to rounding errors by

$$||I-\bar{Q}^T\bar{Q}||_2 \leq f(n,j)\eta||A||_E||\bar{R}^{-1}||_2 ,$$

where $\bar{Q}$ and $\bar{R}$ denote computed quantities, f(n,j) is a low degree polynomial, and η is the rounding unit. We see that $||R^{-1}||$ does not grow until Bv_j is nearly linearly dependent to V_j. We can monitor this growth by using the recurrence

$$R_{j+1} = \begin{bmatrix} e_1 & H_{j+1j} \end{bmatrix} = \begin{bmatrix} 1 & H_{jj-1} & h_j \\ 0 & 0 & h_{j+1j} \end{bmatrix}$$

$$||R_j^{-1}||_2 \leq \rho_j \quad \text{where}$$

$$\rho_1 = 1$$

$$\rho_{j+1} = \rho_j(1+||h_j|| \, |h_{j+1j}|^{-1}) + |h_{j+1j}|^{-1}$$

or use a method similar to the LINPACK condition estimator [3] to get a smaller estimate. When orthogonality finally gets lost, we can stop the algorithm, since then <u>all</u> eigenvalues have converged. In the symmetric Lanczos algorithm, orthogonality gets lost as soon as <u>one</u> eigenvalue has converged.

The main deficiency of the Arnoldi algorithm is the fact that it does not yield a good approximation to the left eigenvector. Assume that the process is run to completion (which is always possible in theory), so that

$$V = (V_j, V_{n-j})$$

is an orthonormal matrix. Then

$$V^H BV = \begin{bmatrix} H_{jj} & H_{jn-j} \\ {}_0{}^{h_{j+1j}} & H_{n-jn-j} \end{bmatrix}$$

where the right block column contains elements that are yet unknown after j steps. The best left vector we can get is obtained from a left vector of H_{jj} ,

$$t^H H_{jj} = \mu t^H, \quad z^H = t^H V_j^H ,$$

but it has the residual,

$$\begin{aligned} q^H &= z^H B - \mu z^H \\ &= t^H V_j^H B - t^H H_{jj} V_j^H \\ &= t^H H_{j\,n-j} V_{n-j}^H \end{aligned}$$

which is not small in norm, unless the entire $H_{j\,n-j}$ block contains small elements, or is rank deficient as in the symmetric case.

The nonsymmetric Lanczos algorithm gets both left and right vectors by building up bases of both left and right Krylov spaces. It is formulated as

Algorithm L (Nonsymmetric Lanczos)

1. Start with q_1 and p_1^H satisfying $p_1^H q_1 = 1$. $\beta_1 = \gamma_1 = 0$

2. For j=1,2,..., do

   (1) $r := Bq_j - q_{j-1}\gamma_j$ (Operate right vector)

   (2) $r := r - q_j \times (\alpha_j := p_j^H r)$ (Biorthogonalize)

   (3) $s^H := p_j^H B - \beta_j p_{j-1}^H - \alpha_j p_j^H$ (Operate left vector)

   (4) Determine $\beta_{j+1}, \gamma_{j+1}$ such that $\beta_{j+1}\gamma_{j+1} = s^H r$

   (5) $q_{j+1} := r/\beta_{j+1}$, $p_{j+1}^H := s^H/\gamma_{j+1}$ (Normalize)

The $\alpha_j \beta_j \gamma_j$ build up a tridiagonal matrix,

$$J_j = \begin{bmatrix} \alpha_1 & \gamma_2 & 0 & & 0 \\ \beta_2 & \alpha_2 & \gamma_3 & & 0 \\ 0 & \beta_3 & \alpha_3 & & \\ & & & & \gamma_j \\ 0 & 0 & & \beta_j & \alpha_j \end{bmatrix}$$

and we have,

$$BQ_j - Q_j J_j = \beta_{j+1} q_{j+1} e_j^T$$

$$P_j^H B - J_j P_j^H = \gamma_{j+1} e_j p_{j+1}^H \ .$$

Now both left and right approximative eigenvectors can be obtained. If

$$J_j s = s\mu, \ y = Q_j s$$

then the residual

$$r = By - y\mu = \beta_{j+1} s_j q_{j+1}$$

$$||r|| = |\beta_{j+1} s_j| \ ||q_{j+1}|| \ ,$$

and if

$$t^H J_j = \mu t^H, \ z^H = t^H P^H$$

then the left residual

$$u^H = z^H B - \mu z^H = \gamma_{j+1} t_j p_{j+1}^H$$

$$||u^H|| = |\gamma_{j+1} t_j| \ ||p_{j+1}^H|| \ .$$

Note that β_{j+1}/γ_{j+1} can be chosen freely, but that neither $||y||$, $||z^H||$, $||q_{j+1}||$, nor $||p_{j+1}{}^H||$ can be set to unity. If both norms $||r||$ and $||u^H||$ are small, we can bound the perturbation of B having μ as exact eigenvalue (1.4), see [5].

The amount of work to perform j steps is now, under the same assumptions as leading to (2.3),

$$nj\{2r_B+8\} \tag{2.4}$$

multiplicative operations, if no reorthogonalization is performed. If one set of vectors is reorthogonalized, we need

$$nj\{2r_B+j+9\} \ . \tag{2.5}$$

If no reorthogonalization is performed, we lose orthogonality as soon as one vector converges. On the other hand, we need only reorthogonalize either of the sets Q_j and $P_j{}^H$, not both. See the analysis by Saad [9].

As is well known, the algorithm may break down in step 2.4, due to that s^Hr may become zero. This does not depend on lack of orthogonality, but on an ill-fated choice of starting vectors. The remedy is to use the look ahead device by Parlett and Taylor [7]. Note that this look ahead device does not obviate the need for reorthogonalization. It makes a marginal addition of less than 5n multiplicative operations each step.

We have seen that the main deficiency of the Arnoldi algorithm is the fact that it does not give any good approximations to left vectors. On the other hand, nonsymmetric Lanczos is a waste of effort when B is close to normal, since then it should be enough with j matrix vector operations, not 2j. The two sided Arnoldi finds both left and right vectors and performs only as many left operations as necessary.

<u>Algorithm TSA</u> (Two sided Arnoldi)

1. Perform algorithm A (Arnoldi) until (2.1) indicates that right vector has converged, giving

$$BV_j - V_jH_{jj} = h_{j+1j}v_{j+1}e_j{}^T$$

2. Compute left eigenvector approximation.

$$t^H H_{jj} = \mu t^H, \; z^H = t^H V_j^H$$

3. Perform algorithm A from the left

(1) $w_1^H = z^H$

(2) Compute W_k^H, orthonormal and K_{kk} lower Hessenberg, such that

$$W_k^H B - K_{kk} W_k^H = k_{kk+1} e_k w_{k+1}^H$$

(3) Stop when left vector has converged indicated by

$$t^H K_{kk} = \nu t^H$$

$|t_k \; k_{kk+1}|$ small

4. Adjust vectors so that adjusted residuals v_{j+1}' and w_{k+1}' satisfy

$$w_{k+1}'^H V_j = 0 \qquad W_k^H v_{j+1}' = 0$$

if that is possible.

We see that if algorithm TSA is applied to a Hermitian matrix, it will stop at the first step in part 3, since then $k_{11}=\mu$, and $k_{12}=||z^H B-\mu z^H||$ $= ||r||$, the same quantity (2.1) that stopped part 1. If it is applied to a normal matrix, the same thing would happen, if we had chosen the right eigenvector of H_{jj} in part 2, note that H_{jj} is not normal even if B is. We anyhow retain the left vector, since it will be a better starting vector in non-normal cases.

If more than one eigenvalue is sought, we could either make several starts of part 3, with different left vectors, or take a linear combination in part 2, and make a run until all have converged in part 3.

Note that neither algorithms A nor TSA may break down in the way algorithm L can do. However, the deficiency that led to breakdown will put up its head in a slightly milder form in the adjustment phase as we will soon explain.

3. Adjustment of Residuals

We notice that the result from parts 1-3 of algorithm TSA leaves something to be desired. We get two different eigenvalue approximations, μ and ν, from the right and left iteration. Moreover, the residuals are orthogonal to the wrong subspaces.

For μ to be a Rayleigh quotient

$$\rho(q,p^H) = p^H Bq/p^H q,$$

it is necessary that the right residual is orthogonal to the left vector and vice versa

$$p^H(Bq-q\mu) = 0$$

After algorithm TSA, p^H is a vector in the subspace spanned by W_k^H, while the right residual is orthogonal to V_j^H.

Let us see if we can make all right residuals orthogonal to all left vectors i.e. to the entire left subspace W_k^H.

We see from (2.1) that an appropriate way to go about this is to change v_{j+1} by subtracting a vector in the V_j subspace, orthogonalizing it against W_k^H,

$$\begin{aligned} v'_{j+1} &= v_{j+1} - V_j c \\ &= (I - V_j (W_k^H V_j)^- W_k^H) v_{j+1} \quad . \end{aligned}$$

The reader is referred to the appendix for a detailed description of an algorithm for such an orthogonalization by oblique projection.

It is necessary for this orthogonalization to be possible, that $W_k^H V_j$ has linearly independent rows, this is equivalent to the condition that the nonsymmetric Lanczos should not break down. In this algorithm we do not get a breakdown; we only get orthogonalization to a subspace $W_{k'}^H$ with $k'<k$. However, we see that

$$||v'_{j+1}|| > ||v_{j+1}||,$$

since $v_j^H v_{j+1} = 0$, and the growth may be large if $W_k^H V_j$ is close to being rank deficient.

The adjustment of v_{j+1} corresponds to changing H_{jj}. Subtracting $h_{j+1j} V_j c e_j^T$ from both sides of (2.1), we get

$$BV_j - V_j\{H_{jj} + h_{j+1j} c e_j^T\} = h_{j+1j}\{v_{j+1} - V_j c\} e_j^T$$

$$BV_j - V_j H'_{jj} = h_{j+1j} v'_{j+1} \; e_j^T \quad ,$$

and we see that the last column of H_{jj} has to be changed. This gives a new eigenvalue and eigenvector approximation μ' and s', and now taking

$$y' = V_j s', \; r' = By' - y'\mu' \quad ,$$

we have gotten the residual orthogonal

$$W_k^H, \; r' = 0$$

and μ' a Rayleigh quotient

$$\mu' = z^H By'$$

We can perform a corresponding adjustment of the left residual s^H and K_{kk}. One thing is important to notice, however, namely that already the starting vector of part 3 has its residual orthogonal to V_j. For each step we destroy this orthogonality towards one dimension of V_j, and after k steps we only need to adjust it towards

$$V_j (V_j^H W_k)$$

which has k columns.

4. A Numerical Example

We have as of now only very scant numerical experience of the algorithm gathered from running an experimental program on artificial test examples. The following runs were performed with a FORTRAN program using EISPACK and LINPACK routines whenever applicable, on an IBM 3081 computer at Stanford Linear Accelerator Center. All computations were performed in double precision.

The Arnoldi algorithm is independent of orthogonal transformations; we therefore have chosen to test it on a triangular matrix B having elements

$$b_{ij} = \begin{Bmatrix} q & i<j \\ 1/i & i=j \\ 0 & i>j \end{Bmatrix}$$

The eigenvalue distribution is representative of what one gets after a spectral transformation of a matrix with real eigenvalues. The quantity q is used to vary the departure from normality.

Iterations are started with a random starting vector consisting of normally distributed random numbers. (Note that uniformly (rectangularly) distributed random numbers does not yield a random direction, a fact not always appreciated but explained in [4]).

Though the theory of Arnoldi runs parallel to that of symmetric Lanczos, several useful observations do not carry over. Even though B has real eigenvalues only, it is the rule rather than the exception that H_{jj} has complex pairs in those positions that have not yet converged. There is hardly a trace left of the interlacing property of the symmetric submatrix eigenvalues, quite to the opposite, when the departure from normality is large the low order approximations H_{jj} generally have large eigenvalues compared to B. We have plotted the eigenvalues of a typical case in figure 1.

A disturbing fact was that it was often impossible to complete the OMGSA algorithm due to linearly dependent rows in $W_k^H V_j$. Moreover, the simple algorithm described in the appendix often gave rise to an L matrix with only moderately small diagonal elements, even though $W_k^H V_j$ was very close to singular. We therefore used a variant where W_k^H was permuted, in order to get a normalized L. Then L always got its smallest diagonal element last and the rest was well conditioned. When OMGSA did not succeed, we used the Rayleigh quotient of the adjusted eigenvectors as a refined eigenvalue approximation.

A series of runs for different values of the departure from normality q is summarized in Table 1. The tolerance used to stop in (2.2) was set to 10^{-5}. All matrices were of order n=50; the convergence behavior did not depend much on n.

Acknowledgements: Beresford Parlett and Yocef Saad have generously supplied me with yet unpublished material. The last part of the work was performed at Stanford, where I enjoyed the hospitality of Gene Golub and his group, notably Mr. Douglas Baxter, who introduced me to the intricacies of the computer facilities there.

TABLE 1

Summary of results of test runs for different departure from normality q

Order n = 50 in all cases

Departure	$\lambda_1 = 1.000$			$\lambda_3 = 0.3333$		
	Condition	Iterations		Condition	Iterations	
q	$\frac{1}{\lvert y_1^H x_1 \rvert}$	j	k	$\frac{1}{\lvert y_3^H x_3 \rvert}$	j	k
0	1	10	1	1	15	3
0.001	1	10	5	1	15	10
0.01	1	10	5	1.2	15	10
0.1	35	15	10	$3 \cdot 10^4$	20	15
0.2	$5 \cdot 10^3$	20	10			
0.5	$7 \cdot 10^6$	20	15			

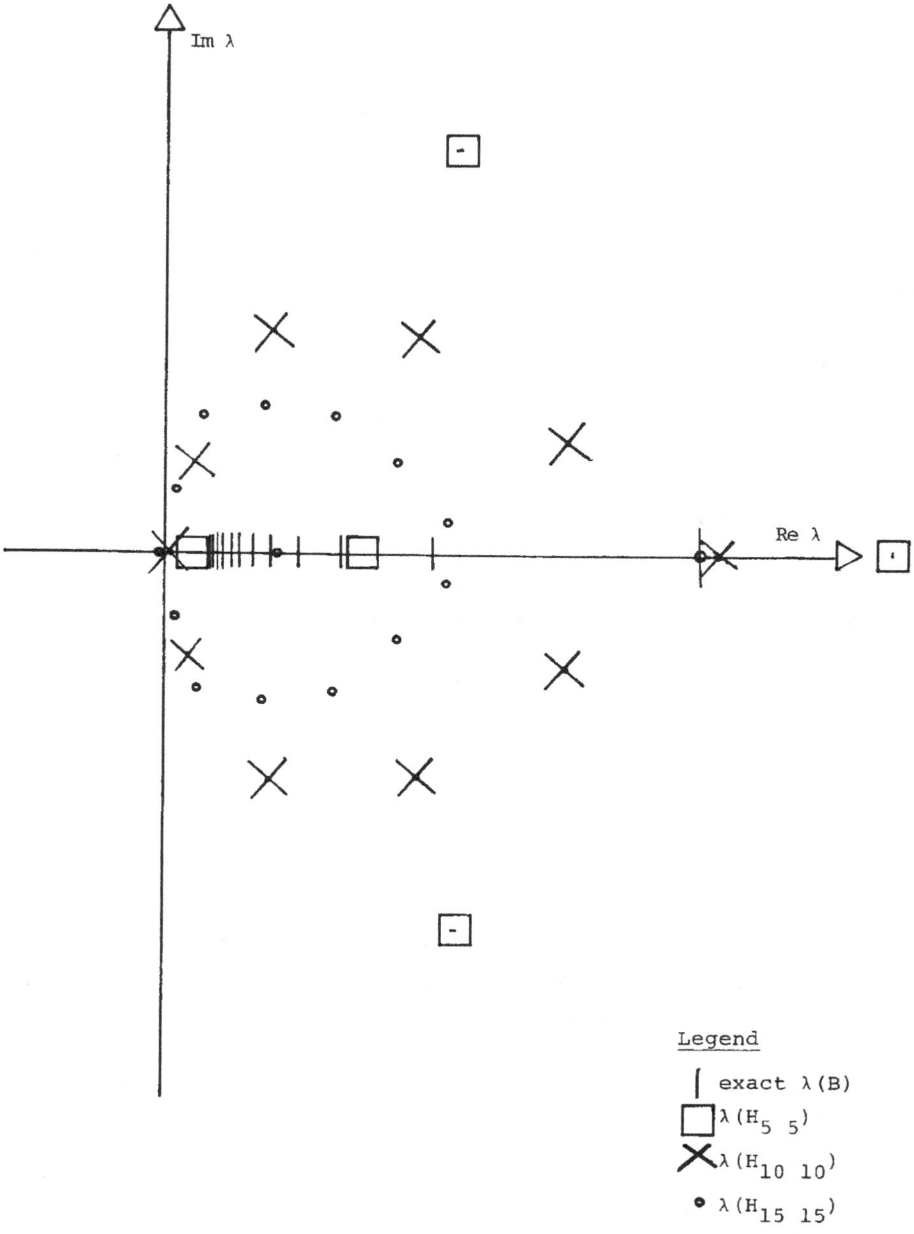

Figure 1 - Eigenvalues for H_{jj} for different j compared to those of B. Test example with n = 50 and departure from normality q = 0.1.

APPENDIX

The Modified Gram Schmidt algorithm for orthogonalization by oblique projections

The aim is to make a given vector y orthogonal to a subspace spanned by the columns of W_k, an $n \times k$ matrix, by means of subtracting a suitable linear combination of the columns of V_j, another $n \times j$ matrix. We see immediately that the new y is,

$$y' = (I - V_j(W_k^H V_j)^- W_k^H)y$$

with X^- denothing a generalized right inverse of X satisfying $XX^- = I$, and that we get a minimal distance between y and y' if the pseudo-inverse is chosen. The orthogonalization is not possible to perform if $W_k^H V_j$ has linearly dependent rows, especially we have to assume that $k \leq j$.

Assume that we perform the Gram Schmidt LQ decomposition

$$W_k^H V_j = LQ^H, \quad \begin{array}{l} L \text{ lower triangular} \\ Q \text{ unitary} \end{array}$$

we get

$$y' = (I - V_j Q L^{-1} W_k^H)y \ .$$

The classical Gram Schmidt algorithm now computes y' this way, by first forming $W_k^H y$, then a forward substitution to get $L^{-1}W_k^H y$, then multiply by Q, and last making a linear combination of V_j. We also note that Q and L can be formed one column at a time for increasing k.

The notable fact now is that the modified Gram Schmidt algorithm does not need a forward substitution and makes no explicit use of the non-diagonal elements of L. Then y is successively orthogonalized to w_1, w_2,....until w_k. If $w_1^H y = \ldots . w_{i-1}^H y = 0$ then $W_i^H y = (0,\ldots,w_i^H y)^T$, and $L^{-1}W_i^H y = (0,\ldots,l_{ii}^{-1} w_i^H y)^T$, and finally

$$V_j Q L^{-1} W_i^H y = V_j q_i \cdot l_{ii}^{-1} w_i^H y,$$

only using the i:th column of Q and W_k. This is a considerable simpl fication, compared to the straightforward CGSA approach and is also likely to be more stable numerically.

In algorithmic notation we get:

Algorithm OMGS (Oblique Modified Gram Schmidt)

1. Start $y^{(0)}:=y$
2. for $i:=1$ to k do

   (1) $q_i:=V_j^H w_i$

   (2) for $\ell:=1$ to $i-1$ do

      1. $q_i:=q_i-q_\ell \times (l_{i\ell}:=q_\ell^H q_i)$

   (3) $q_i:=q_i/(l_{ii}:=||q_i||)$

   (4) $y^{(i)}:=y^{(i-1)}-V_j q_i l_{ii}^{-1} w_i^H y^{(i-1)}$
3. Now $y^{(k)}$ satisfies $W_k^H y^{(k)}=0$

Breakdown may occur in step 2.(3) if $W_i^H V_j$ does not have linearly independent rows.

REFERENCES

1. W. E. ARNOLDI, The principle of minimized iteration in the solution of the matrix eigenvalue problem, Quart. Appl. Math. 9:17-29, (1951).

2. Å. BJÖRCK, Solving linear least squares problems by Gram-Schmidt orthogonalization, BIT 7,1-21 (1967).

3. A. K. CLINE, C. B. MOLER, G. W. STEWART and J. H. WILKINSON, An estimate for the condition number of a matrix, SIAM J. Numer. Anal. 16,368-375 (1979).

4. T. ERICSSON and A. RUHE, The spectral transformation Lanczos method for the numerical solution of large sparse generalized symmetric eigenvalue problems, Math Comp. 35, 1251-1268 (1980).

5. W. KAHAN, B. N. PARLETT and E. JIANG, Residual bounds on approximate eigensystems of nonnormal matrices, SIAM J. Numer. Anal. 19, 470-484(1982).

6. C. LANCZOS, An iteration method for the solution of the eigenvalue problem of linear differential and integral operators, J. Res. Nat. Bur. of Standards 45:255-282 (1950).

7. B. N. PARLETT and D. TAYLOR, A look ahead Lanczos algorithm for unsymmetric matrices, Technical Report PAM-43, Center for Pure and Applied Mathematics, Berkeley (1981).

8. Y. SAAD, Variations on Arnoldi's method for computing eigenelements of large unsymmetric matrices, Linear Algebra and its Applications 34:269-295 (1980).

9. Y. SAAD, The Lanczos biorthogonalization algorithm and other oblique projection methods for solving large unsymmetric systems, SIAM J. Numer. Anal. 19, 485-506(1982).

10. J. H. WILKINSON, The algebraic eigenvalue problem, Clarendon Press, Oxford (1965).

PROJECTION METHODS FOR SOLVING LARGE SPARSE EIGENVALUE PROBLEMS

Youcef Saad
Computer Science Department,
Yale University
New Haven, CT. 06520, USA[1]

Abstract.

We present a unified approach to several methods for computing eigenvalues and eigenvectors of large sparse matrices. The methods considered are projection methods, i.e. Galerkin type methods, and include the most commonly used algorithms for solving large sparse eigenproblems like the Lanczos algorithm, Arnoldi's method and the subspace iteration. We first derive some a priori error bounds for general projection methods, in terms of the distance of the exact eigenvector from the subspace of approximation. Then this distance is estimated for some typical methods, particularly those for unsymmetric problems.

1. Introduction. In the previous few years a fairly important effort has been devoted to solving large sparse eigenvalue problems. Although more attention has been directed towards symmetric eigenvalue problems, many applications are now encountered where one requires the eigenvalues of a large unsymmetric matrix.

The purpose of this paper is to attempt to present a unified view of the most commonly used algorithms for solving large sparse eigenvalue problems. We will start by reviewing the general framework of projection methods and describe orthogonal as well as oblique projection methods. A projection method consists in approximating the exact eigenvector u, by a vector $\tilde{u}$ belonging to some subspace K, referred to as the right subspace, by requiring that the residual vector of $\tilde{u}$ satisfies the Petrov-Galerkin condition that it is orthogonal to some subspace L, possibly different from K, often referred to as the left subspace. When L=K we have an orthogonal projection method otherwise we say that the method is an oblique projection method. As it turns out most methods for solving large sparse eigensystems can be formulated in terms of projection methods . As will be seen, the

[1] This work was supported in part by the U.S. Office of Naval Research under grant N000014-76-C-0277 and in part by NSF Grant MCS-8104874

common feature which makes these methods work, is that the exact eigenvector is well approximated by some vector of the subspace K. It then becomes important to analyse the distance between the exact eigenvector and the subspace of approximation. This will be done for several methods with emphasis on those for solving unsymmetric problems, including the method of subspace iteration and the method of Arnoldi.

Concerning the subspace iteration method we will see that Chebyshev acceleration can also be efficiently used in the unsymmetric case and we will derive some estimates of the convergence factor.

Throughout the paper, the norm $\| \ \|$ represents the Euclidean norm. The spectrum of a matrix is denoted by $\sigma(A)$. The matrices treated may be complex and the transpose conjugate of A is denoted by A^H.

2.__General projection methods for matrix eigenvalue problems.__ In this section we present the general projection methods which provide a unified approach to many methods for computing eigenvalues and eigenvectors of large matrices.

2.1. Orthogonal projection methods. The material of this subsection summarizes part of the previous paper [11]. Let A be an NxN complex matrix and K be an m-dimensional subspace of $\mathbb{C}^N$. We will make the notational convention of representing by the same symbol A the matrix and the linear operator represented by the matrix A.

Consider the eigenvalue problem

$$A\,u = \lambda\,u \tag{2.1}$$

An orthogonal projection method on the subspace K seeks an approximate eigenpair $\tilde{\lambda}$, $\tilde{u}$ to problem (2.1), which belongs to the subspace K and such that the following Galerkin condition is satisfied.

$$A\,\tilde{u} - \tilde{\lambda}\,\tilde{u} \perp K \tag{2.2}$$

In terms of projection operators, if P represents the orthogonal projector onto the subsapce K, then the above Galerkin condition (2.2) can be rewritten as

$$P\,(\,A\,\tilde{u} - \tilde{\lambda}\,\tilde{u}\,) = 0 \tag{2.3}$$

We will refer to (2.2) or (2.3) as the approximate problem. Assuming that we have an orthonormal basis $V = [v_1, v_2, \ldots v_m]$ of K, we can solve the approximate problem (2.2) by expressing the approximation $\tilde{u}$ in the basis V as

$$\tilde{u} = V y \tag{2.4}$$

in which case $\tilde{\lambda}$ and y constitute an eigenpair of the m dimensional eigenproblem derived from (2.2):

$$B_m y = \tilde{\lambda} y \tag{2.5}$$

with

$$B_m = V^H A V \tag{2.6}$$

We will denote by A_m the <u>linear application of rank m, defined by A_m=PAP</u> . Note that the restriction of A_m to K is represented by the matrix B_m with respect to the basis V. An important quantity for the convergence properties of the method is the distance $\|(I-P)u\|$ of the exact eigenvector u, which is supposed to be of norm 1, from the subspace K. First it is clear that the eigenvector u cannot be well approximated from K if $\|(I-P)u\|$ is not small , because

$$\|u-\tilde{u}\| \geq \|(I-P)u\|$$

The fundamental quantity $\|(I-P)u\|$ can also be interpreted as the sine of the acute angle between the eigenvector u and the subspace of approximation K. In [11] it was shown that if we consider the exact eigenvector u as an approximate eigenvector of A_m then the corresponding residual vector satisfies the inequality:

$$\| (A_m - \lambda I) u \| \leq (|\lambda|^2 + \gamma^2)^{1/2} \| (I-P) u \| \tag{2.7}$$

where

$$\gamma = \| PA(I-P)\|$$

Let us illustrate the above notation with Arnoldi's method which is a typical orthogonal projection method [1, 11]. Assuming the algorithm does not break down before the final step N, it will reduce A to Hessenberg form by a simpe algorithm, so we get $B=Z^H AZ$, where B is N by N and Hessenberg, Z is N by N and orthonormal. The projection technique described above consists in taking as approximation to the eigenvalues of A, the eigenvalues of the mxm principal submatrix B_m of B, which clearly is equal to $V^H AV$, where $V=[v_1, v_2, \ldots v_m]$ consists of the first m vectors of

Z. In Arnoldi' method, it is known that V is an orthonormal basis of the m-th Krylov subspace, i.e. the subspace spanned by the system $[v_1, Av_1, .. A^{m-1}v_1]$.

Thus the situation can be schematically described as follows.

$$A \approx B = \begin{vmatrix} B_m & C_m \\ E_m & D_m \end{vmatrix}$$

in which $\approx$ denotes the similarity relation. The operator A_m is represented by the N by N matrix VB_mV^H of rank m. Its restriction to K, here equal to the m-th Krylov subspace K_m, is represented by B_m with respect the basis V. This illustrates the relation (2.6). The orthogonal projector P is represented by the N by N matrix VV^H of rank m. The quantity γ of (2.7) is nothing but the norm of the m by (N-m) matrix C_m which represents the linear operator PA(I-P).

Note that γ can be bounded by $\|A\|$, and this indicates that we can obtain a good approximation provided that the distance $\|(I-P)u\|$ is small. Among the methods which are of the type described above let us mention the symmetric Lanczos method (see e.g. [7]) , some of the Subspace Iteration methods [7, 16], the method of Arnoldi [1, 11].

2.2. Oblique projection methods. Several iterative methods suitable for large matrices can be interpreted in terms of oblique projection methods, or Petrov-Galerkin methods. In these methods we are given a second subspace L which may be different from K, and we seek an approximation $\tilde{u}$ belonging to K and satisfying the Petrov Galerkin condition:

$$A\tilde{u} - \tilde{\lambda}\tilde{u} \perp L \tag{2.8}$$

The subsapce K is often referred to as the right subspace and L as the left subspace. The two subspaces L and K are assumed to be of the same dimension, throughout the paper. In order to interpret the above condition in terms of operators we will need the obliquэ projector Q onto K and orthogonal to L which is defined by:

$Qx \in K$

$$x - Qx \perp L.$$

Note that the vector Qx is uniquely defined only under the assumption that <u>no vector of L is orthogonal to K</u>. It is easy to see that this fundamental assumption is equivalent to:

Assumption: For any two bases V and W of K and L respectively we have

$$\mathrm{Det}(\, W^H V \,) \neq 0. \tag{2.9}$$

The projector Q is illustrated in Figure 2-1.

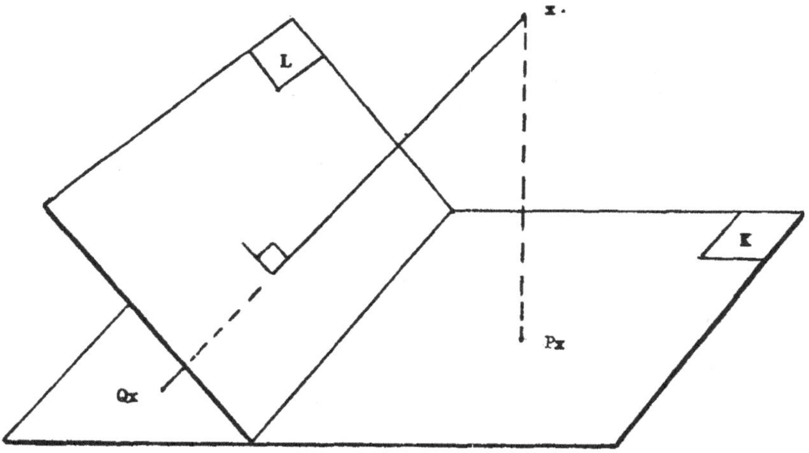

Figure 2-1: The oblique projector Q

We can rewrite equation (2.8) as

$$Q(A\,\tilde{u} - \tilde{\lambda}\,\tilde{u}) = 0 \tag{2.10}$$

which again can be translated matricially by expressing the approximate eigenvector $\tilde{u}$ in an appropriate basis of K. Assume that a basis V of K and a basis W of L can be found such that V and W form a biorthogonal pair i.e. such that

$$V^H W = I$$

where I is the identity matrix. Then if we write $\tilde{u}=Vy$, the Petrov-Galerkin condition (2.8) gives the same approximate problem as (2.5) except that the matrix B_m is this time defined by:

$$B_m = W^H A V$$

We should however emphasize that in order for a biorthogonal pair V, W to exist we must make the above assumption that no vector of L is orthogonal to K.

We can establish the following theorem which generalizes the result (2.7) of [11].

<u>Theorem 2.1</u>: Let $\gamma=\| Q(A-\lambda I)(I-P) \|$. Then the following two inequalities hold:

1) $$\| (A_m - \lambda I) Pu \| \leq \gamma \| (I-P) u \| \tag{2.11}$$

2) $$\| (A_m - \lambda I) u \| \leq \sqrt{|\lambda|^2 + \gamma^2} \| (I-P) u \| \tag{2.12}$$

Proof:

1) Since Py belongs to K for any vector y, we have QP=P and

$$\begin{aligned}(A_m - \lambda I) Pu &= Q (A - \lambda I) Pu \\ &= Q(A-\lambda I)(Pu - u) \\ &= - Q(A-\lambda I)(I-P)u \end{aligned} \tag{2.13}$$

and since (I-P) is a projector

$$(A_m - \lambda I) Pu = - Q(A-\lambda I)(I-P).(I-P)u$$

Taking the Euclidean norms of both sides we immediately obtain the result (2.11)

2) We have

$$\begin{aligned}(A_m - \lambda I)u &= (A_m - \lambda I)[Pu + (I-P)u] \\ &= (A_m - \lambda I)Pu + (A_m - \lambda I)(I-P)u\end{aligned}$$

Noticing that $A_m(I-P)=0$, this becomes:

$$(A_m - \lambda I)u = (A_m - \lambda I)Pu -\lambda(I-P)u$$

Since the two terms of the right hand side are orthogonal we obtain

$$\|(A_m - \lambda I)u\|^2 = \|(A_m - \lambda I)Pu\|^2 + \|\lambda(I-P)u\|^2$$

Using inequality (2.11) this immediatly gives the result (2.12). Q.E.D.

Note that (2.12) is a simple generalization of (2.7). In the case of orthogonal projection methods we have $\|Q\|=1$ and γ may be bounded by $\|A\|$. It may seem that since we obtain very similar error bounds for both the orthogonal projection method and the oblique projection method we are likely to get similar errors when we use the same subspace K. This is unfortunately not the case because unlike in the orthogonal projection method, the scalar γ can no longer be bounded by $\|A\|$, since we have $\|Q\|\geq 1$ and $\|Q\|$ is unknown in general. In fact the constant γ might very well be a large number even if A is a matrix with a moderate norm. Besides γ we also have the difficulty that we do not have any information about the condition number of the approximate problem.

2.3. Application: orthogonal projection methods for symmetric problems. We will now restrict ourself to the case where an orthogonal projection method is applied to a symmetric problem and will establish a few consequences of theorem 2.1. First notice the important fact that the first inequality can be matricially translated in the subspace K by expressing the vector Pu and the operator A_m in a certain basis $V:=\{v_1,v_2,..v_m\}$ of K. Assuming V is orthonormal let us set $Pu/\|Pu\|=Vy$ and define B_m as in (2.6). Then inequality (2.11) translates itself into:

$$\|(B_m - \lambda I)y\| \leq \gamma \|(I-P)u\|/\|Pu\| \tag{2.14}$$

where B_m is the representation of the restriction of A_m into K as defined by (2.6).

As mentioned above when A is unsymmetric and Q is oblique then we may encounter some difficulties because we have little information on the conditioning of the approximate eigenvalue problem and because the constant γ may be large. However, in the symmetric case and when L=K, the situation is somewhat simpler, and we are able to give more precise bounds. We would like to apply the following well-known result , in conjunction with our inequality (2.11).

<u>Theorem 2.2</u>: Let B be any symmetric matrix, and y a vector of norm 1. Consider the Rayleigh quotient $\mu=(By,y)$ and let ρ be the norm of the residual vector associated with y, i.e. $\rho=\|(B-\mu I)y\|$. Then there exists an eigenvalue $\tilde{\lambda}$ of B with associated eigenvector $\tilde{u}$ such that:

$$| \mu - \tilde{\lambda} | \leq \rho^2 \qquad (2.15)$$

$$\sin[\theta(u,\tilde{u})] \leq \rho / \delta \qquad (2.16)$$

where δ is the distance between μ and $\sigma(B)-\{\tilde{\lambda}\}$ i.e.

$$\delta = \min\{ |\mu - \tilde{\lambda}'|, \tilde{\lambda}' \in \sigma(B), \tilde{\lambda}' \neq \tilde{\lambda} \}.$$

and $\theta(u,\tilde{u})$ denotes the acute angle between u and $\tilde{u}$.

For the proof see e.g. [7] or [2]. The above fundamental result shows in particular the well known fact that the error for the Rayleigh quotient as an eigenvalue is of order the square of the residual norm. Our next objective is naturally to show that the Rayleigh quotient of Pu is not too different from the exact eigenvalue λ when $\|(I-P)u\|$ is small.

<u>Lemma 2.3</u>: Let u be an eigenvector of A associated with the eigenvalue λ. Then the Rayleigh quotient $\dot{\mu}$ of the vector Pu satisfies the inequality

$$| \lambda - \dot{\mu}| \leq \| A -\lambda I \| \, \| (I-P)u \|^2 / \| Pu \|^2 \qquad (2.17)$$

Proof: From the identity

$$(A-\lambda I)Pu = (A-\lambda I)(u-(I-P)u) = -(A-\lambda I)(I-P)u$$

and the fact that A is Hermitian we get

$$| \lambda - \dot{\mu} | = \left| \frac{((A-\lambda I)P\,u,\ P\,u)}{(P\,u,\ P\,u)} \right| = \left| \frac{((A-\lambda I)(I-P)u\ ,\ (I-P)u)}{(Pu\ ,\ Pu)} \right| .$$

Hence the result by use of the Cauchy Schwartz inequality. Q.E.D.

With the help of the above two results we can prove:

<u>Theorem 2.4</u>: Let λ be an eigenvalue of A, with associated eigenvector u of norm 1 and let

$$\varepsilon = \| (I-P)u \| / \| Pu \|$$

Then there exists an eigenvalue $\tilde{\lambda}$ of the approximate problem satisfying:

$$| \lambda - \tilde{\lambda} | \leq \tau \varepsilon^2 \qquad (2.18)$$

in which $\tau = \| A-\lambda I \| + \gamma^2/\delta$, with γ defined in theorem 2.1 and where δ is the distance from $\dot{\mu}$, the Rayleigh quotient of Pu, to the approximate eigenvalues $\tilde{\lambda}'$ different from $\tilde{\lambda}$.

Proof: From inequality (2.14) and theorem 2.2, there exists $\tilde{\lambda}$ an eigenvalue of B_m, i.e. an approximate eigenvalue of A, such that

$$|\tilde{\lambda}-\dot{\mu}| \leq [\gamma\ \varepsilon]^2 / \delta$$

Then the result follows from the triangle inequality

$$|\lambda - \tilde{\lambda}| \leq |\lambda - \dot{\mu}| + |\dot{\mu} - \tilde{\lambda}|$$

and lemma 2.3. Q.E.D.

The above inequality expresses the error in the eigenvalue directly in terms of he distance between the exact eigenvector u and the subspace of approximants and shows that this error is of order the square of this distance. A somewhat weaker but more general inequality can be derived from the following well known a posteriori error bound [7]:

__Lemma 2.5__: Let B be any symmetric matrix, y a vector of norm 1 and λ any scalar. Let ρ be the norm of the residual vector associated with y and λ i.e.

$$\rho = \| (B-\lambda I)y \|$$

Then there exists an eigenvalue $\tilde{\lambda}$ of B such that:

$$|\lambda - \tilde{\lambda}| \leq \rho \qquad (2.19)$$

This with inequality (2.14) immediately proves the theorem:

__Theorem 2.6__: Let λ be an eigenvalue of A. Then there exists an eigenvalue $\tilde{\lambda}$ of the approximate problem such that :

$$|\lambda - \tilde{\lambda}| \leq \gamma\ \varepsilon \qquad (2.20)$$

where γ is defined in theorem 2.1, and ε is as in theorem 2.4.

Clearly, inequality (2.20) is weaker than the previous result (2.18). It has, however, its own importance. Most projection methods use a sequence of subspaces K_m

for which the orthogonal projectors P_m onto them are such that $\| (I-P_m)u\|$ converges to zero when m tends to infinity. Thus (2.20) proves the convergence of an approximate eigenvalue towards the exact eigenvalue λ in those situations. As an example for the symmetric Lanczos algorithm it is known that $\|(I-P_m)u\|$ converges to zero under the assumption that the initial vector in the Lanczos algorithm is not deficient in the direction u , see [12]. Therefore, there will be at least one sequence of eigenvalues converging towards the exact eigenvalue. Note however that in this particular case there are alternative a priori error bounds which are more accurate, see [12]

We now turn to the problem of estimating the error in the eigenvector.

<u>Theorem 2.7</u>: Let u be a unit eigenvector of A and let $\theta(u,K)$ be the acute angle between u and the subspace K, defined by

$$\sin[\theta(u,K)] = \| (I-P)u \| \tag{2.21}$$

Then there exists an eigenvector $\tilde{u}$ of the approximate problem such that

$$\sin[\theta(u,\tilde{u})] \leq \sqrt{1 + \gamma^2/d^2} \quad \sin[\theta(u,K)] \tag{2.22}$$

where γ is defined in theorem 2.1 and where d is the distance between λ and the set of approximate eigenvalues other than $\tilde{\lambda}$.

Proof: This result was shown in [12] in the context of the Lanczos algorithm , i.e. K is a Krylov subspace. See also [7] for an elegant presentation. It is transparent from the proofs presented there that we do not make use at any time of the fact that the subspace of approximation is a Krylov subspace. The result is therefore true for any orthogonal projection method. Q.E.D.

These results do not extend immediately to unsymmetric problems or to oblique projection methods. The main reason is that we do not have at our disposal the powerful theorem 2.2. In a recent article, Kahan, Parlett and Jiang have derived alternative error bounds generalizing (2.2) but using a residual norm of the right and the left approximate eigenvectors [4]. Their idea may be used in our context and this remains to be done. Without the knowledge of the left eigenvector , the best we can hope for is some partial information whereby the bound contains parameters from the approximate problem which are not known a priori. As an example the inequality (2.20) becomes in the unsymmetric case, assuming B_m diagonalizable [18]:

$$| \lambda - \tilde{\lambda} | \leq \gamma \| X \| \| X^{-1} \| \| (I-P)u \|$$

where X is a matrix which diagonalizes B_m. Clearly, the global condition number $\| X \| . \| X^{-1} \|$ is not known a priori and neither is γ. Note that even in the symmetric case many a priori error bounds use some a posteriori knowledge on the eigenvalues like for example the distance δ in theorem (2.2). The result (2.20) is an exception. Finally we mention that some analysis of the norm of the projector Q is proposed in [14].

3.__The subspace iteration for unsymmetric eigenvalue problems.__ In this class of methods the space of approximants is a subspace S_m spanned by a system S_m given by $S_m = A^m S_o$ where S_o is some initial system of r linearly independent vectors. There are two main versions of the method, both originally due to Bauer. The first uses two subspaces and is an oblique projection method [3], known originally under the name of bi-iteration. The second uses one subspace and is an orthogonal projection method, originally named treppeniteration. We will restrict ourself to this second class of methods which require less work in general. An efficient way of implementing the method has been presented by Stewart [16], and an analysis of the convergence was made by Parlett and Poole [9], and by Stewart [16]. We would like to show how our results of section 1 can be applied here.

We will denote by P_m the orthogonal projector onto the subspace S_m and will assume that the eigenvalues of A are labelled in decreasing order of magnitude and that $|\lambda_{r+1}|<|\lambda_r|$ i.e.

$$|\lambda_1| \geq |\lambda_2| \geq |\lambda_3| \ldots\ldots \geq |\lambda_r| > |\lambda_{r+1}| \geq \ldots \geq \lambda_N$$

Again u_i denotes an eigenvector of A of norm unity associated with λ_i. The spectral projector associated with the invariant subspace corresponding to $\lambda_1,\ldots\lambda_r$ will be denoted by π. We can then establish the following theorem concerning the distance $\|(I-P_m)u_i\|$

<u>Theorem 3.1</u>: Let $S_o = \{x_1, x_2, \ldots x_r\}$ and assume that S_o is such that the system of vectors $\{\pi x_i\}_{i=1,\ldots r}$ is linearly independent . Then for each u_i, $i=1,2..r$ there exists at least one vector s_i in the subspace S_o = span{ S_o } such that $\pi s_i = u_i$. Moreover the following inequality is satisfied:

$$\| (I-P_m)u_i \| \leq \| u_i - s_i \| \left(| \lambda_{r+1} / \lambda_i | + \varepsilon_m \right)^m \tag{3.1}$$

where ε_m tends to zero as m tends to infinity.

Proof: Let us write any vector s of S_o as

$$s = \sum_{j=1}^{r} \xi_j x_j$$

Projecting this onto the invariant subspace associated with $\lambda_1, \ldots \lambda_r$ we get

$$\pi s = \sum_{j=1}^{r} \xi_j \pi x_j$$

Since the πx_j's are assumed linearly independent and since u_i belongs to the invariant subspace associated with $\lambda_1, \ldots \lambda_r$, there exists at least one vector s_i such that $\pi s_i = u_i$. Then the vector s_i is such that

$$s_i = u_i + w \tag{3.2}$$

where $w=(I-\pi)s_i$. Note that s_i is not unique, since adding to s_i any vector of the intersection of S_o and $(I-\pi)\mathbb{C}^N$ would still give a vector satifying the requirement $\pi s_i = u_i$. Next consider the vector y of S_m defined by $y=(1/\lambda_i)^m A^m s_i$. We have from (3.2) that

$$u_i - y = (1/\lambda_i)^m A^m w \tag{3.3}$$

Denoting by W the invariant subspace corresponding to the remaining eigenvalues $\lambda_{r+1}, \ldots \lambda_N$, and noticing that w belongs to W, we clearly have

$$u_i - y = (1/\lambda_i)^m [A_{|W}]^m w$$

Hence

$$\| u_i - y \| \leq \| [\lambda_i^{-1} A_{|W}]^m \| \, \| w \| \tag{3.4}$$

Since the eigenvalues of $A_{|W}$ are λ_j with $j>r$, the spectral radius of $\lambda_i^{-1} A_{|W}$ is simply $|\lambda_{r+1}/\lambda_i|$ and from a well known result [17] we have

$$\| [\lambda_i^{-1} A_{|W}]^m \| = [\, |\lambda_{r+1}/\lambda_i| + \varepsilon_m]^m \tag{3.5}$$

where ε_m tends to zero as $m \to \infty$. Using the fact that

$$\| (I-P_m)u_i \| = \min \{ \| s - u_i \|, \; s \in S_m \} \; ,$$

and (3.4), (3.5) we obtain the desired result (3.1). Q.E.D.

A few remarks are in order. First notice that we can take advantage of the nonuniqueness of the vector s_i to improve the bound (3.1) by replacing s_i in the theorem by the vector $\bar{s}_i$ satisfying $\pi s_i = u_i$, for which the norm $\| s_i - u_i \|$ is minimum.

A second remark concerns the sequence ε_m for which we can be more specific by using the more precise bound for $\| B^m \|$ given in [17]

$$\| B^m \| \leq \alpha \, \rho^m \, m^{\eta-1} \tag{3.6}$$

where B is any matrix, ρ represents the spectral radius of B, η is the dimension of its largest Jordan block, and α is some constant <u>independent</u> <u>of</u> <u>m</u>. We may assume without loss of generality that $\alpha \geq 1$ (otherwise we can consider the weaker bound obtained by replacing α by $\alpha'=1$). Thus in the case where A is diagonalizable[2] we have $\eta=1$, and after taking the m-th root of both sides of (3.6) the above inequality (3.1) becomes simply

$$\| (I-P_m)u_i \| \leq \alpha \, \| u_i - s_i \| \, (\, | \, \lambda_{r+1} \, / \, \lambda_i | \,)^m \tag{3.7}$$

In the diagonalizable case s_i is a vector having the eigenexpansion

$$s_i = u_i + \sum_{j=r+1}^{N} \xi_j \, u_j$$

and letting $\beta = \sum_{j=r+1}^{N} | \, \xi_j |$, the proof of theorem 3.1 can be repeated to yield the inequality

$$\| (I-P_m)u_i \| \leq \beta \, | \, \lambda_{r+1} \, / \, \lambda_i |^m$$

In case $A_{|W}$ is not diagonalizable, then from the result (3.6) we can majorize ε_m as follows:

$$\varepsilon_m \leq |\lambda_i / \lambda_{r+1}| \, [\alpha^{1/m} \, m^{(\eta-1)/m} - 1 \,]$$

which tends to zero as m tends to infinity.

Finally we would like to interpret the assumption of the theorem. It can easily be shown that the assumption that $\{\pi x_i\}$ is a linearly independent system, is

---

[2] In fact we only need that the restriction of A to W defined in the above proof to be diagonalizable.

equivalent to the condition

$$\det(U^H S_o) \neq 0$$

in which U is any basis of the invariant subspace. Clearly this is a generalization of a similar condition required for the convergence of the power method.

4.**Chebyshev Acceleration of the Unsymmetric Subspace Iteration.** An experiment described in [11] indicated that even in the unsymmetric case the Chebyshev polynomials can be efficiently used to accelerate the convergence of the subspace iteration.

Let us assume that we can find an ellipse of center d and focii d+e, d−e which contains all the eigenvalues of A except the r dominant ones $\lambda_1,\ldots\lambda_r$, see Figure 4-1. Then Rutishauser's symmetric subspace iteration can be generalized to unsymmetric problems by simply replacing the subspace S_m of Section 3, by the better subspace R_m defined by

$$R_m = \text{span}\{\ C_m(A)S_0\ \}$$

in which C_m is the shifted Chebyshev polynomial

$$C_m(z) = T_m[(z-d)/e]$$

Assuming that A is diagonalizable and denoting by P_m the orthogonal projector onto R_m, then theorem 3.1 can be extended as follows.

__Theorem 4.1__: Let $S_o = \{x_1, x_2,\ldots x_r\}$ be such that the vectors $\{\pi x_j\ ,\ j=1,..r\}$ are linearly independent . Assume that the eigenvalues $\lambda_{r+1}, \lambda_{r+2},..\lambda_N$ are contained in an ellipse E of center d, focii d+e, d−e and major semi axis a. Then for each u_i i=1,2..r there exists at least one vector s_i in the subspace R_0=span$\{S_o\}$ such that

$$s_i = u_i + \sum_{j=r+1}^{N} \xi_j\, u_j \tag{4.1}$$

Moreover letting $\beta = \sum_{j=r+1}^{N} |\ \xi_j\ |$, the following inequality is satisfied:

$$\|\ (I-P_m)u_i\ \| \leq \beta\ \frac{T_m(a/e)}{|T_m[(\lambda_i-d)/e]\ |} \tag{4.2}$$

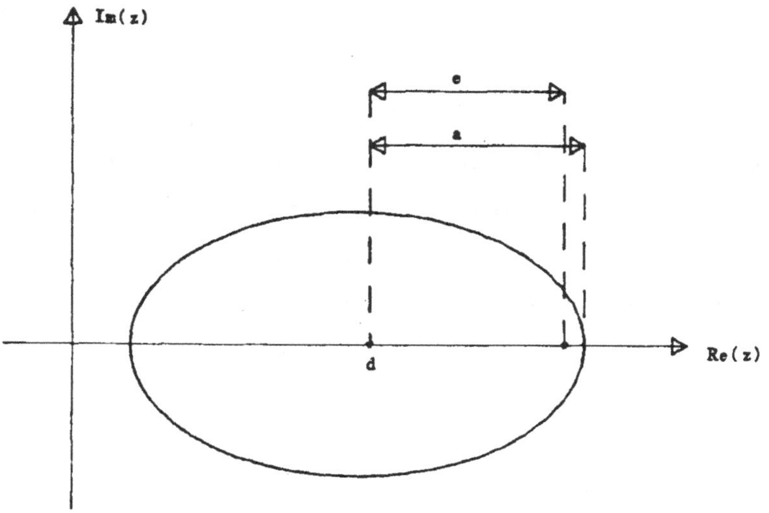

Figure 4-1: Optimal ellipse for the unsymmetric subspace iteration

Proof: The existence of s_i defined by (4.1) can be proved in the same way as for theorem 3.1. Proceeding as in the proof of theorem 3.1, consider the vector y of R_m defined by

$$y=(C_m(A)/C_m(\lambda_i)\ )\ s_i$$

Then it is clear that

$$y-u_i = (1/C_m(\lambda_i)) \sum_{j=r+1}^{N} C_m(\lambda_j)\ )\xi_j\ u_j$$

Taking the norm of both sides we obtain the bound

$$\| y-u_i \| \leq |1/C_m(\lambda_i)| \sum_{j=r+1}^{N} |\ C_m(\lambda_j)\ |\ \ |\xi_j| \qquad (4.3)$$
$$\leq \max_{z \in E} |C_m(z)/C_m(\lambda_i)| \sum_{j=r+1}^{N} |\xi_j|$$

where E is the ellipse containing the eigenvalues $\lambda_{r+1},\ldots\lambda_N$. It is easy to show that the above maximum is achieved for m different points on the boundary of the ellipse, including the point d+a of the major axis. Replacing this in (4.3), the proof can be completed in the same way as for theorem 3.1 Q.E.D.

Again as in theorem 3.1, because of the nonuniqueness of s_i, the constant β can be replaced by the smallest possible β:

$$\bar{\beta} = \min\{ \sum_{j=r+1}^{N} | \xi_j | \; ; \text{ all } \xi_j \text{ such that } u_i + \sum_{j=r+1}^{N} \xi_j u_j \in R_m \}.$$

The above bound is a generalization of Rutishauser's result. In the case where all eigenvalues of A are real we can take for E the degenerate ellipse which has eccentricity e=a, i.e. E is the interval [d-a,d+a] and a=e. In this case, assuming the eigenvalues are labelled in increasing order, the numerator of (4.2) becomes one and the denominator can be written as:

$$T_m(1+2\gamma_r)$$

with

$$\gamma_r=(\lambda_i-\lambda_{r+1})/(\lambda_{r+1}-\lambda_N)$$

This is precisely the result obtained by Rutishauser in the symmetric case, see [7]. Note that generally the result is more difficult to interpret in the presence of complex eigenvalues . It can be shown that the right hand side of (4.2) always tends to zero, see [19, 6]. When the ellipse is flat along the real axis, i.e. when the eigenvalues have small imaginary parts , then the convergence will be faster because a/e is closer to one. The ideal case is when all eigenvalues are real.

5.**Methods using Krylov subspaces.** There are several techniques which realize a projection method on Krylov subspaces of the form K_m = span$\{v_1,Av_1,..A^{m-1}v_1\}$. Unlike in the subspace iteration method where the dimension of the subspace of approximation is fixed during the iteration, here the dimension of K_m increases by one at every step.

Among the methods which use Krylov subspaces, we mention the following: . The symmetric Lanczos algorithm, see e.g. [7].

. The method of Arnoldi for unsymmetric systems, [1, 11].:

. The unsymmetric Lanczos method [5, 8]

. The method of incomplete orthogonalization [11, 14].

The first two methods are orthogonal projection methods while the last two are oblique projection methods.

We now show a characteristic property for all techniques which realize an orthogonal projection method onto the Krylov subspace K_m. It will be assumed throughout that $\dim(K_m)=m$.

<u>Theorem 5.1</u>: Assume that an orthogonal projection technique is applied to A, using the subspace K_m and let $\bar{p}_m(t)$ be the characteristic polynomial of the approximate problem. Then $\bar{p}_m$ minimizes the norm $\| p(A)\, v_1 \|$ over all monic polynomials p of degree m.

Proof: By Cayley Hamilton's theorem, we have $\bar{p}_m(A_m) = 0$, so that clearly

$$(\bar{p}_m(A_m)v_1, v) = 0, \quad \text{for any vector } v \text{ in } K_m \tag{5.1}$$

It can easily shown by induction that for $k \le m$ we have the property

$$(A_m)^k\, v_1 = P_m\, A^k\, v_1 \tag{5.2}$$

Therefore (5.1) becomes

$$(P_m\, \bar{p}_m(A)v_1\,,\, v\,) = 0, \quad \forall\ v \text{ in } K_m$$

or

$$(\bar{p}_m(A)v_1\,, P_m\ v\,) = 0, \quad \forall\ v \text{ in } K_m$$

which is equivalent to

$$(\bar{p}_m(A)v_1\,,\, v\,) = 0, \quad \forall\ v \text{ in } K_m$$

Now writing $\bar{p}_m(t)$ as $\bar{p}_m(t) = t^m - q(t)$, where q is a polynomial of degree less than m, we obtain the equality

$$(A^m v_1 - q(A)v_1\,,\, v\,) = 0 \quad \forall\ v \text{ in } K_m \tag{5.3}$$

which is equivalent to

$$(A^m v_1 - q(A)v_1\,,\, A^j v_1\,) = 0 \quad j=0,1,2....m-1$$

In the above system of equations we recognize the normal equations for minimizing the Euclidean norm of $A^m v_1 - s(A)v_1$ over all polynomials s of degree $\le m-1$ and the result is proved. Q.E.D.

The above characteristic property was shown in the particular context of the Lanczos algorithm for symmetric problems in [15]. What we have just shown is that it holds for any orthogonal projection method onto a Krylov subspace K_m and that it is independent of the particular algorithm applied. It indicates that the method can be regarded as an optimization process whereby we attempt to mininize some norm of the minimal polynomial of v_1. Indeed under the assumption that the minimal polynomial of v_1 is of degree at least m then $\| p(A) v_1 \|$ can be regarded as a discrete norm on the set of polynomials of degree not exceeding m-1.

Let us now consider the distance of a particular exact eigenvector u_i from the subspace of approximation K_m. It is simplifying to assume that A is diagonalizable and to denote by $\varepsilon_i^{(m)}$, the quantity

$$\varepsilon_i^{(m)} = \min_{p \in P_{m-1}} \max_{\lambda \in \sigma(A)-\{\lambda_i\}} |p(\lambda)|$$

where P_{m-1} represents the set of all polynomials p of degree not exceeding m-1 such that $p(\lambda_i)=1$.

It can easily be shown that $\| (I-P_m)u_i \|$ is related to $\varepsilon_i^{(m)}$ by the following inequality, see [11]

$$\| (I-P_m)u_i \| \leq \| v_1 \|_1 \varepsilon_i^{(m)}$$

where $\| x \|_1$ is the norm defined as the sum of the absolute values of the components of x in the eigenbasis, assuming the eigenvectors are all of norm unity. This means that we will obtain an estimate for $\| (I-P_m)u_i \|$ from an estimate of $\varepsilon_i^{(m)}$.

Without loss of generality we assume that i=1. In [13] a result similar to the following one was stated without proof:

<u>Theorem 5.2</u>: Let m<N. Then there exists m eigenvalues which can be numbered $\lambda_2, \ldots \lambda_{m+1}$, such that:

$$\varepsilon_1^{(m)} = \left[\sum_{j=2}^{m+1} \prod_{\substack{k=2 \\ k\neq j}}^{m+1} \frac{|\lambda_k - \lambda_1|}{|\lambda_k - \lambda_j|} \right]^{-1} \qquad (5.4)$$

A proof of this equality is proposed in the appendix.

Let us suppose that all of the eigenvalues except λ_1, lie inside a certain circle, see Figure 5-1. Then as m increases $\varepsilon_1^{(m)}$ will decay rapidly. This can be

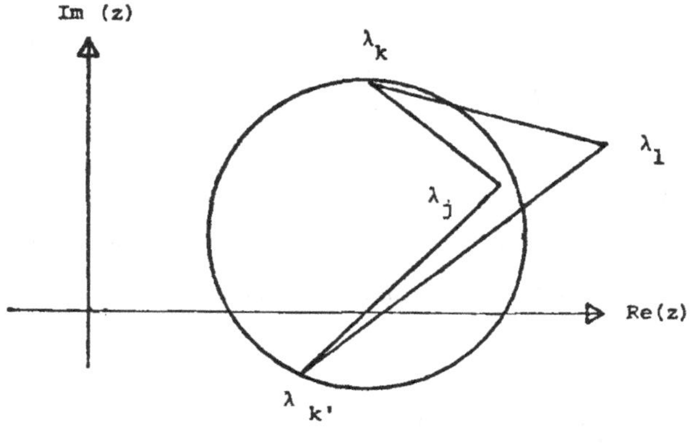

Figure 5-1: Illustration of theorem 5.2

made clear by only considering the product term in (5.4) associated with the eigenvalue nearest to λ_1, called λ_j in Figure 5-1. As is seen in fig. 5-1, we will have in general $|\lambda_1-\lambda_k| > |\lambda_j-\lambda_k|$ and therefore the product

$$\prod_{k=2}^{m+1} \frac{|\lambda_k - \lambda_i|}{|\lambda_k - \lambda_j|}$$

in (5.4) will be large in general, thus showing that $s_1^{(m)}$ will be a small quantity. Unfortunately, we do not know in general what the eigenvalues $\lambda_2, \ldots \lambda_{m+1}$ are. The interesting indication provided by the theorem is that the convergence of $s_1^{(m)}$ towards zero is fastest when λ_1 is in the outest part of the spectrum. We now propose a few illustrations.

<u>Example</u> <u>1</u>. Assume that $\lambda_k = (k-1)/(N-1)$, $k=1,2,..N$ (Uniform distribution on the real line), and take $m=N-1$. Then $s_1^{(m)} = 1/(2^m - 1)$

Proof: Since $m=N-1$ there is no choice for the λ_j's but the remaining eigenvalues

$\lambda_2,\ldots\lambda_N$. We have from (5.4)

$$[\varepsilon_1^{(m)}]^{-1} = \sum_{j=2}^{m+1} \prod_{k=2:m+1,\ k\neq j} |k-1|/|k-j|$$

$$= \sum_{j=2}^{m+1} \frac{m!}{(j-1)!(m+1-j)!} = \sum_{l=1}^{m} \binom{1}{m} = 2^m - 1$$

__Example 2.__ Let again m=N-1 and assume the following distribution of the eigenvalues: $\lambda_k = e^{i\,2(k-1)\pi/N}$, k=1,2..N. (Uniform distribution on the cercle). Then $\varepsilon_1^{(m)} = 1/m$.

Proof: Letting $w=e^{i2\pi/N}$ each product term in (5.4) can be written as

$$\prod_{k=2:m+1,\ k\neq j} \frac{|w^{k-1}-1|}{|w^{k-1}-w^{j-1}|} = \prod_{k=1:m,\ k\neq j} \frac{|w^k-1|}{|w^k-w^j|}$$

$$= [\,|w^j-1| \prod_{k=1:m,\ k\neq j} |w^k-1|\,] \;/\; [\,|w^j-1| \prod_{k=1:m,\ k\neq j} |w^k-w^j|\,]$$

$$= [\prod_{k=1}^{m} |w^k-1|\,] \;/\; [\,|w^j-1| \;.\; \prod_{k=1:m,\ k\neq j} |w^{k-j}-1|\,]$$

Recalling that the w^k's are the k-th powers of the N-th root of unity, a simple renumbering of the products in the denominator of the above expression shows that this expression has modulus one. Thus taking the sum over the m different eigenvalues yields the result Q.E.D.

The above examples show a striking difference between the complex uniform distribution and the real uniform distribution, and indicate that the approximation can be very poor in some complex cases.

In [13] we have shown a number of ways of bounding a quantity similar to $\varepsilon_1^{(m)}$ which occurs when solving a system of linear equations by projection methods onto Krylov subspaces. Similar results hold for the dominant eigenvalue λ_1, in the context of the eigenvalue problem. For example we can prove the following proposition.

__Proposition 5.3__: Assume that all the eigenvalues of A except λ_1 lie inside the ellipse having center d, focii d+e,d-e and major semi axis a. Then

$$\varepsilon_1^{(m)} \leq \frac{T_{m-1}(a/e)}{|T_{m-1}[(\lambda_1-d)/e]\,|}$$

where T_{m-1} is the Chebyshev polynomial of degree m-1 of the first kind.

Note that a/e is a real positive number although for generality e and a may be complex , in case the main axis of the ellipse is not on the real line.

6.__APPENDIX: Proof of theorem 5.2.__ In this appendix we propose a proof of theorem 5.2. We need the following lemma from approximation theory, see [10].

__Lemma 6.1__: Let $\tilde{q}$ be the best uniform approximation of a function f by a set of n polynomials satisfying the Haar conditions, on a compact set σ consisting of at least n+1 points. Then there exist at least n+1 points $\lambda_0,..\lambda_n$ of σ, such that the error $\tilde{e}(z)=f(z)-\tilde{q}(z)$ reaches its maximum at the λ_j's, i.e.

$$|\tilde{e}(\lambda_j)| = \max_{z\in\sigma} |\tilde{e}(z)|$$

Such points are called critical points.

Consider

$$\varepsilon_i^{(n)} = \min_{p\in P_{n-1}} \max_{\lambda\in\sigma(A)-\{\lambda_i\}} |p(\lambda)|$$

Clearly $\varepsilon_i^{(n)}$ represents the smallest possible uniform norm on the set $\sigma(A)$ of polynomials of the form $1-(z-\lambda_1)s(z)$, with s of degree not excceeding n-2. Otherwise stated this means that we seek for the best approximation of the constant function unity over the set $\sigma(A)$ by polynomials of degree $\leq n-1$ which are linear combinations of the polynomials $\omega_1,\omega_2,..\omega_n$ where

$$\omega_j(z)=(z-\lambda_1)\ z^{j-1} \tag{6.1}$$

Since the set of polynomials (6.1) verifies the Haar conditions, from the above lemma there exists at least n+1 critical points, i.e. points where the maximum error is reached. We will denote by $\tilde{p}(z)$ the optimal polynomial $1-\tilde{q}$.

We can easily prove

__Lemma 6.2__: Let $\lambda_2, \lambda_3,...\lambda_{n+2}$ be the n+1 critical points. Then there exists a nontrivial solution to the system of equations:

$$\sum_{i=2}^{n+2} \omega_j(\lambda_i) z_i = 0 , \quad j=1,2\ldots n \tag{6.2}$$

Proof. This is a system of n equations with n+1 unknowns. Because of the Haar conditions when we isolate one unknown, e.g. the last one, then the n by n resulting system is nonsingular Q.E.D.

<u>Lemma 6.3</u>: Let z_j, $j=2,\ldots n+2$ a certain solution of the system (6.2) and let us write $z_k = \delta_k e^{-i\theta_k}$, where δ_k is real and positive. Then the best approximation polynomial $\tilde{p}$ is :

$$\tilde{p}(z) = \sum_{k=2}^{n+2} e^{i\theta_k} l_k(z) \Big/ \sum_{k=2}^{n+2} e^{i\theta_k} l_k(\lambda_1) \tag{6.3}$$

where $l_k(z)$ is the Lagrange polynomial of degree n at the points $\lambda_2, \lambda_3 \ldots \lambda_{n+2}$, taking the value one at λ_k:

$$l_k(z) = \prod_{j=2:n+2,\ j\neq k} \frac{z-\lambda_j}{\lambda_k - \lambda_j}$$

Proof: Because of (6.2) for any v belonging to the space of polynomials $Q_n = \text{span}\{\omega_1, \ldots \omega_n\}$ we have:

$$\sum_{k=2}^{n+2} \delta_k e^{-i\theta_k} v(\lambda_k) = 0 \tag{6.4}$$

Let $\tilde{p}$ be defined by (6.3). We have to show that

$$\| \tilde{p} + v \|_\infty \geq \| \tilde{p} \|_\infty \quad \text{for any v in } Q_n \tag{6.5}$$

where $\|.\|_\infty$ represents the uniform norm on the set σ. Let us set

$$\rho = \Big[\sum_{k=2}^{n+2} e^{i\theta_k} l_k(\lambda_1) \Big]^{-1} \tag{6.6}$$

Notice that $|\rho|$ is precisely the uniform norm of $\tilde{p}$ in σ. From (6.4) it is clear that for some k' we have

$$\text{Re} \,[\rho\, e^{-i\theta_{k'}} v(\lambda_{k'})] \geq 0$$

Therefore

$$\begin{aligned} \| \tilde{p} + v \|_\infty^2 &= \max_{j=2,n+2} | (\tilde{p} + v)(\lambda_j)|^2 \geq |\tilde{p}(\lambda_{k'}) + v(\lambda_{k'})|^2 \\ &= |\tilde{p}(\lambda_{k'}) + v(\lambda_{k'})|^2 = |\rho\, e^{i\theta_{k'}} + v(\lambda_{k'})|^2 \\ &= |\rho|^2 + |v(\lambda_{k'})|^2 + 2 \,\text{Re}\, \{\rho e^{-i\theta_{k'}} v(\lambda_{k'})\} \quad \geq |\rho|^2 \end{aligned}$$

which shows that (6.5) is true and completes the proof of the lemma Q.E.D.

Proof of theorem 5.2. The system (6.2) can be solved by using Cramer's rule and

some Vandermonde determinant equalities. Doing this it is possible to show that one particular solution of (6.2) is $z_k = \bar{l}_k(\lambda_1)$, k=2,..n+2. Hence

$$e^{i\theta_k} = \bar{l}_k(\lambda_1) \;/\; | \; l_k(\lambda_1) \; |$$

replacing this in the expression (6.6) gives the desired result Q.E.D.

REFERENCES

(1) W.E. ARNOLDI. The principle of minimized iteration in the solution of the matrix eigenvalue problem. Quart. Appl. Math. 9:17-29, 1951.

(2) F. CHATELIN. Spectral approximation of linear operators. Academic Press, New York, 1983.

(3) M. CLINT AND A. JENNINGS. The evaluation of eigenvalues and eigenvectors of real symmetric matrices by simultaneous iteration method. J. Inst. Math. Appl. 8:111-121,1971.

(4) W. KAHAN, B.N.PARLETT AND E. JIANG. Residual bounds on approximate eigensystems of nonnormal matrices. SIAM J. Numer. Anal. 19:470-484, 1982. (to appear).

(5) C. LANCZOS. An iteration method for the solution of the eigenvalue problem of linear differential and integral operator. J. Res. Nat. Bur. of Standards 45:255-282, 1950.

(6) T.A. MANTEUFFEL. An iterative method for solving nonsymmetric linear systems with dynamic estimation of parameters. Technical Report UIUCDCS-75-758, University of Illinois at Urbana-Champaign, 1975. Ph.D. dissertation.

(7) B.N. PARLETT. The Symmetric Eigenvalue Problem. Prentice Hall, Englewood Cliffs, 1980.

(8) B.N. PARLETT AND D. TAYLOR. A look ahead Lanczos algorithm for unsymmetric matrices. Technical Report PAM-43, Center for Pure and Applied Mathematics, 1981.

(9) B.N. PARLETT AND W.G. POOLE. A geometric theory of the QR, LU and Power Iterations. SIAM J. of Num. Anal. 10 #2,:389-412, 1973.

(10) RIVLIN T.J. The Chebyshev Polynomials. J. Wiley and Sons Inc., New York, 1976.

(11) Y. SAAD. Variations on Arnoldi's method for Computing Eigenelements of large unsymmetric matrices. Linear Algebra and it's Applications 34:269-295, 1980.

(12) Y. SAAD. On the rates of convergence of the Lanczos and the block Lanczos methods. SIAM J. Numer. Anal. 17:687-706, 1980.

(13) Y. SAAD. Krylov subspace methods for solving large unsymmetric linear systems. Mathematics of Computation 37:105-126, 1981

(14) Y. SAAD. The Lanczos biorthogonalization algorithm and other oblique projection methods for solving large unsymmetric systems. SIAM J. Numer. Anal. 19:470-484, 1982.

(15) Y. SAAD. Computation of eigenvalues of large Hermitian Matrices by Partitioning Techniques. Technical Report, INPG- University of Grenoble, 1974. Dissertation.

(16) G.W. STEWART. Simultaneous iteration for computing invariant subspaces of non-Hermitian matrices. Numer. Mat. 25:123-136, 1976.

(17) R.S. VARGA. Matrix Iterative Analysis. Prentice Hall, Englewood Cliffs, New Jersey, 1962.

(18) J. H. WILKINSON. The Algebraic Eigenvalue Problem. Clarendon Press, Oxford, 1965.

(19) H. E. WRIGLEY. Accelerating the Jacobi method for solving simultaneous equations by Chebyshev extrapolation when the eigenvalues of the Iteration Matrix are complex. Computer Journal 6:169-176, 1963.

SECTION B

SYMMETRIC $(A-\lambda B)$-PENCILS AND APPLICATIONS

THE GENERALIZED EIGENVALUE PROBLEM IN SHIPDESIGN AND OFFSHORE INDUSTRY - A COMPARISON OF TRADITIONAL METHODS WITH THE LANCZOS PROCESS.

Liv Aasland and Petter Bjørstad
Det norske Veritas
Research Division
1322 Høvik, Norway

Abstract:

The solution of generalized eigenvalue problems based on finite element models of ships and offshore structures like North Sea oil drilling platforms, is important in the analysis of design and structural safety. A few examples of such problems and their solution within the framework of SESAM'80, a large commercial finite element package developed by Det norske Veritas [7], is presented and compared with preliminary results using the Lanczos method.

1. INTRODUCTION

The eigenvalue problem:

$$(1.1) \qquad Kx = \omega^2 Mx$$

is frequently encountered in finite element calculations. Here K is the stiffness matrix and M is the corresponding mass matrix. This problem is of fundamental importance in structural engineering. An efficient algorithm for calculating a few of the lowest eigenvalues and corresponding eigenvectors from large finite element models is increasingly more important due to the complex structures that are now being modelled.

At Det norske Veritas eigenvalue calculation is often required for ships and offshore structures. Marine structures and vessels require proper dynamic response to internal machinery vibration and dynamic environmental loading. Eigenvalue calculation is important, for example, to prevent resonance when designing propeller blades for a ship [8]. Also, waves cause vibrations in offshore oil-platforms and thus the need to calculate a few of the platforms eigenvalues to be certain that they do not approach wave frequencies. Offshore structures are also vulnerable to fatigue from vibrations, and eigenvalue calculation may, for example, indicate that the structure should be stiffened. Figures 1 and 2 show two examples of finite element models where eigenvalue calculations have been carried out.

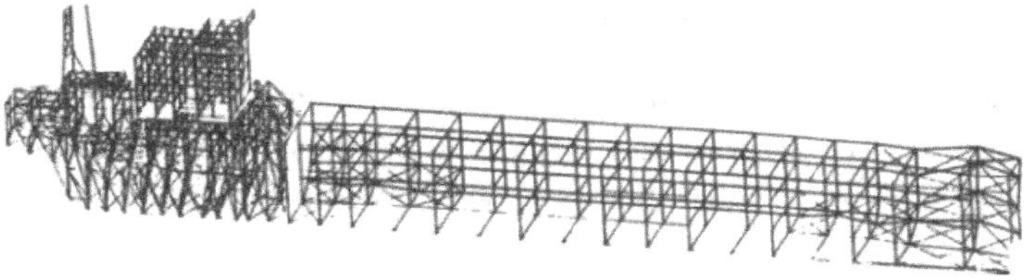

Fig. 1 : Finite element model of a 33000 ton d.w. chemical tanker. 4836 degrees of freedom.

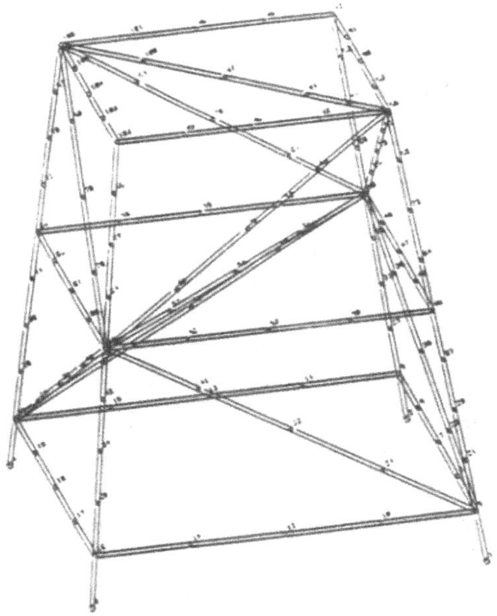

Fig. 2 : Finite element model of offshore oil drilling (jacket) platform. 624 degrees of freedom.

2. SOME METHODS USED IN EIGENVALUE CALCULATION.

The finite element model of a large oil producing platform in the North Sea may have up to 750,000 degrees of freedom. Dynamic calculations for problems of this size are very time consuming and are therefore not performed unless the number of variables can be reduced. Two related and commonly used methods for doing this will be outlined.

2.1 Master-slave reduction

The components of x are split into two sets, $x = \begin{bmatrix} x_1 \\ x_2 \end{bmatrix}$ (masters and slaves) and the matrices K and M are partitioned accordingly. It is now required that a relation $x = Bx_1$ holds for some B [9]. Most often B is chosen as

$$(2.1) \qquad B = \begin{bmatrix} I \\ -K_{22}^{-1} \quad K_{21} \end{bmatrix}$$

giving the reduced eigenvalue problem:

$$(2.2) \qquad B^T K B x_1 = \omega^2 B^T M B x_1$$

Note that the reduced stiffness matrix is a Schur complement in this case. If $M_{21} = M_{22} = 0$ in the original problem, then no approximation has been done. In practice, good engineering skill and some knowledge about the shape of the lowest eigenmodes are required in order to get a satisfactory result. If a component of x has large amplitude and a large contribution to the kinetic energy it should belong to the set x_1, if it has small amplitude and zero or small mass then it should belong to the set x_2 [5].

A calculation of this type has been carried out on the model in Fig. 1. The size of this model was reduced from 4836 degrees of freedom using symmetry and the described reduction technique to a final eigenvalue problem of dimension 426. The reduction process required 25 minutes and the eigen-value calculation 17 minutes of CPU time on a Univac 1110.

2.2 Component mode synthesis.

This approach to the problem of reducing the size of an eigenvalue problem is closely related to the idea of substructures in finite element analysis [3], [4]. Assume that a given problem can be divided into two substructures having variables x_1 and x_2 respectively, and that a separatorset of variables x_3 represents the only coupling between the two substructures. Let Ω^2_{11} and Φ_{11} be the normalized solution of the eigenvalue subproblem

$$(2.3) \qquad K_{11}\Phi_{11} = M_{11}\Phi_{11}\Omega^2_{11}$$

on the first substructure.

Next consider this subproblem with the coupling to the separatorset x_3

$$(2.4) \qquad \begin{bmatrix} K_{11} & K_{13} \\ K_{13}^T & K_{33} \end{bmatrix} \begin{bmatrix} x_1 \\ x_3 \end{bmatrix} = \omega^2 \begin{bmatrix} M_{11} & M_{13} \\ M_{13}^T & M_{33} \end{bmatrix} \begin{bmatrix} x_1 \\ x_3 \end{bmatrix}$$

A change of variables $x = By$ with

$$(2.5) \qquad B = \begin{bmatrix} \Phi_{11} & -K_{11}^{-1} K_{13} \\ 0 & I \end{bmatrix}$$

results in the new problem

$$(2.6) \qquad \begin{bmatrix} I & 0 \\ 0 & K_{33} - K_{13}^T K_{11}^{-1} K_{13} \end{bmatrix} \begin{bmatrix} y_1 \\ y_3 \end{bmatrix} = \omega^2 \begin{bmatrix} \Omega_{11}^{-2} & \overline{M}_{13} \\ \overline{M}_{13}^T & \overline{M}_{33} \end{bmatrix} \begin{bmatrix} y_1 \\ y_3 \end{bmatrix}$$

where $\overline{M}_{13} = \Phi_{11}^T (M_{13} - M_{11} K_{11}^{-1} K_{13})$

$$\overline{M}_{33} = M_{33} - K_{13}^T K_{11}^{-1} M_{13} - M_{13}^T K_{11}^{-1} K_{13} + K_{13}^T K_{11}^{-1} M_{11} K_{11}^{-1} K_{13} .$$

A similar development can be carried out for the second substructure and the matrix for the full problem is then obtained from the two subproblems.

The size of the final problem is not reduced if Φ_{ii} is the complete set of eigenvectors for substructure i. The reduction in size is performed by including only a few of the lowest modes in each Φ_{ii}. Thus the complete structure is represented by the lowest eigenvectors of each substructure plus the degrees of freedom in the separatorsets. The number of eigen-vectors to be included for a given substructure is largely based on experience and engineering skill.

It is our opinion that these techniques and possible alternatives deserve more attention. More automatic procedures are needed and they must be based on more theoretical insight providing estimates for the accuracy obtained.

2.3 Eigenvalue algorithms in a finite element analysis package.

The finite element program system developed by VERITAS (Super Element Structural Analysis Modules) includes programs for static and dynamic analysis of frames, membranes, shells and solids. A multilevel superelement technique is used and the problem size is only limited by CPU-time consumption. All matrices are stored on peripheral storage, whereas all numerical operations are performed on submatrices. The numerical algorithms are therefore block algorithms and sparsity is exploited only on the block level. The block size is in principle arbitrary, but it must be small enough so that a few blocks can be kept in main memory simultaneously. Large block sizes also reduce the number of zero blocks in the block matrix. On the other hand a small block size leads to excessive I/O and inefficient computation [2].

The eigenvalue algorithms that have been adapted to this framework are the EISPACK path based on Householdder tridiagonalization, bisection and inverse iteration and an implementation of the simultaneous iteration technique due to Rutishauser. In practice these algorithms have been used successfully on problems of dimension less than one thousand.

3. A COMPARISON OF THE SIMULTANEOUS ITERATION METHOD AND THE LANCZOS ALGORITHM

A preliminary implementation of the block Lanczos algorithm written by David Scott [6], DNLASO, has been tested within the framework of the SESAM'80 finite element analysis package. The results have been compared with simultaneous iteration. The simultaneous iteration routine, SSIT25, is quite similar to the algorithm described by Bathe and Wilson [1]. This method is designed for the generalized eigenvalue problem. The tests revealed that the performance of the method depends quite strongly on the number of iteration vectors used. This number was chosen as $\min\{2p, p+8\}$ for the computations shown in Figures 3 and 4. (p is the number of eigenvalues computed.) In particular note the reduction in time when finding 10 eigenvalues using 18 vectors compared to finding 8 eigenvalues with 16 vectors in Figure 4.

In order to use the block Lanczos algorithm written by David Scott, the generalized eigenvalue problem was first transformed to standard form using a Cholesky decomposition of the matrix M. A Lanczos blocksize of 2 was used, but other tests confirmed that the efficiency of the method is quite insensitive to small variations in the parameters. All tests were carried out on a VAX 11/780 computer.

The first test example is a membrane model problem (the Laplace operator) using square elements and bilinear basisfunctions. The results of this test are given in Figure 3.

The second test examle is an eigenvalue analysis of the jacket oil drilling platform modelled in Figure 2. The results of this example are given in Figure 4.

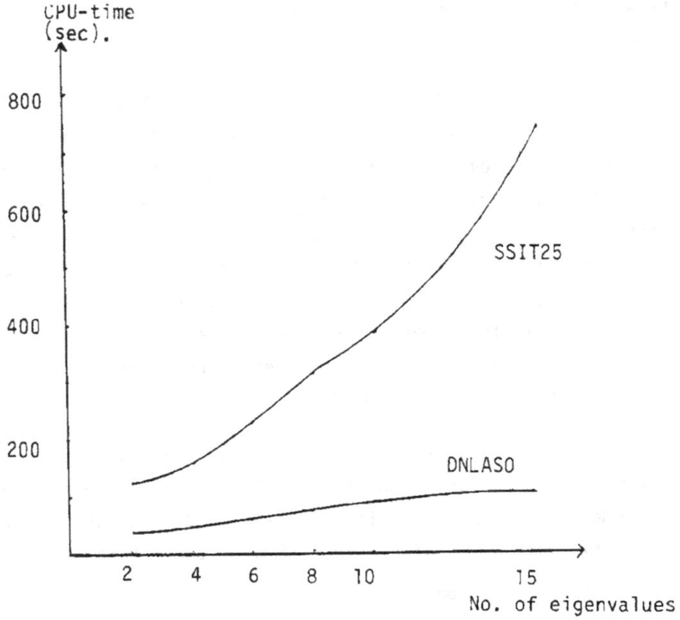

Fig. 3a: CPU-time as a function of eigenvalues found, for membrane problem with 108 degrees of freedom.

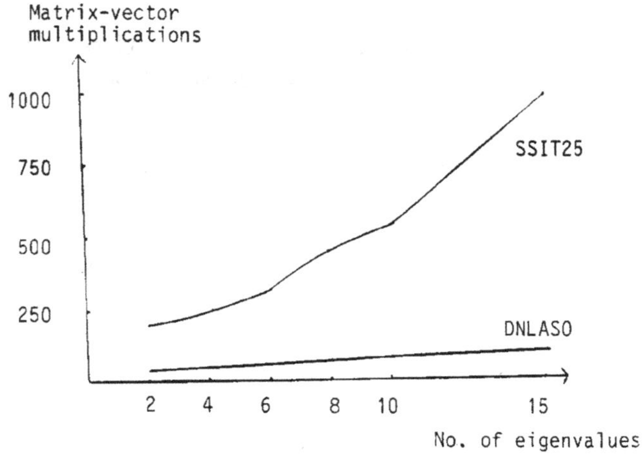

Fig. 3b: Matrix-vector product as a function of eigenvalues found, for a membrane problem with 108 degrees of freedom.

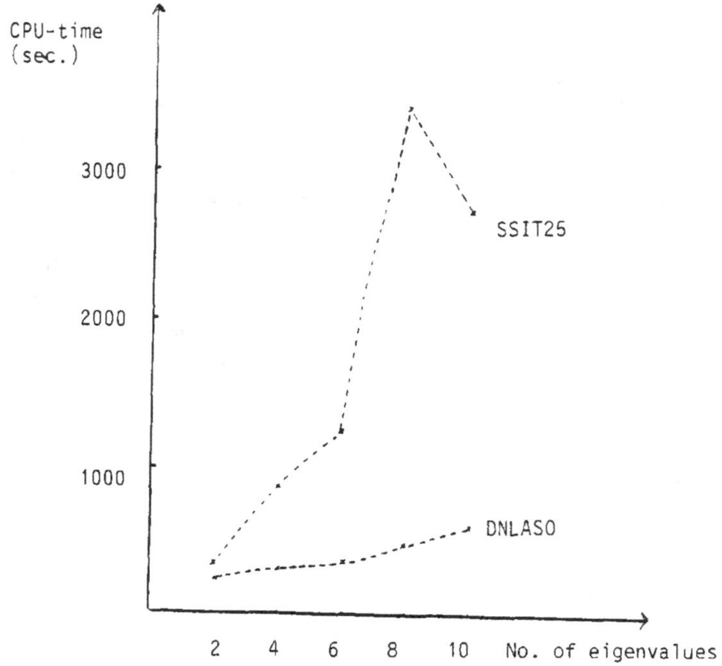

Fig. 4a: CPU-time as a function of eigenvalues found, for an oil-platform with 624 degrees of freedom.

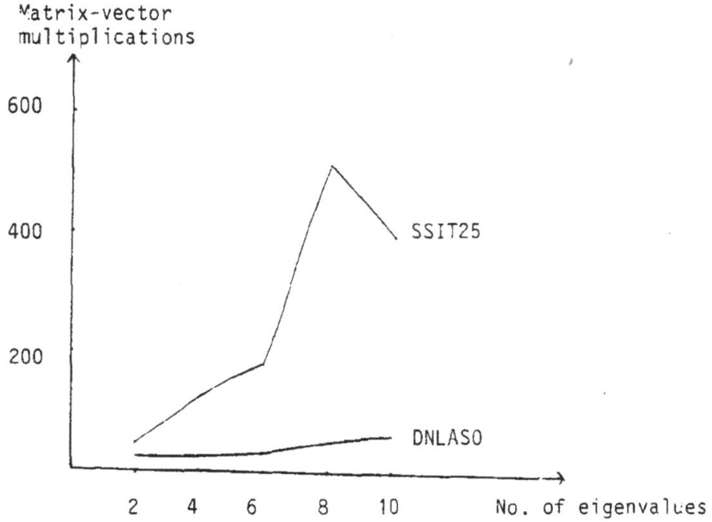

Fig. 4b: Matrix-vector multiplications as a function of eigenvalues found, for a jacket platform with 624 degrees of freedom.

4. CONCLUSION

The tests show that the Lanczos algorithm can be used successfully within a general finite element analysis package. It outperforms previously known methods that are widely used in structural engineering today. This does not mean that the various techniques for reducing the size of large finite element models are obsolete. Rather, it is our view that these and alternative methods deserve more investigation. A more solid theoretical basis for reduction techniques would help when developing more automatic algorithms for large scale eigenvalue calculations. In such procedures the reduction need not be carried as far as it is today, since it can be coupled with efficient implementations of the Lanczos algorithm.

REFERENCES

[1] BATHE, K.J. and WILSON, E.L.:
"Numerical methods in finite element analysis."
Prentice Hall, 1976.

[2] BELL, K., HATLESTAD, B., HANSTEEN, O.E. and ARALDSEN, P.O.:
"NORSAM User's Manual, Part 1 - General Description."
Trondheim 1973.

[3] CALLAGHER, R.M.:
"Approximation procedures, reduced stiffness and mass matrices. Substructuring. Component mode synthesis."
Course on advanced topics in finite element analysis, St. Margherita, Italy, June 3-7, 1974. Lecture series N.1/7, International centre for computer aided design, Genova.

[4] CRAIG, R.R. and Bampton, M.C.C.:
"Coupling of substructures for dynamic analysis."
AIAA Journal, Vol. 6, No. 7, 1968.

[5] HENSHELL, R.D. and ONG, I.M.:
"Automatic masters for eigenvalue economization."
Earhtq. Engng. Struct. Dyn., Vol. 3, 375-383, 1975.

[6] SCOTT, D.S.:
"Block Lanczos software for symmetric eigenvalue problems."
Computer Sciences Division, Oak Ridge National Laboratory, Nov.1979.

[7] SESAM'80, Project Description. October 1981.
Computas, 1322 Høvik, Norway.

[8] SKAAR, K.T. and CARLSEN, C.A.:
"Modelling aspects for finite element analysis of ship vibration."
Computer and Structures, Vol. 12, 409-419, 1980.

[9] ZIENKIEWICZ, O.C.:
"The finite element method". Chapter 20.5, third edition.
McGraw Hill 1977.

ON THE PRACTICAL USE OF THE LANCZOS ALGORITHM IN FINITE ELEMENT APPLICATIONS TO VIBRATION AND BIFURCATION PROBLEMS

by

E.G. CARNOY * and M. GERADIN **

Aerospace Laboratory, University of Liège, Belgium

ABSTRACT

Vibration and bifurcation analyses of structures modeled by finite elements yield a linear eigenvalue problem, $Kq = \lambda Bq$, where K and B are symmetric matrices of large dimension in practical applications. An iterative reduction of the matrix size is attained by the biorthogonal Lanczos algorithm which allows extraction of the lower eigenvalue spectrum. For solving the problem when coincident eigenvalues occur, a restart procedure is implemented so that further iterations can be performed from a new arbitrary vector, yielding thus to modifications in the interaction eigenvalue problem.

In addition, practical suggestions for the implementation of the method are made and efficiency of the proposed approach is demonstrated through several numerical examples.

1. SOME EIGENVALUE PROBLEMS IN STRUCTURAL ANALYSIS

Structural engineers are often faced whith problems such as determination of natural vibration frequencies, critical buckling loads, critical speeds, ... which take the form of a generalized eigenvalue problem.

For a conservative system the so-called tangent stiffness matrix K_T is symmetric. If the fundamental state described by the vector of degrees of freedom q_o corresponds to a stable state of equilibrium, this matrix $K_T(q_o)$ is also positive semi-definite. The possible zero energy modes correspond either to rigid body motions or to mechanisms of the structure.

In vibration analysis, the mass matrix M is symmetric and positive semi-definite in most cases. Zero mass modes can arise in simplified finite element models such as shell elements where rotatory inertia terms are neglected. The natural frequencies ω are solutions of the eigenvalue problem

$$K_T(q_o)q = \omega^2 M q \qquad (1.1)$$

Negative values for ω^2 would indicate that the fundamental state of equilibrium is unstable.

For critical rotation speed analysis, the centrifugal stiffness matrix K is the mass matrix of the same structure in which the inertia terms associated with axial displacement degrees of freedom have been suppressed. The critical speed ω corresponds

---

* Chargé de Recherche of the Belgian National Foundation of Scientific Research

** Professor

to the minimum eigenvalue of the problem

$$K_T(q_0)q = \omega^2 K_\omega q \qquad (1.2)$$

In stability analysis, the structure is initially assumed in a stable equilibrium state and is submitted to a proportional load increment with the load factor λ. The structural response can be obtained by a perturbation method as a power series of the load factor

$$q(\lambda) = q_0 + \lambda q_1 + \lambda^2 q_2 + \ldots \qquad (1.3)$$

This relation defines the fundamental path in the space (q, λ). For a conservative system, the fundamental path becomes unstable when the tangent stiffness matrix is no longer positive definite for increasing values of λ. Using (1.3), we can expand the tangent stiffness matrix in a power series of the load factor

$$K_T\left(q(\lambda)\right) = K_0 + \lambda K_1 + \lambda^2 K_2 + \ldots \qquad (1.4)$$

and the critical buckling load is the minimum positive eigenvalue of the bifurcation problem

$$(K_0 + \lambda K_1 + \lambda^2 K_2 + \ldots) q = 0 \qquad (1.5)$$

When both expansions (1.3) and (1.4) are limited to the first-order term the relation (1.5) defines a linear eigenvalue problem

$$K_0 q = \lambda S q \qquad (1.6)$$

with the stability matrix, $S = - K_1$. This gives a first approximation to the buckling load. Improvement of this approximation could be obtained at larger computer expense by solving the matrix pencil (1.5). A more efficient approach based on Koiter's theory of elastic stability [16] which uses the first few eigenmodes of (1.6) as a modal basis to reduce the system of equations (1.5) has been presented elsewhere [1, 2].

An alternative way to improve the first approximation of (1.5) consists in applying (1.6) in an incremental manner [20] which consists to perform the bifurcation analysis from successive deformed equilibrium configurations of the structure instead of the unstressed configuration ($q_0 = 0$) used in the initial bifurcation analysis.

In both cases the tangent stiffness matrix K_0 is symmetric positive definite while the stability matrix S is simply symmetric. The stability matrix is the sum of four terms

$$S = K_\sigma + K_u + K_p + K_M \qquad (1.7)$$

K_σ, the geometrical stiffness matrix, is the stability matrix of the Euler bifurcation problem, it is linear in the initial stresses associated with q_1 in (1.3). K_u is the initial displacement stiffness matrix, and is linear in the rotations associated with q_1 ; it takes into account the change of geometry due to the applied load [3]. K_p is the load stiffness matrix and is associated with a *lively* load such as a fluid pressure. For a conservative pressure, this matrix is also symmetric [17].

The last term K_M denotes a material stiffness matrix which accounts for the change of Young tangent modulus in the case of an elasto-plastic material [4]. For a proportional loading system, the finite theory of plasticity (deformation theory) is generally verified up to the buckling load. Then, the material stiffness matrix comes from the first-order term of the expansion of the elasto-plastic Hooke matrix in terms of the load factor.

The generalized eigenvalue problems to be solved takes thus the general form

$$K\,q = \lambda\,B\,q \tag{1.8}$$

where K is a symmetric and positive semi-definite stiffness matrix. The matrix B stands alternatively for M, K_ω and S and is thus also symmetric, but indefinite.

Remark

1) In some problems, it is convenient to introduce linear constraints between degrees of freedom

$$C\,q = 0 \tag{1.9}$$

by means of Lagrange multipliers ℓ. The extended eigenvalue problem takes the form

$$\begin{bmatrix} K & C^T \\ C & 0 \end{bmatrix} \begin{bmatrix} q \\ \ell \end{bmatrix} = \mu \begin{bmatrix} B & 0 \\ 0 & 0 \end{bmatrix} \begin{bmatrix} q \\ \ell \end{bmatrix} \tag{1.10}$$

in which case the extended stiffness matrix is no longer positive semi-definite.

2) As it will be seen, difficulties in the solution of (1.8) by the Lanczos method arise from the fact that the norm $q^t\,B\,q$ is not positive definite. Solving (1.8) in the inverse form

$$B\,q = \frac{1}{\lambda}\,K\,q \tag{1.11}$$

instead would avoid this problem, but would be unsatisfactory from engineering point of view, as it would yield higher eigenvalues first.
Another reason for rejecting the inverse form above is that B may contain a large number of singularities.

3) Also fo the latter reason, the alternate form

$$B\,q = \lambda\,B K^{-1}\,B\,q \tag{1.12}$$

is generally discarded.

4) The most adequate procedure to obtain a positive definite norm is to transform (1.8) in the squared form

$$K\,q = \lambda^2\,B\,K^{-1}\,B\,q \tag{1.13}$$

The signature of the eigenvalue spectrum has then to be recuperated afterwards.

2. THE BIORTHOGONAL LANCZOS ALGORITHM FOR SYMMETRIC MATRICES

The eigenvalue problem (1.8) involves symmetric matrices and only its lower eigenvalues are generally useful for engineering purpose.

The Lanczos method is a variant of the power method where the successive iterates are obtained according to

$$\begin{aligned} p_r &= B\,q_r \\ q_{r+1} &= A^{-1}\,p_r \end{aligned} \tag{2.1}$$

The matrix A^{-1} is the inverse of K or possibly, its pseudo-inverse. Biorthogonality is maintained between both sequences

$$q_{r+1}^{t}\, p_i = 0 \qquad p_{r+1}^{t}\, q_i = 0 \qquad i \leqslant r \tag{2.2}$$

by modifying (2.1) into

$$q_{r+1}^{*} = A^{-1}\, p_r - \alpha_r\, q_r - \beta_{r-1}\, q_{r-1} \tag{2.3}$$

$$p_{r+1}^{*} = B\, q_{r+1}^{*}$$

followed by a normalization of the new iterates

$$\gamma_{r+1} = \left| p_{r+1}^{*t}\, q_{r+1}^{*} \right|^{1/2} \tag{2.4}$$

$$q_{r+1} = q_{r+1}^{*} / \gamma_{r+1} \quad ; \quad p_{r+1} = p_{r+1}^{*} / \gamma_{r+1}$$

Care must be taken of the non positive definite property of the B matrix and the sign of the scalar product (2.4) is denoted by

$$\varepsilon_{r+1} = p_{r+1}^{t}\, q_{r+1} = \pm 1 \tag{2.5}$$

The coefficients in (2.3) are obtained as follows

$$\alpha_r = \varepsilon_r\, p_r^{t}\, A^{-1}\, p_r \tag{2.6}$$

$$\beta_{r-1} = \varepsilon_{r-1}\, p_{r-1}^{t}\, A^{-1}\, p_r = \varepsilon_{r-1}\, q_r^{*t}\, p_r = \varepsilon_{r-1}\, \varepsilon_r\, \gamma_r$$

The recurrence relations (2.3) can be written in the matrix form

$$A^{-1}\, B\, \{q_0\, q_1\, \dots\, q_r\} = \{q_0\, q_1\, \dots\, q_r\}\, T_r + \{0\, \dots\, q_{r+1}^{*}\} \tag{2.7}$$

with the tridiagonal matrix

$$T_r = \begin{bmatrix} \alpha_0 & \beta_0 & & & & \\ \gamma_1 & \alpha_1 & \beta_1 & & & 0 \\ & \gamma_2 & \alpha_2 & \cdot & & \\ & & \cdot & \cdot & \cdot & \\ & 0 & & \cdot & \cdot & \beta_{r-1} \\ & & & & \gamma_r & \alpha_r \end{bmatrix} \tag{2.8}$$

In order to obtain the interaction eigenvalue problem we premultiply equation (2.7) at step r by the orthogonal sequence $\{p_0\, p_1\, \dots\, p_r\}^t$. If use is made of (2.2) we obtain the matrix equation

$$\{p_0\, p_1\, \dots\, p_r\}^t\, A^{-1}\, B\, \{q_0\, q_1\, \dots\, q_r\} = \{p_0\, p_1\, \dots\, p_r\}^t\, \{q_0\, q_1\, \dots\, q_r\}\, T_r \tag{2.9}$$

Let us next denote by a an eigenvector of T_r associated with the eigenvalue μ.

With the definition of the two successive iterates

$$v_0 = \{ q_0\ q_1\ \dots\ q_r \}\ a \qquad\qquad v_1 = A^{-1}\ B\ v_0$$

equation (2.9) yields

$$\mu = \frac{v_1^t\ B\ v_0}{v_0^t\ B\ v_0} = \frac{1}{\lambda} \tag{2.10}$$

Therefore the eigenvalues of T_r are the Schwartz quotients [7] that can be constructed in the orthogonal subspace spanned by the vectors $\{q_0\ \dots\ q_r\}$. They are thus expected to converge rapidly to the eigenvalues of (1.8).

3. RESTART PROCEDURE

When coincident eigenvalues occur, the foregoing iteration scheme gives only one eigenmode for each multiple eigenvalue. Indeed, in the initial trial vector q_0, the eigenmodes associated with the same eigenvalue are represented through a linear combination, the coefficients of which are multiplied by the same factor in each iteration. Thus, there is only one combination of these eigenmodes in the sequence $\{ q_0\ q_1\ \dots\ q_r\}$ and their separation can not be performed.

In order to avoid this drawback, Golub et al. [12] proposed to replace the trial vector q_0 by a set of independent trial vectors. Then, each iteration (2.3) yields a set of vectors and orthogonality is kept between the vectors of each iteration and also between the vectors of the same iteration. This process yields to a block tridiagonal matrix.

A more economical solution consists in developing a restart procedure when a given number n of iterations have been performed on the initial trial vector q_0. A new trial vector u is chosen orthogonal to the first sequence

$$u_0^t\ p_i = 0 \qquad\qquad i = 1, \dots n \tag{3.1}$$

and the recurrence relation (2.3) is modified into

$$\begin{aligned} u^*_{r+1} &= A^{-1}\ v_r - \alpha_{n+r}\ u_r - \beta_{n+r-1}\ u_{r-1} - \sum_0^n \zeta_i^r\ q_i \\ v^*_{r+1} &= B\ u^*_{r+1} \\ \gamma_{n+r+1} &= | u^{*t}_{r+1}\ v^*_{r+1} |^{1/2} \quad ; \quad u_{r+1} = \frac{u^*_{r+1}}{\gamma_{n+r+1}} \quad ; \quad v_{r+1} = \frac{v^*_{r+1}}{\gamma_{n+r+1}} \end{aligned} \tag{3.2}$$

which ensures the biorthogonality relations

$$\begin{aligned} u_r^t\ p_i &= v_r^t\ q_i = 0 \qquad\qquad i = 1, \dots n \\ u_r^t\ v_s &= u_s^t\ v_r = 0 \qquad\qquad r < s \end{aligned} \tag{3.3}$$

The coefficients of (3.2) are obtained in the same way as (2.5) and (2.6) while the ζ's are given by

$$\zeta_i^r = \varepsilon_i\ p_i^t\ A^{-1}\ v_r$$

which take a non zero value due to the existence of a remaining term q^*_{n+1} in (2.7).

Using (2.3) and the orthogonality property (3.3), we have

$$\zeta_i^r = \varepsilon_n \, v_r^t \, q_{n+1}^* \, \delta_{ni}$$

where δ_{ni} denotes the Kronecker symbol. This relation is transformed with (2.3) into

$$\zeta_i^r = \varepsilon_n \, v_r^t \, A^{-1} \, p_n \, \delta_{ni} \tag{3.4}$$

and allows rewriting the recurrence relation (3.2) in the simpler form

$$\gamma_{n+r+1} \, u_{r+1} = A^{-1} \, v_r - \alpha_{n+r} \, u_r - \beta_{n+r-1} \, u_{r-1} - \zeta_n^r \, q_n \tag{3.5}$$

Let us next gather the recurrence relations (2.3) and (3.2) in a matrix form and denote by $X_{n+r} = \{ q_0 \ldots q_n \; u_0 \ldots u_r \}$ the set of vectors at the iteration $(n + r)$

$$A^{-1} \, B \, X_{n+r} = X_{n+r} + \{ 0 \ldots q_{n+1}^* \; 0 \ldots u_{r+1}^* \} \tag{3.6}$$

The matrix H_{n+r} is tridiagonal except for one line containing the ζ's

$$H_{n+r} = \left[\begin{array}{c|c} T_n & 0 \\ \hline 0 & T_r \end{array} \right] \leftarrow \{ \zeta_n^r \} \tag{3.7}$$

In order to obtain the interaction matrix, we premultiply (3.6) by $X_{n+r}^t \, B$

$$X_{n+r}^t \, B A^{-1} \, B X_{n+r} = X_{n+r}^t \, B X_{n+r} \, H_{n+r} + \{ p_0 \ldots p_n \; v_0 \ldots v_r \}^t \{ 0 \ldots q_{n+1}^* \; 0 \ldots u_{r+1}^* \} \tag{3.8}$$

The last term of this relation yields a null matrix except for the column n which corresponds to

$$\{ p_0 \ldots p_n \; v_0 \ldots v_r \}^t \, q_{n+1}^* = \{ 0 \ldots 0 \; v_0 \ldots v_r \}^t \, q_{n+1}^* \tag{3.9}$$

Taking into account that $X_{n+r}^t \, B \, X_{n+r} = \mathrm{diag}\,(\varepsilon_i)$ we obtain

$$X_{n+r}^t \, B \, A^{-1} \, B \, X_{n+r} = X_{n+r}^t \, B \, X_{n+r} \, \mathcal{H}_{n+r} \tag{3.10}$$

where the interaction matrix $\mathcal{H}_{n+r}$ is the sum of two matrices

$$\mathcal{H}_{n+r} = \left[\begin{array}{c|c} T_n & 0 \\ \hline 0 & T_r \end{array} \right]_{\leftarrow \zeta\text{'s}} + \left[\begin{array}{c|c} 0 & 0 \\ \hline 0 & 0 \end{array} \right]_{\xi\text{'s}} = \left[\begin{array}{c|c} T_n & 0 \\ \hline 0 & T_r \end{array} \right] \tag{3.11}$$

The elements of the last one are denoted ξ_n^r and take the values

$$\xi_n^r = \varepsilon_{n+r} \, v_r^t \, q_{n+1}^* = \varepsilon_{n+r} \, v_r^t \, A^{-1} \, p_n$$

or

$$\xi_n^r = \varepsilon_n \, \varepsilon_{n+r} \, \zeta_n^r \tag{3.12}$$

This restart procedure can be employed several times and the foregoing developments are easily generalized to give the following recurrence relation and interaction matrix

$$w_{r+1}^* = A^{-1} \, B \, w_r - \alpha_{n+m+r} \, w_r - \beta_{n+m+r-1} \, w_{r-1} - \zeta_n^{m+r} \, q_n - \zeta_{n+m}^r \, u_m \tag{3.13}$$

$$
H_{n+m+r} = \left[\begin{array}{c|c|c} T_n & 0 & 0 \\ \hline 0 & T_m & 0 \\ \hline 0 & 0 & T_r \end{array}\right] \begin{array}{l} \leftarrow \zeta_n \\ \leftarrow \zeta_{n+m} \\ \\ \end{array} \qquad \xi_n \nearrow \quad \xi_{n+m} \nearrow \tag{3.14}
$$

To extract the eigenvalues of H, the interaction matrix is first transformed into an upper Hessenberg matrix and then solved by the QR method (procedures PREHQR and HQR, ref. 22).

4. REORTHOGONALIZATION

When the foregoing method is applied in its crude form, a rapid loss of orthogonality is observed between the two sequences of vectors q_r and p_r . As observed by Golub et al. [13] , departure from orthogonality is the result of cancellation when computing q_{r+1} and p_{r+1} from (2.3) and not the result of accumulation of rounding errors.

In order to be certain of obtaining the full set of eigensolutions it is necessary to ensure that the computed q_r are orthogonal to working accuracy. The conventional way of restoring orthogonality with all previously computed vectors is the well-known Schmidt process. The relation (2.3) is then transformed into

$$
\begin{aligned}
q^0_{r+1} &= A^{-1}\, p_r \\
p^0_{r+1} &= B\, q^0_{r+1}
\end{aligned} \tag{4.1}
$$

with the recurrence relations

$$
\begin{aligned}
q^i_{r+1} &= q^{i-1}_{r+1} - (q^{i-1}_{r+1})^t\, p_{i-1}\, \varepsilon_{i-1}\, q_{i-1} \\
p^i_{r+1} &= p^{i-1}_{r+1} - (q^{i-1}_{r+1})^t\, p_{i-1}\, \varepsilon_{i-1}\, p_{i-1}
\end{aligned} \tag{4.2}
$$

for $i = 1, \ldots, r+1$

and finally

$$
\begin{aligned}
\gamma^2_{r+1} &= \left| (q^{r+1}_{r+1})^t\, p^{r+1}_{r+1} \right| & \varepsilon_{r+1} &= (q^{r+1}_{r+1})^t\, p^{r+1}_{r+1} \,/\, \gamma^2_{r+1} \\
q_{r+1} &= q^{r+1}_{r+1} \,/\, \gamma_{r+1} & p_{r+1} &= p^{r+1}_{r+1} \,/\, \gamma_{r+1}
\end{aligned} \tag{4.3}
$$

Check of orthogonality can be performed by comparing the coefficients of the interaction matrix which are obtained before and after reorthogonalization, namely

$$
\begin{aligned}
\alpha^e_r &= \varepsilon_r\, p^t_r\, q^0_{r+1} & \alpha_r &= \varepsilon_r\, p^t_r\, q^r_{r+1} \\
\beta^e_r &= \varepsilon_{r-1}\, p^t_{r-1}\, q^{r-1}_{r+1} & \beta_r &= \varepsilon_r\, \varepsilon_{r+1}\, \gamma_{r+1}
\end{aligned} \tag{4.4}
$$

where $(\ )^e$ denotes a first estimation. In some problems involving an even small number of degrees of freedom it was found that the relative error could be larger than the required accuracy. This is related to the occurrence in a previous iteration of a difference of magnitude orders in the absolute values of the coefficients

$$\gamma_{r+1} , \alpha_r \text{ and } \beta_{r-1}$$

in (2.3). In this case, it is advisable to reorthogonalize the new vectors q^0_{r+1} and p^0_{r+1} twice by using (4.2). This procedure has appeared to be sufficient and the iterative Schmidt process proposed by Ojalvo and Newman [18] does not seem necessary.

An alternative to the Schmidt process has been proposed by Golub et al. [13] in the symmetric case and generalized to the unsymmetric case by Geradin [8] . Orthogonalization is then performed by using elementary hermitian matrices. These matrices allow transformation of the iteration vectors $\{ q_0 , \dots q_r \}$ into $\{\alpha_0 e_0 , \dots \alpha_r e_r \}$ where e_i denotes a unit vector. The choice of the direction e_{r+1} is based on the largest product of the corresponding elements of q_{r+1} and p_{r+1} after orthogonalization to the previously computed base vectors, and such that this product has the same sign as ε_{r+1} . However, this second procedure may lead to difficulties in the presence of linear constraints in the stiffness matrix since all the unit vectors e_i can no longer be considered as independent.

5. COMPUTATIONAL IMPLEMENTATION

Several features of the Lanczos algorithm implemented in the finite element software SAMCEF [21] are worth noting and are described in the following.

5.1 Factorization of the stiffness matrix

The most significant time-consuming step in the foregoing process is contained in the equations (4.1). The B matrix is not assembled and calculation of p^0_{r+1} is performed by accumulation of the contributions of each finite element

$$p^0_{r+1} = \sum_1^{n_e} L_e \, B_e \, L_e^t \, q^0_{r+1} \tag{5.1}$$

n_e denotes the number of finite elements and L_e , B_e are the incidence matrix and the B matrix of an element, respectively. The other step is equivalent to the solution of the linear system

$$K \, q^0_{r+1} = p_r \tag{5.2}$$

The equation solver uses a Gauss elimination method with a maximum pivot strategy. As shown in reference [9] , it can be implemented even in positive semi-definite and non positive definite cases without any artificial transformation of the initial matrix K such as frequency shifting [19] . A frontal technique organized in substructures is employed which allows factorization of the stiffness matrix

$$K = L \, D \, L^t \tag{5.3}$$

where D denotes a block diagonal matrix and L is a lower triangular matrix with unit diagonal blocks. For two substructures, for example, we have the following decomposition

$$K = \begin{bmatrix} K_{11} & K_{12} \\ K_{21} & K_{22} \end{bmatrix} \qquad D = \begin{bmatrix} K_{11} & 0 \\ 0 & K^*_{22} \end{bmatrix} \qquad L = \begin{bmatrix} I & 0 \\ K_{21} K^{-1}_{11} & I \end{bmatrix} \tag{5.4}$$

with $K^*_{22} = K_{22} - K_{21} \, K^{-1}_{11} \, K_{12}$.

Inversion of the stiffness matrix is straighforward by

$$K^{-1} = L^{-t} \, D^{-1} \, L^{-1}$$

$$\text{with} \qquad L^{-1} = \begin{bmatrix} I & 0 \\ -K_{21}\,K_{11}^{-1} & I \end{bmatrix} \quad \text{and} \quad D^{-1} = \begin{bmatrix} K_{11}^{-1} & 0 \\ 0 & K_{22}^{*-1} \end{bmatrix} \tag{5.5}$$

These relations are easily generalized to a large number of substructures in an iterative way such that the only elements K_{11}^{-1} and $K_{21}\,K_{11}^{-1}$ of D^{-1} and L respectively are to be stored for each substructure. Then, the solution of (5.2) is performed in two steps, namely the condensation

$$x = D^{-1}\,L^{-1}\,p_r \tag{5.6}$$

and the restitution

$$q^{0}_{r+1} = L^{-t}\,x \tag{5.7}$$

which involves the reading of the elements $K_{21}\,K_{11}^{-1}$ in the reverse order of substructures. As advocated in [19] , the backward solution (5.7) is much more time-consuming than the forward solution unless we define the reverse matrix $L_{Rev} = (L^{-t})\text{reverse}$ such that the elements $K_{21}\,K_{11}^{-1}$ are stored in the reverse order of substructures. Then the file containing the factorized stiffness matrix is organized as follows

$$(K_{11}^{-1}\,,\,K_{21}\,K_{11}^{-1})\,\ldots\,(K^{*-1}_{n-1,n-1}\,,\,K_{n,n-1}\,K^{*-1}_{n-1,n-1})\,(K^{*-1}_{n,n})\,(K_{n,n-1}\,K^{*-1}_{n-1,n-1})\,\ldots\,(K_{21}\,K_{11}^{-1}) \tag{5.8}$$

where n denotes the number of substructures. The solution of the linear system (5.2) only requires reading (5.8) in sequence.

5.2 Occurence of kinematical modes

The case of a singular stiffness matrix can be included in the previous scheme of factorization provided that the kinematical modes are contained in the stiffness matrix of the last substructure. If these modes are known a priori it is always possible to satisfy this condition by retaining appropriate degrees of freedom up to the last substructure. In this case the reduced stiffness matrix K^*_{nn} is singular to working accuracy and Gauss'elimination with a maximum pivot strategy yields the following matrix

$$\begin{bmatrix} K_{cc}^{-1} & -\,K_{cc}^{-1}\,K_{cr} \\ -\,K_{rc}\,K_{cc}^{-1} & 0 \end{bmatrix} \tag{5.9}$$

with a zero diagonal block. The pseudo-inverse matrix of K^*_{nn} is defined as

$$(K^*_{nn})^{-1} = \begin{bmatrix} K_{cc}^{-1} & 0 \\ 0 & 0 \end{bmatrix} \tag{5.10}$$

which gives the matrix A^{-1} by (5.5). The kinematical modes in the last substructure are defined by the columns of the matrix

$$x = \begin{bmatrix} -K_{cc}^{-1}\,K_{cr} \\ I \end{bmatrix} \tag{5.11}$$

The backward solution

$$q_R = L^{-t}\,x \tag{5.12}$$

yields the kinematical modes in the complete system. These modes are then orthonormalized with respect to the B matrix

$$p_i = B\, q_i \quad ; \quad p_i^t\, q_j = \delta_{ij} \quad ; \quad i,j = -r_0\, , -r_0 + 1\, , \ldots \tag{5.13}$$

where r_0 denotes the number of kinematical modes. In some cases, these kinematical modes may correspond to a zero eigenvalue of the B matrix, too, and may then be simply ignored in the solution of the eigenvalue problem (1.8). Otherwise, these modes are to be taken into account in the reorthogonalization process such that the index i in (4.2) takes the values $-(r_0 - 1)$, ... 1, ..., r + 1 with

$$q_{r+1}^{-r_0} = A^{-1}\, p_r \quad ; \quad p_{r+1}^{-r_0} = B\, q_{r+1}^{-r_0} \tag{5.14}$$

The rigid body modes occur in the case of free vibration analysis. These modes have a global pattern and are generally well represented in the degrees of freedom of the last substructure. However, for particular geometries, it is not the case and loss of accuracy may occur in the Lanczos algorithm. Experience has shown that adding a few degrees of freedom randomly distributed to the front of equations of the last substructure is generally sufficient to restore the accuracy. When a shift is performed this problem may become more critical. The modified eigenvalue problem takes the form

$$(K - \lambda_1\, B)\, q = \lambda^*\, B\, q \tag{5.15}$$

and the pseudo inverse matrix

$$A^{-1} = (K - \lambda_1\, B)^{-1} = L^{-t}\, D^{-1}\, L^{-1} \tag{5.16}$$

may have a kinematical mode if λ_1 is solution of the initial eigenvalue problem. However, this mode may correspond to a local mode located far from the last substructure or poorly represented in this substructure. In this case, the choice of suitable degrees of freedom to be retained in the last substructure is quite more difficult. On the other hand, the number of negative pivots in D^{-1} (5.16), allows determination of the number of eigenvalues contained between zero and λ_1 by the difference of this number and the one corresponding to D^{-1} in (5.5). This property has been widely used by Ericsson and Ruhe [5] in their algorithm.

5.3 Choice of starting vectors

As pointed out in [19] , the number of independent modes in the system (1.8) is limited by the rank of the B matrix, and a random starting vector may not be included in the exact eigenvector space. Moreover, the tangent stiffness matrix may involve linear constraints which will not be satisfied by an arbitrary vector. Therefore, in order to restrict the complete space to the appropriate one, the arbitrary vector q_0^* , is transformed into

$$q_0^{-r_0} = A^{-1}\, B\, q_0^* \tag{5.17}$$

$$p_0^{-r_0} = B\, q_0$$

which are then orthogonalized with respect to the possible kinematical modes. The same procedure is employed for each restarting vector (3.1). If only the eigenvalues are required and not the eigenmodes, this starting procedure is not necessary since it does not affect the coefficients of the interaction matrix.

5.4 Convergence strategy

In the classical Lanczos algorithm the sequence (2.3) is continued until $p_r^t\, q_r = 0$ (breakdown) or $r = N - 1$ (normal termination) where N is the dimension of the effective space. In case of breakdown the coefficient γ_r is zero with respect to the eigenvalues of the interaction matrix. The pattern of the tridiagonal matrix (2.8) then shows that a complete subspace of vectors $\{ q_0 \; \dots \; q_{r-1} \}$ has been isolated if either p_r or q_r or both vectors are zero.

The dead end breakdown corresponding to $(p_r^t \, q_r = 0, \; p_r \neq 0, \; q_r \neq 0$ for $r < N - 1)$ may only occur if the B matrix is not positive semi-definite as it is generally the case in bifurcation analysis. According to Faddeev and Faddeeva [6] , dead end breakdown is very unlikely to occur. In any case of breakdown the restart procedure (3.5) can be employed and except for dead end breakdown the coupling terms ζ are zero which simplifies the search of the eigenvalues of the interaction matrix. If breakdown still occurs with the restarting vector, we conclude that all the vectors of the effective space N have been found and the problem is solved. Generally the number of required eigenvalues s is much lower than N and the Lanczos sequence is truncated. Convergence criterion is based on the eigenvalues of the interaction matrix

$$H \, a = \mu \, a \tag{5.18}$$

Let $| \mu_1^{(r)} | \geqslant | \mu_2^{(r)} | \geqslant \dots \geqslant | \mu_r^{(r)} |$ and $\{ a_{(1)} \; \dots \; a_{(r)} \}$ be the eigensolutions of (5.18) at step r. Convergence of the Lanczos sequence is reached when

$$| \mu_k^{(r+1)} - \mu_k^{(r)} | < \varepsilon \, | \mu_k^{(r)} | \qquad k = 1, \dots s \tag{5.19}$$

where ε denotes the precision required on the eigenvalues. This simple criterion which does not require the calculation of the first s eigenmodes $a_{(i)}$ at each step has been proved to be efficient. A more elaborated criterion can be found in the literature [5, 19] which employs the eigenmodes $a_{(i)}$. Let us assume that the criterion (5.19) is satisfied after n_1 iterations (2.3). The restart procedure is then employed in order to investigate the possibility of coincident eigenvalues. The iteration process (3.5) is continued until the criterion (5.19) is satisfied for an additional eigenvalue (s + 1). If this eigenvalue is coincident the number s is incremented by one and the same procedure is repeated with possibly a third starting vector and so on. Otherwise, the Lanczos sequence is stopped and the eigenmodes are calculated as described below. For each starting vector, a maximum number of iterations n_1 , n_2 , ... can be defined. The following choice is usually made

$$n_1 = 3 \, s + 5 \quad ; \quad n_i = 5 \qquad i > 2 \tag{5.20}$$

For some ill-conditioned problems, the number n of iterations may be not sufficient to ensure satisfaction of the convergence criterion (5.19). As in the case of nearly coincident eigenvalues, the change of starting vector is expected to improve the rate of convergence of the algorithm.

5.5 Eigenmodes and error analysis

Once the eigenvalues $(\mu_1 \; , \; \dots \; \mu_s)$ of H have been determined, inverse iteration provides an efficient algorithm for computing the corresponding eigenvectors $a_{(1)}$, ... $a_{(s)}$ (procedure INVIT, ref. 22). To restore the eigenvectors of the initial problem (1.8), we return to equation (3.10) from which we deduce that the approximation to $q_{(k)}$ contained in the subspace X is

$$q_{(k)} = X \, a_{(k)} \tag{5.21}$$

If error bounds to the initial eigenproblem (1.8) are needed they can be obtained from the bracketing algorithm [10] . The error analysis is rendered straightforward by the fact that the Lanczos algorithm provides directly the Schwartz quotients (2.10) associated to the approximate eigenvectors

$$\rho_1 = \lambda_k = \frac{q_{(k)}^t \; B \; q_{(k)}}{q_{(k)}^t \; B \; A^{-1} \; B \; q_{(k)}} \tag{5.22}$$

Error bounds can be computed with the associated Rayleigh quotients

$$\rho_0 = \frac{q^t_{(k)}\ K\ q_{(k)}}{q^t_{(k)}\ B\ q_{(k)}} \tag{5.23}$$

For a positive semi-definite stiffness matrix, the positive coefficient

$$\sigma^2 = \frac{\rho_0}{\rho_1} - 1 \tag{5.24}$$

can be regarded as an "error measure coefficient" associated with the approximate modal shape $q_{(k)}$, since it vanishes when $q_{(k)} = \lambda_{(k)}\ A^{-1}\ B\ q_{(k)}$. Convergence to the eigenvalue is generally sufficient if $\sigma < 10^{-2}$.

6. NUMERICAL EXAMPLES

6.1 Free vibration analysis of the planet arm of a planetary gear train

The structure represented in fig. 1 has been modeled by 288 plane shell finite elements yielding to a total of 3912 degrees of freedom. The finite element discretization is shown for one eighth of the structure. The factorization of the stiffness matrix is performed with 183 substructures, with a maximum front width of 276. The last substructure contains 27 d.o.f. corresponding to one of the triangular elements along the edge of the plate. Five eigenvalues are requested with the accuracy $\varepsilon = 10^{-8}$ in (5.19). After scaling of the stiffness matrix, the factorization yields the six rigid body modes with a 10^{-9} accuracy. 17 iterations are needed with two starting vectors. The first seven natural frequencies are obtained as follows

ω_1	3.710128
ω_2 , ω_3	3.990583 , 3.990639
ω_4 , ω_5	4.819108 , 4.819165
ω_6 , ω_7	5.645672 , 5.645744

and the error measure coefficients (5.24) are lower than 5.10^{-8} for the five requested eigenvalues. The computer time consumed for the extraction of the eigenvalues is about 60 % of the time needed for the factorization process. In addition to the consistent mass matrix, several concentrated masses have been introduced next to make a second free vibration analysis of the same structure. 16 eigenvalues are requested with a 10^{-8} precision. To this end, 40 iterations are performed with three starting vectors, which consume 112 % of the factorization computer time. The first twenty natural frequencies are

ω_1 , ω_2	1.891263 , 1.891276	ω_{10} , ω_{11}	6.398325 , 6.398511
ω_3	2.248469	ω_{12} , ω_{13}	8.311802 , 8.311869
ω_4 , ω_5	2.635696 , 2.635712	ω_{14} , ω_{15}	10.90717 , 10.90932
ω_6 , ω_7	3.347863 , 3.347888	ω_{16} , ω_{17}	10.92499 , 10.92750
ω_8	3.373715	ω_{18} , ω_{19}	11.33196 , 10.34597
ω_9	5.150341	ω_{20}	11.44350

Since eigenvalues 16 and 17 are coupled, the number of requested eigenvalues is put equal to 17. All the error coefficients of these eigenvalues are lower than 5.10^{-8} except the following ones

$\sigma_{14} = 4.10^{-7}$, $\sigma_{15} = 3.10^{-2}$, $\sigma_{16} = 7.10^{-7}$, $\sigma_{17} = 6.10^{-2}$

It was concluded that the poor approximation of the null pivots corresponding to the rigid body modes was the reason for the occurence of large error measure coefficients. It was decided to retain in the last substructure the three displacement components of five nodes, three fo which were located on the sun gear and the two others on the planetary. The largest null pivot was then equal to $1.8\ 10^{-11}$ and convergence of the first 17 eigenvalues was achieved with $\sigma = 10^{-8}$ after 50 iterations and using three starting vectors.

6.2 Plastic buckling of a clamped spherical cap

The next example consists in the plastic bifurcation analysis of the clamped spherical cap shown in fig. 2. This very simple structure is submitted to a uniform external pressure. A complete non linear analysis of this structure can be found in [15] and the corresponding buckling pressure is $\lambda_{ref} = 2\,060$ psi.

The structure is modeled by 5 isoparametric volume finite elements which employ an uncoupled Hooke matrix for the normal stress and a reduced integration technique [11]. This yields to a very small eigenvalue problem of 98 degrees of freedom. We have chosen this example in order to show the very different nature of the bifurcation eigenvalue problem as a function of the stability matrix. Three analyses have been performed as described below.

- For the Euler bifurcation problem, the stability matrix is reduced to the geometrical stiffness matrix. Indeed, the load stiffness matrix does not have any effect in the case of a shallow shell. The eigenvalue problem is well-conditioned and 10 iterations with 2 starting vectors are required to obtain the first two bifurcation loads with an error measure coefficient less than $3\ 10^{-11}$. The first buckling loads are in [10^3 psi]

$$\lambda_1 = 14.10006$$
$$\lambda_2 = 26.13725$$
$$\lambda_3 = 39.70582$$

- For the elastic bifurcation analysis, the stability matrix is the sum of the geometrical stiffness matrix and the displacement stiffness matrix. The eigenvalue spectrum is a slightly more clustered and 14 iterations are needed with 2 starting vectors to obtain the first three buckling loads with an error measure coefficient $\sigma < 5\ 10^{-10}$, namely

$$\lambda_1 = 7.903000$$
$$\lambda_2 = 11.39269$$
$$\lambda_3 = 17.81012$$
$$\lambda_4 = 26.42743$$

- Finally, for the plastic bifurcation analysis, the material stiffness matrix is added (see appendix) to the precedent stability matrix. As could be expected, the eigenvalue problem becomes ill-conditioned with a very clustered eigenvalue spectrum. This is the important feature of the plastic bifurcation analysis, especially when the stress state of the linear approximation of the fundamental path is rather uniform. Three eigenvalues were requested with a 10^{-8} accuracy but after 24 iterations with 3 starting vectors, the convergence criterion (5.19) was not satisfied. The eigenvalues are obtained as follows

$\lambda_1 = 1.812456$	$\lambda_5 = 2.425991$	$\lambda_9 = 2.855203$
$\lambda_2 = 1.875896$	$\lambda_6 = 2.535616$	$\sigma_1 \quad 6.7\ 10^{-4}$
$\lambda_3 = 2.121697$	$\lambda_7 = 2.585674$	$\sigma_2 \quad 7.6\ 10^{-5}$
$\lambda_4 = 2.322662$	$\lambda_8 = 2.752342$	$\sigma_3 \quad 1.3\ 10^{-2}$

The approximation λ_1 of the plastic buckling load is 12 % lower than the reference buckling load. This rather small discrepancy is explained by the linearization of the fundamental path which neglects the important stress redistribution that occurs in the plastic zones. On the other hand, it is worth noting that the bifurcation analysis is much less time consuming than a complete non linear analysis. Moreover, this result could be improved by an incremental bifurcation analysis [4] .

- Now, considering the solution of the eigenvalue problem, we can expect improvement of the algorithm efficiency by transforming the initial eigenvalue problem into

$$K\ q = \lambda^2\ B\ K^{-1}\ B\ q \qquad (6.1)$$

When the eigenvalues have been extracted, the sign of these can be found by calculation of the corresponding Rayleigh quotient of the initial problem (5.23). On one hand, the eigenvalue problem (6.1) is a little more well-conditioned than (1.8) by working with the square of the eigenvalues and by using a positive semi-definite norm $B^* = B\ K^{-1}\ B$ instead of the non definite matrix B.
On the other hand, the eigenvalue spectrum of (6.1) may become more dense in presence of negative eigenvalues in (1.8). Moreover, the computer time consumed for each iteration of (6.1) is about twice larger than the one corresponding to the initial eigenvalue problem. In the present problem of plastic bifurcation, the lower part of the eigenvalue spectrum of (1.8) corresponds to positive eigenvalues so that the only positive effect is expected to occur by working with (6.1) instead of (1.8). Indeed, 20 iterations with 3 starting vectors are required to obtain the first three buckling loads with a 10^{-8} precision for (5.19).

$\lambda_1 = 1.812411$	$\lambda_5 = 2.425257$	$\lambda_9 = 2.797088$
$\lambda_2 = 1.875890$	$\lambda_6 = 2.517797$	$\sigma_1 = 1.3\ 10^{-9}$
$\lambda_3 = 2.120176$	$\lambda_7 = 2.558439$	$\sigma_2 = 3.9\ 10^{-10}$
$\lambda_4 = 2.32091$	$\lambda_8 = 2.685792$	$\sigma_3 = 2.8\ 10^{-6}$

It should be noted that the first two buckling loads are stabilized with 7 digits after 14 iterations.

6.3 Elastic buckling of a tilted cylindrical shell with fluid

A tilted cylindrical shell is subjected to internal hydrostatic loading, as shown fig. 3. Hughes et al. [14] have performed a complete non linear analysis with a rather crude mesh of 9 × 6 HS2 elements. They defined an effective specific gravity $\gamma_{eff} = \gamma\ g$ and buckling occured for the value $g = 1.8$. In the following, the load factor will be defined by $\gamma_{eff} = 1.8\ \gamma\ \lambda$.

As presented en [14] , the prebuckling circumferential stress is positive everywhere except in a very small neighborhood of the clamped edge. The axial stress is negative for $\phi = 0$ and positive at $\phi = \pi$. Thus, it is expected that the buckling mode will have a small number of waves in the circumferential direction and a large number of waves in the axial direction. Therefore, the finite element model will be more refined in the axial direction than in the circumferential direction.

For the bifurcation analysis, the stability matrix is the following

$$S = (K_\sigma + K_u + K_p)$$

and a first crude mesh of 9 × 6 volume elements [11] is employed in order to have some idea of the eigenvalue spectrum. We have 1008 degrees of freedom and only the first eigenvalue is requested. We obtained the following load factors after 8 iterations

$\lambda_1 = 0.1142$	$\lambda_2 = -\ 0.1529$	$\lambda_3 = -\ 0.2274$
$\lambda_4 = -\ 0.3997$	$\lambda_5 = -\ 0.5700$	$\lambda_6 = 0.756$

The first buckling load is parasitic with a $\sigma_1 = 10^2$. Such a phenomenon may occur when the matrices K and S are not both positive semi-definite since the Schwartz inequality is not satisfied in this case. Indeed, the Schwartz quotient may be much lower than the lowest eigenvalue in absolute value although the Rayleigh quotient is always larger than this eigenvalue for K positive definite.

The next four eigenvalues are negative and do not matter. The order of magnitude of the first positive eigenvalue of interest corresponds to λ_6. Since the lowest part of the eigenvalue spectrum is more dense in the negative eigenvalues, we have to perform a positive shift λ_s to make the convergence to the positive eigenvalues easier.

$$(K - \lambda_s \; S)\; q = \eta\; S\; q \qquad (6.2)$$

and $$\lambda = \eta + \lambda_s \qquad (6.3)$$

For the next analysis, a finer mesh of 20 × 6 (fig. 3) elements is employed yielding to a dimension of 2240 degrees of freedom. The factorization of the shifted stiffness matrix (λ_s = 0.5) involves 115 substructures (maximum size 110) and takes 547" CPU time on a IBM 370/158. The first five eigenvalues are requested with a 10^{-8} precision. Convergence was not ensured after 30 iterations with 3 starting vectors (822" CPU time).

η_1 = 0.183928	λ_1 = 0.683928	σ_1 = $1.6\ 10^{-3}$
η_2 = 0.361752	λ_2 = 0.861752	σ_2 = $2.9\ 10^{-3}$
η_3 = 0.473357	λ_3 = 0.973357	σ_3 = $6.2\ 10^{-2}$
η_4 = - 0.645686	λ_4 = - 0.145686	σ_4 = 4.10^{-5}
η_5 = - 0.695221	λ_5 = - 0.195221	σ_5 = 2.10^{-2}
η_6 = - 0.722967		
η_7 = 0.746913		
η_8 = - 0.795698		

The first three eigenvalues are stabilized with 7 digits after 13 iterations. The reason for the poor convergence of the algorithm in this example are not known.

Transformation of the eigenvalue problem (6.2) into

$$(K - \lambda_s \; S)\; q = \eta^2 \; S\; (K - \lambda_s \; S)^{-1} \; S\; q \qquad (6.4)$$

does not much improve the situation. Indeed, we requested 5 eigenvalues with a 10^{-6} accuracy. 18 iterations with 2 starting vectors were sufficient to satisfy the convergence criterium and took 768" CPU time.

η_1 = 0.183928	λ_1 = 0.683928	σ_1 = $1.6\ 10^{-3}$
η_2 = 0.361752	λ_2 = 0.861752	σ_2 = $2.9\ 10^{-3}$
η_3 = 0.473357	λ_3 = 0.973357	σ_3 = $6.2\ 10^{-2}$
η_4 = - 0.645686	λ_4 = - 0.145686	σ_4 = 2.10^{-5}
η_5 = - 0.695228	λ_5 = - 0.195228	σ_5 = $1.8\ 10^{-4}$
$\vert\eta_6\vert$ = 0.722984		
$\vert\eta_7\vert$ = 0.794388		
$\vert\eta_8\vert$ = 0.85574		

We obtain a small improvement for the fifth eigenvalue only, while the first three eigenvalues were stabilized with 7 digits after 9 iterations. It was then decided to perform a larger shift λ_s = 1 in (6.2), in which case we obtained two negative pivots and one kinematical mode, which means λ_3 = 1. We were lucky because the deformation of the associated eigenmode was importantly located in the last finite element to be assembled so that the kinematical mode was well represented in the last substructure. We requested one eigenvalue and we obtained convergence after only 7 iterations

η_1 = - 0.137426	η_3 = - 0.316146	λ = 0.731182	σ_1 = $2\ 10^{-4}$
η_2 = - 0.268818	λ_1 = 0.683854	λ_2 = 0.862574	

From the sequence of eigenvalues at each iteration, it can be concluded that the second eigenvalue η_2 is parasitic. This last analysis is quite satisfying so far as the eigenvalue problem is concerned. However, in the experimental test performed by Babcock at Caltech, it was found that the actual buckling load was much higher than the present one ($\lambda_1 = 0.684$). Looking at the first three eigenmodes, we can observe that the deformation is mainly located in the neighborhood of the free edge of the cylinder (fig. 4). This was not the case for the actual buckling mode since some stiffening of the upper edge of the cylinder was realized in the experimental apparutus such that the cross-section of the free edge remained circular.

Therefore, we made a last analysis in which the radial component of the displacement field was fixed at the free edge. Using (6.2) with a shift $\lambda_s = 1$, one eigenvalue was requested which was stabilized with 7 digits at the sixth iteration

$$\eta_1 = 0.7295909 \qquad \eta_3 = -1.417247 \qquad \lambda_1 = 1.7296$$
$$\eta_2 = -1.161637 \qquad \eta_4 = 1.75055 \qquad \sigma_1 = 5\ 10^{-6}$$

The final bifurcation load is in excellent agreement with the experimental result and the corresponding buckling mode is shown in figure 5.

7. CONCLUSIONS

A highly efficient algorithm has been described for the computation of eigenvalues and eigenvectors of large symmetric matrices. Care is taken of coincident or nearly coincident eigenvalues by means of a restart procedure within the iteration process. In vibration analysis a good representation of the kinematical modes allow improvement of the algorithm convergence. In bifurcation analysis the solution of the eigenvalue problem seems to be a little more critical depending upon the nature of the B matrix. In some bifurcation problems in which the lowest part of the eigenvalue spectrum correspond to positive eigenvalues it has been shown that efficiency can be slightly improved by working with $B^* = B\ K^{-1}\ B$ instead of B.

Use of a shifted stiffness matrix may also improve convergence to the desired eigenvalues. However, further researches are needed along the lines developed in [5] in order to optimize the computer time between factorizations and iterations respectively.

REFERENCES

[1] E. CARNOY, *Asymptotic study of the elastic postbuckling behavior of structure by the finite element method*, Compt. Meth. in App. Mech. Engng, 29 (1981), 147-173

[2] E. CARNOY, G. SANDER, *Stability and Postbuckling analysis of non-linear structures*, Comp. Meth. in App. Mech. Engng, 30 (1982)

[3] E. CARNOY, *Etude de la stabilité élastique des coques par éléments finis*, Ph. D. thesis, Univ. of Liège, June (1980)

[4] E. CARNOY, *Extended bifurcation analysis of elasto-plastic structures*, in preparation

[5] T. ERICSSON, A. RUHE, *The spectral transformation Lanczos method for the numerical solution of large sparse generalized symmetric eigenvalue problems*, Mathematics of Computation, 35 (1980), n° 152, 1251-1268

[6] D.K. FADDEEV, V.N. FADDEEVA, *Computational methods of linear algebra*, W.H. Freeman and Co (1963)

[7] B.M. FRAEIJS de VEUBEKE, M. GERADIN, A. HUCK, *Structural dynamics*, CISM Lectures n° 126, Springer Verlag (1972)

[8] M. GERADIN, *Application of the biorthogonal Lanczos algorithm*, LTAS, report VA-17, Univ. of Liège (1977)

[9] M. GERADIN, *Eigenvalue analysis by matrix iteration in the presence of kinematical modes*, Shock and Vibration Digest, vol. 6, n° 3 (1974)

[10] M. GERADIN, E. CARNOY, *On the practical use of eigenvalue bracketing in finite element applications to vibration and stability problems*, Euromech 112, Budapest, Hungarian Academy of Sciences (1979), 151-171

[11] A. GODINAS, P. JETTEUR, G. LASCHET, *Développement d'un élément fini tridimensionnel spécialisé pour l'étude des coques dans le domaine non linéaire*, LTAS, Univ. of Liège, report SF-101 (1981)

[12] G.H. GOLUB, R. UNDERWOOD, *The block Lanczos method for computing eigenvalues*, Mathematical Software 3, Academic Press, New York, 361-377 (1977)

[13] G.H. GOLUB, R. UNDERWOOD, J.H. WILKINSON, *The Lanczos algorithm for the symmetric* AX = λ B x *problem*, Stanford Univ., Computer Science Dpt, STAN-CS 72-270 (1972)

[14] T.J.R. HUGHES, W.K. LIU, I. LEVIT, *Nonlinear dynamic finite element analysis of shells* , in Nonlinear Finite Element Analysis in Structural Mechanics, Springer Verlag, Berlin (1981), 151-168

[15] R. KAO, *Large deformation elastic-plastic buckling analysis of spherical caps with initial imperfections*, Computer and Structures, 11 (1980), 609-619

[16] W.T. KOITER, *The stability of elastic equilibrium*, AFFDL-TR-70-25 (1970) Ph. D. Thesis (1945)

[17] H.A. MANG, R.H. GALLAGHER, *Finite element analysis of thin shells of general form for displacement dependent loads*, in Nonlinear Finite Element Analysis of Plates and Shell, ASME, AMD 48 (1981), 65-82

[18] I.U. OJALVO, M. NEWMAN, *Vibration modes of large structures by an automatic reduction method*, AIAA Jnl, vol. 8, n° 7 (1970), 1234-1239

[19] I.U. OJALVO, *ALARM - a highly efficient eigenvalue extraction routine for very large matrices*, Shock and Vibration Digest, vol. 7, n° 12 (1975)

[20] E. RAMM, *Nonlinear finite element stability analysis*, Univ. of Stuttgart, Lecture at the Euromech Coll. 128, Delft (1980)

[21] SAMCEF, *Système d'Analyse des Milieux Continus par Eléments Finis*, Manuels LTAS, Univ. of Liège

[22] J.H. WILKINSON, C. REINSCH, *Handbook for automatic computation*, vol. 2 Linear Algebra, Springer Verlag (1971)

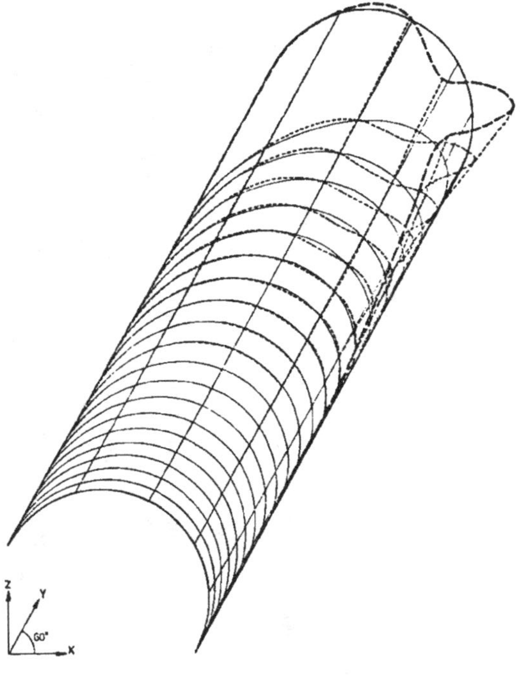

CYLINDER WITH FLUID (inside view) CRITICAL LOAD λ=.6839

FIG. 4

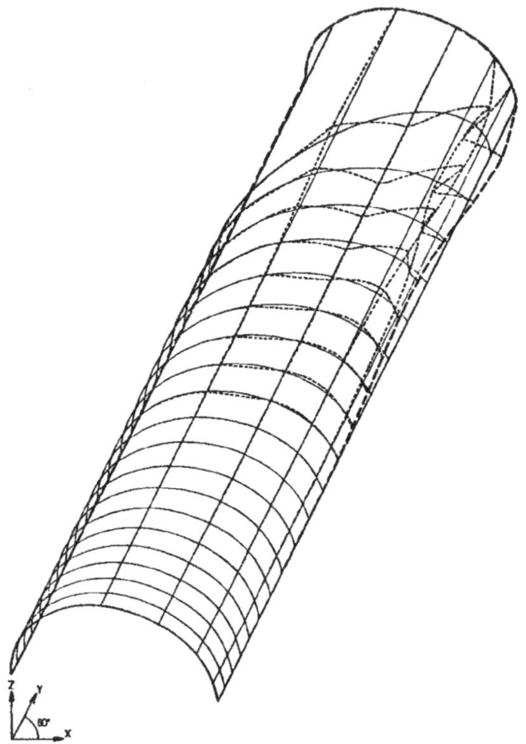

CYLINDER WITH FLUID: UPPER EDGE RESTRAINED BUCKLING LOAD 1.7296

FIG. 5

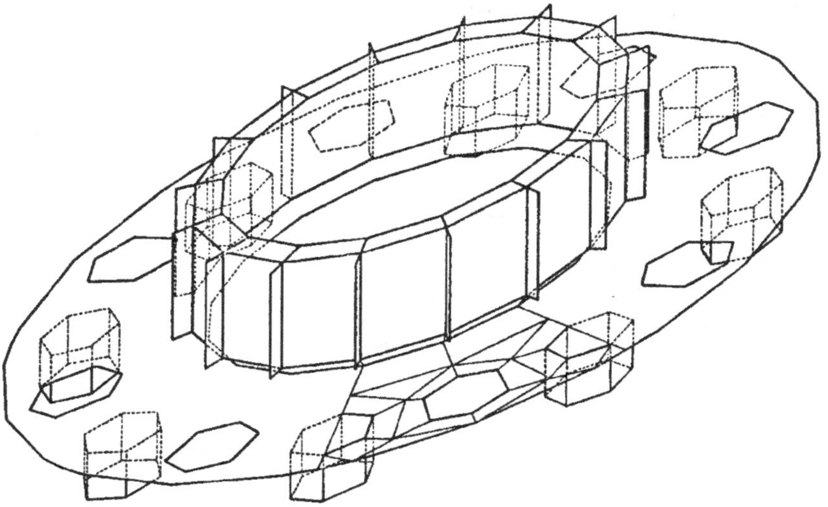

FIG. 1   PLANET ARM OF A PLANETARY GEAR TRAIN

MATERIAL DATA

V = .3
E = 30.10⁶ psi
$\sigma_y/E = .002$
$E_t/E = .1$

GEOMETRICAL DATA

$\alpha = 16.26°$
R = 25 in
H = 1 in
$\lambda = 2[3(1-\nu^2)]^{1/4} \cdot \left(\frac{H}{h}\right)^{1/2} = 4$

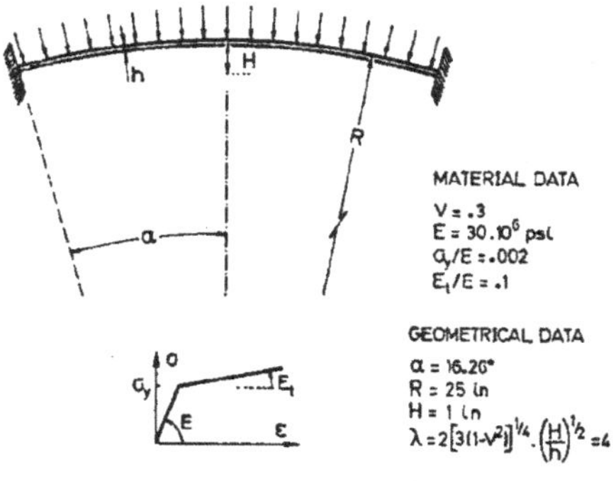

CLAMPED SPHERICAL CAP
UNDER UNIFORM EXTERNAL PRESSURE

FIG. 2

MATERIAL
DATA:

V = .3
E = 7.35 10⁵ lb/in²
Y = .0361 lb/in²

GEOMETRICAL
DATA:

R = 4 in
L = 20 in
h = .01 in
D = 16 in
$\alpha = 30°$

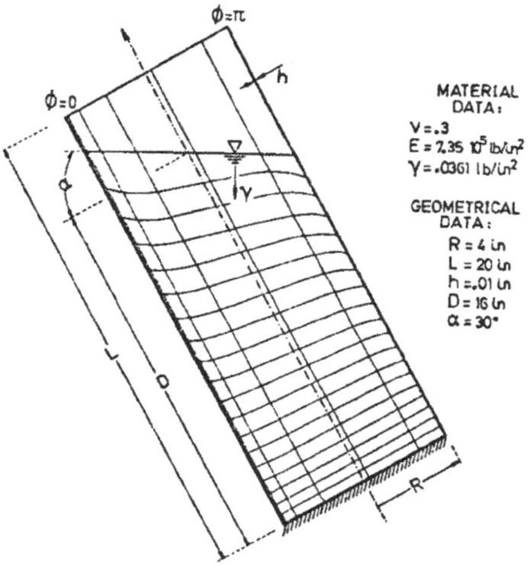

TILTED CYLINDRICAL SHELL CONTAINING A FLUID

FIG. 3

APPENDIX : the material stiffness matrix K_M

Denoting σ and ε the stress tensor and the strain tensor

$$\sigma^T = \{ \sigma_{xx} \ \sigma_{yy} \ \sigma_{zz} \ \sigma_{xy} \ \sigma_{yz} \ \sigma_{zx} \}$$
$$\varepsilon^T = \{ \varepsilon_{xx} \ \varepsilon_{yy} \ \varepsilon_{zz} \ \varepsilon_{xy} \ \varepsilon_{yz} \ \varepsilon_{zx} \} \qquad (A.1)$$

respectively, we relate these tensors by the constitutive law

$$\varepsilon = H^{-1} \sigma + \varepsilon_p (\sigma) \qquad (A.2)$$

where ε_p is the plastic part of the deformation.

In the initial bifurcation analysis, the fundamental path is approximated by the linear solution of the static problem

$$q (\lambda) = \lambda q_1 + 0 (\lambda^2) \qquad (A.3)$$

for which the constitutive law is reduced to the classical Hooke matrix H

$$\sigma_1 = D (0) \varepsilon_1 = H \varepsilon_1 \qquad (A.4)$$

Introducing a small perturbation to the fundamental solution, we obtain (A.2) in the form [4]

$$\Delta \sigma = D (\lambda \sigma_1) \Delta \varepsilon \qquad (A.5)$$

Assuming an isotropic hardening rule based on von Mises yield criterion, we have the classical result

$$D (\sigma_y) = H - \frac{H a a^t H}{a^t H a + \frac{4}{9} E_p \bar{\sigma}^2} \qquad (A.6)$$

with

σ_y	yield stress
E_p	plastic modulus $= \dfrac{E E_t}{E - E_t}$
$\bar{\sigma}$	von Mises equivalent stress
$f = \bar{\sigma}^2 - \sigma_y^2$	yield criterion

and $a = \dfrac{df}{d\sigma}$

The following approximation is used for (A.5)

$$D (\lambda \sigma_1) = D (0) + \frac{\lambda \sigma_1}{\sigma_y} \left(D (\sigma_y) - D (0) \right) \qquad (A.7)$$

or

$$D (\lambda \sigma_1) = H - \frac{\lambda \sigma_1}{\sigma_y} \frac{H a a^t H}{a^t H a + \frac{4}{9} E_p \bar{\sigma}^2} \qquad (A.8)$$

The first term of this relation gives the linear stiffness matrix while the second one yields the material stiffness matrix. Justification of (A.8) can be found in [4] .

IMPLEMENTATION AND APPLICATIONS OF THE SPECTRAL TRANSFORMATION LANCZOS ALGORITHM

Thomas Ericsson
Institute of Information Processing
Dept of Numerical Analysis
University of Umeå
S-901 87 UMEÅ, SWEDEN

Abstract

This paper gives an orientation on some practical details of the program package STLM (=Spectral Transformation Lanczos Method). STLM is a FORTRAN implementation of an algorithm for computing some eigenpairs to large, sparse, symmetric, and generalized eigenproblems. Some portability and flexibility aspects of the package are also discussed.

1. The Problem

Compute some of the eigenpairs (λ_i, x_i) of

$$Kx_i = \lambda_i M x_i, \quad i=1,2,\dots,n$$

$$\lambda_1 \leq \lambda_2 \leq \dots \leq \lambda_n$$

where K and M are real and symmetric n by n matrices. K may be indefinite and M must be positive semidefinite. Further K and M are supposed to have non intersecting nullspaces, i.e. $N(K) \cap N(M) = \{0\}$.

In practice we are interested in the cases where K and M are large and sparse.

The notation K and M comes from vibration mode analysis in FEM computations, where K is the stiffness matrix and M is the massmatrix. But, the matrices may have other origin too.

Limits

Eigenvalues can be computed in any region of the spectrum, and regardless of the signs of the eigenvalues. Formulations like the following are thus possible:

compute all eigenpairs (λ_i, x_i) with $\lambda_i \in (a,b)$, $a < b$

compute ℓ consecutive λ_i (and corresponding x_i) starting at a.

The figure below illustrates three possible cases:

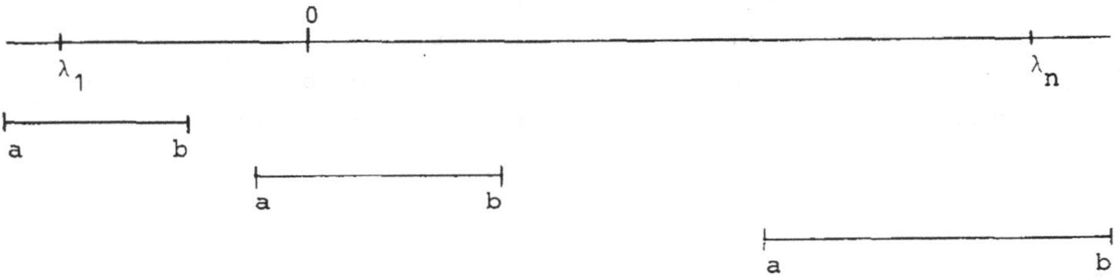

Multiple eigenvalues are permitted, if the multiplicity is not too high. A multiplicity of five of an eigenvalue should not cause any major problems.

For the dimension of the problem, n, we have the bound

$$n \geq 1,$$

the upper bound depends on available memory and CPU-time, number of routines supplied by the user (see below).

An Example with $a \neq 0$

In FEM computations one is often interested only in some of the smallest eigenvalues, $\lambda_1, \lambda_2, \ldots, \lambda_m$, where $\lambda_i = \omega_i^2 > 0$, so the varying a (in (a,b)) does not seem to be too useful. Leaving the world of FEM computations there are of course problems that require eigenvalues in, for example, the middle of the spectrum, and which have negative eigenvalues But even in FEM computations there may be an advantage of an $a \neq 0$, as in the following example (computation on a wetwell in a nuclear reactor, from Dr. Christer Gustafsson, at the time employed by ASEA-ATOM):

$$n = 245, \ \lambda_1 = \lambda_2 = \dots = \lambda_{95} = 0, \ \lambda_{96} \approx 3.97 \cdot 10^3$$

When using the company's standard package (which only could start at zero or use a shift to the left of zero for the special case $\lambda_1 = 0$) it stopped with a registerdump while working on the multiple eigenvalue.

STLM could hardly have computed the eigenvectors to the multiple eigenvalue either (due to loss of orthogonality between the x_i), though it would have ended in a nicer way. When they used (the first version of) STLM with a positive shift (a) everything went fine. This was possible since the engineers were not interested in $x_1, \dots, x_{95}$.

2. The STLM Algorithm

We will only give a sketch of the algorithm in this paper, for more details, see [3], where the first version of STLM is described.

The basic algorithm is to compute a sequence of shifts μ_i (μ_1=a), and then for each shift apply the Lanczos algorithm on the symmetric matrix $M(K-\mu_i M)^{-1}M$, for more details see below.

The idea is illustred in the following snapshots:

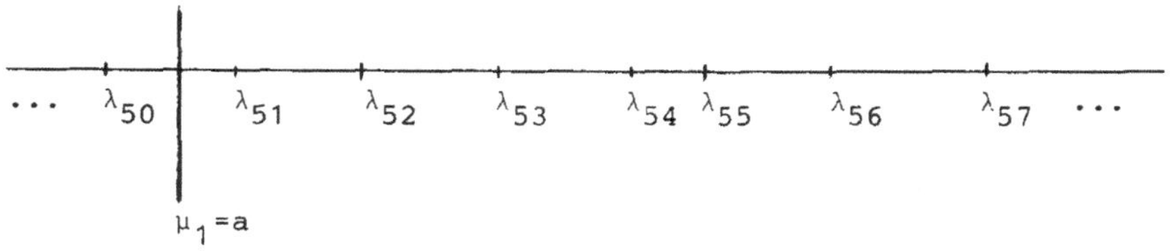

$(\lambda_{50}, x_{50}), (\lambda_{51}, x_{51})$, and (λ_{52}, x_{52}) are computed (the number of computed eigenpairs for one shift depends for example on the distribution of eigenvalues and the number of Lanczos steps).

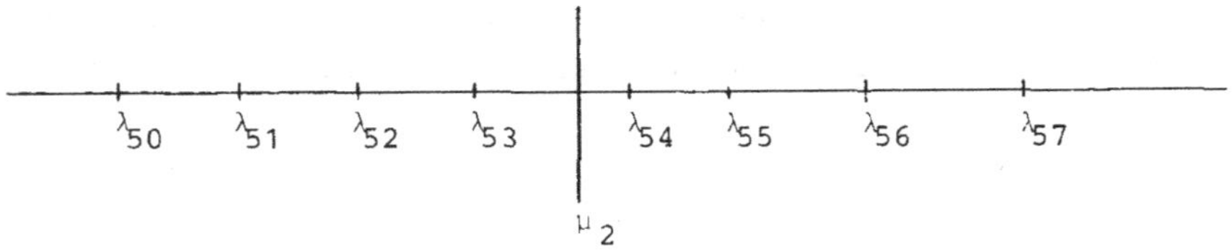

(λ_{53},x_{53}), (λ_{54},x_{54}), (λ_{55},x_{55}), and perhaps (λ_{56},x_{56}) are computed. To avoid to get the eigenpairs computed for μ_1, we orthogonalize the startingvector to the Lanczos routine against x_{51} and x_{52}.

We can now continue in this way with a shift μ_3 till we have computed the requested pairs. Should eigenvalues be missing to the left of a shift μ_{i+1} after we have used Lanczos on it, we probably have a multiple eigenvalue in (μ_i,μ_{i+1}). We then run Lanczos again (the same shift) with a startingvector that is orthogonal against the eigenvectors having eigenvalues in (μ_i,μ_{i+1}). This extra run may have to be repeated more than once.

The Lanczos routine is almost a standard Lanczos with full reorthogonalization. We do however everything in M inner products, see [10]. For more details, see for example [2], [7], and [10]. The Lanczos routine produces a tridiagonal subproblem, $Ts_i=\nu_i s_i$, $i=1,2,\ldots,p$, of much lower dimension, e.g., if $n = 1000$, p may be 40. Some of the ν's are good approximations to some of the extreme eigenvalues of the big problem $M(K-\mu M)^{-1}Mx=(\lambda-\mu)^{-1}Mx$. We have:

$$Kx=\lambda Mx, \text{ and if } (K-\mu M)^{-1} \text{ exists then}$$

$$M(K-\mu M)^{-1}Mx=(\lambda-\mu)^{-1}Mx$$

(one M will cancel in the Lanczos routine). $(K-\mu M)^{-1}$ is <u>not</u> computed explicitly.

We see that the extreme eigenvalues of this problem correspond to the λ's nearest to μ.

To know where our shift is placed in the spectrum, we use the triangular decomposition $L_iD_iL_i^T=K-\mu_iM$ (L_i=lower triangular, D_i=diagonal). With r_i=number of negative elements in D_i, then if M is positive definite

r_i = number of $\lambda < \mu_i$. If M is singular we only get a relative measure as in the following example:

$$K = \begin{pmatrix} -1 & & 0 \\ & 1 & \\ 0 & & 2 \end{pmatrix}, M = \begin{pmatrix} 0 & & 0 \\ & 0.5 & \\ 0 & & 0 \end{pmatrix}$$

$\mu_1=1$ gives $r_1 = 1$, i.e. $\lambda_1 < 1$ ($\lambda_1=$ "$-1/0$")

$\mu_2=10$ gives $r_2 = 2$, i.e. $\lambda_1, \lambda_2 < 10$

Note that we always get a correct count for the number of eigenvalues in the interval (μ_1, μ_2) by subtracting r_2-r_1, in this example $r_2-r_1=1$.

3. The FORTRAN Routines

The main routine in the package is called STLM and has the following structure:

> μ: = a
> initiate
> CALL INITU(...)
> REPEAT
>
> > $LDL^T := K-\mu M$
> > Run Lanczos on the shifted and inverted problem, producing the tridiagonal matrix T. Compute the eigenpairs (ν_i, s_i) of T and then the corresponding (λ_i, x_i) ($\lambda=\mu+1/\nu, x=Vs$ + a bit of the next v-vector. V is the transformation matrix, with orthogonal columns produced by the Lanczos algorithm, see [3]).
> >
> > Compute a new shift μ.
>
> UNTIL TERMIN(...)
> terminate.

The routines INITU and TERMIN must be supplied by the user. In INITU the user may set up the problem (this can also be done before referencing STLM). TERMIN is a logical function that gives the termination criteria, e.g. b in (a,b), or whatever the user supplies.

There are three additional routines that must be given by the user, namely:

SECOND - should return the elapsed CPU-time.

RANF - should return a randomnumber.

IO - the number of tasks depends on which additional routines the user writes, see below, but IO needs never do more than:

store and retrieve n-vectors (e.g. x_i)
save λ_i
retrieve K

Usually IO communicates between a direct accessfile and primary storage. It is also possible to utilize virtual memory. K may be stored on a sequential file.

Using maximum default (also regarding routines written by the user) the following input must be given to STLM:

n = the dimension of the problem
DIAGM = TRUE if M is diagonal
MEQI = TRUE if M equals the identity
a = the first shift

two vectors (the only primary storage the user must supply):

W (real) that stores K and M, and has an additional $2n+5p+p^2$ elements as a work area.

IW (integer) that stores a pointer vector P and an additional p+1 elemets.

p is the maximum number of Lanczos steps. If n=1000, p may be 40. The program determines itself how many steps that are optimal to perform in an actual run, p is just a limit set on available storage.

In the default case K and M are stored according to the method described in [5] (page 97), i.e. for each column all the elements from the first nonzero to the diagonal are stored. These parts of the columns are store consecutively in a vector. The pointer vector P above points to the diagonal elements in the vector. The set that consists of the indices of the first nonzero elements is called the profile of the matrix.

We permit three cases:

profile (K) = profile (M) = arbitrary
profile (K) = arbitrary, M is diagonal but not equal to the identity.
profile (K) = arbitrary and M is equal to the identity, in which case M is not stored (i.e. this is the standard eigenvalue problem).

It is of course possible to change the message level and other parameters of interest.

4. Flexibility

There are some drawbacks with the default method, for example: If n is big (non virtual memory) we can not hold K and M in primary storage, in which case the default storage method will not work. Suppose that we have a problem with $K_{ij} = 0$ everywhere except on the diagonal and in the first row and column. Using the profile method we would have to store the whole upper triangle (unless we made a reordering of the matrix, see [4]).

Although the profile method is often used, it is not the only way. In the ASKA package (see [9]) for example, partitioned matrices in three levels are used.

Not all problems have profile (K) = profile (M), or M diagonal.

In order to get the package flexible enough it is thus essential to give the user the opportunity to choose the storage method he finds most suitable.

The reasons for having the default version are:

- the method is often used in practice
- we think it is important with an easy to use version of the package
- the default routines may serve as models for how the alternative

routines may be written.

If the user wishes to change storing method, he must supply the three routines making the following:

- compute $LDL^T:=K-\mu M$
- solve $(K-\mu M)x=b$ (b known, x unknown) using LDL^T above.
- given x, compute $y:=Mx$

W and IW need of course not store K and M now, since the user handles that (he may still use W, IW or blank common for K and M). W has thus a length of $2n + 5p + p^2$, and IW has p+1 elements.

We have still the dependence on n in the length of W. This may cause trouble on a computer with a small primary storage if n is big. Since only two n-vectors in primary storage give rise to some I/O between primary and secondary storage, it would also be nice if one could lessen this transfer. That is for example possible if the machine has a large primary storage or a virtual memory.

We have thus incorprated the possibility of the user supplying the n-vector handling routines himself, in which case he must write the following four routines: (α,β, and γ are scalars. x and y are vectors).

- $\alpha := x^Ty$
- $y := y-\beta x$
- $y := \gamma x$
- $y :=$ randomvector

W has now a length of $5p + p^2$ elements and IW contains p+1 elements. Since $p = p(n)$ is a very slowly increasing function we have essentially an independence of n.

5. Portability

In addition to the flexibility aspects above, which can be said to be a part of the portability, the package is written in a subset of FORTRAN66 (ANSI X3.9-1966). We have also tested it with the PFORT-verifier [8] with no remarks, and we have taken most of the recommendations in [1] into consideration.

The package will run as a FORTRAN77-program (in one routine two lines may have to be changed). The user can then utilize the ANSI direct access file handling.

To increase portability we have tried to make the package selfcontained (i.e. we do not demand the existence of certain packages at the user's computer). The only routine that is not of our own production is a slightly modified EISPACK routine IMTQL2 [6].

The package is equipped with some auxiliary routines of which one can convert it to DOUBLE PRECISION, INTEGER * 2, LOGICAL * 1, or whatever the user wishes.

A detailed manual will be available in the late spring of 1982. To control that the user has installed the package correctly, we have supplied a self running testpackage.

6. Experience

We do not have much experience with the new version of the program. The first version [3] has however been used extensively on practical problems, and since the major differences between the versions are in portability and flexibility, we are convinced that the new version will work well too.

The only problem with the first version was the routine computing $LDL^T := K - \mu M$. Since $K - \mu M$ is factored without any sort of pivoting we will get into trouble whenever μ lies near an eigenvalue to any of the subproblems. Should this happen the program tries with a new, slightly different, shift. Should it go wrong the second time too, the program writes an error message and returns to the users program. In the new version we have a better criterion for discarding a decomposition ([7] pages 43-46).

We have not tested the new version on any real problems so far, but only on constructed test problems. This we have done on two computers, a CDC Cyber 730 (precision ≈ 14 decimals) and a PRIME(virtual memory) with about half the precision.

On the Cyber (using both FORTRAN 66, 77, single and double precision) we have not had any difficulties. When installing the package on the PRIME (which was easy) the difficulties were of system limit character, e.g. limited record length on a direct accessfile, a vector, unless it lies in common, may not consist of more than 128 kbytes.

Due to the lower precision we also had to change two tolerances (checking the size of x^TMx) to be able to run the testpackage without warning messages.

We have also converted the package to double precision and compiled and run it with no problems on the PRIME.

7. An Example on the PRIME

We end with an example on the PRIME (FORTRAN 77, single precision and using the virtual memory).

n=2000
profile (K) = profile (M) = bandmatrix with halfband width = 45.
(i.e. K and M needed 90965 elements each to be stored).

Two shifts were used and the Lanczos algorithm took 44 and 34 steps. We got 24 eigenpairs. One addition and multiplication took about 40 µs, Some other times of interest are:

$LDL^T:=K-\mu M$	took	101	CPU-seconds each
$(LDL^T)x:=b$	-"-	7.5	-"-
$y:=Mx$	-"-	7.2	-"-

To answer one question we sometimes get:
Let T_i be the tridiagonal matrix produced by the Lanczos algorithm in step i, we do (p=44 and 34 in this example) for each shift:

```
FOR i = 1 TO p DO
   | compute top and bottom elements in the
   | eigenvectors to T_i
   | Compute the eigenvalues of T_i

Compute all the eigenpairs of T_p.
```

Is this not very time consuming?
No, not for a n > 500 say, in this example it took 2 % of the total time.

The total run time (2151 CPU-seconds) can be divided as follows (in % of the total run time)

Factorization, $LDL^T := K - \mu M$,	9%
Lanczos	77%, of which
Solution, $(LDL^T)x = b$,	28%
$y := Mx$ and recursion ,	28%
Reorthogonalization	21%
Compute T_p:s eigenpairs (IMTQL2)	2%
Compute x_i-vectors from T_p:s eigenvectors	10%
Initiation	2%

References

1. A. COLIN DAY, Compatible Fortran, Cambridge U. P., 1978.

2. IAIN S. DUFF (ed), Sparse Matrices and their Uses, Academic Press, 1981.

3. T. ERICSSON AND A. RUHE, The Spectral Transformation Lanczos Method for the Numerical Solution of Large Sparse Generalized Symmetric Eigenvalue Problems, Math of Comp., vol 35, no 152, 1251-1268.

4. ALAN GEORGE AND JOSEPH W. LIU, Computer Solution of Large Sparse Positive Definite Systems, Prentice Hall, 1981.

5. ALAN JENNINGS, Matrix Computation for Engineers and Scientists, John Wiley, 1978.

6. Matrix Eigensystem Routines - EISPACK Guide (1974)(1977), B. T. SMITH et al., Springer Lecture Notes in Computer Science 6 and 51

7. BERESFORD N. PARLETT, The Symmetric Eigenvalue Problem, Prentice Hall, 1980.

8. B. G. RYDER, The PFORT Verifier, Software - Practice and Experience, vol. 6 no. 4, 473-486.

9. E. SCHREM, Computer Implementation of the Finite Element Procedure, ISD-Report no. 111, University of Stuttgart, 1975.

0. D. S. SCOTT, The Advantages of Inverted Operators in Rayleigh-Ritz Approximations, Computer Sciences Division at Oak Ridge National Laboratory, 1980.

PRECONDITIONED ITERATIVE METHODS FOR THE GENERALIZED EIGENVALUE PROBLEM

D.J. Evans
Department of Computer Studies
Loughborough University of Technology
Loughborough, Leicestershire,
U.K.

ABSTRACT

In this paper a preconditioned iterative method suitable for the solution of the generalized eigenvalue problem is presented. The proposed method is suitable for the determination of the extreme eigenvalues and their corresponding eigenvectors of the large sparse matrices derived from finite element/difference discretisation of partial differential equations. The new strategy when coupled with the conjugate gradient algorithm yields a powerful method for this class of problems.

1. INTRODUCTION

In this paper we describe a preconditioned algorithm based on the theory of convergent splittings for the solution of large sparse eigenvalue problems by the minimisation of the Rayleigh quotient corresponding to the generalized eigenvalue problem,

(1.1) $$Ax = \lambda Bx \ ,$$

where A and B are symmetric matrices with B positive definite.

Briefly, a general iterative procedure for the solution of an eigenvalue problem generates a sequence of vectors $x_1, x_2, \ldots, x_i, \ldots$ which in turn produces a sequence of eigenvalue approximations μ_i defined as,

(1.2) $$\mu_i = \mu(x_i) = x_i^t A x_i / x_i^t B x_i \ .$$

In general, we consider the splitting of the matrix C_i defined as,

(1.3) $$C_i = A - \lambda_i B = V_i - H_i \ .$$

It can be seen that the sequence of vectors x_i satisfy the relation

(1.4) $$x_{i+1} = x_i - p_i \ , \quad i=1,2,\ldots \ ,$$

where, (1.5) $$p_i = V_i^{-1} C_i x_i \ .$$

In the case of convergence, the vectors x_i and their corresponding values μ_i, will converge to an eigenvector and its corresponding eigenvalue of the system (1.1). It can be seen that in this case, the residual vector r_i defined as,

(1.6) $$r_i = r(x_i) = (A - \mu_i B) x_i \ ,$$

will converge to zero.

2. FORMULATION OF THE PRECONDITIONED ALGORITHM

The basic iteration formula used to evaluate the consecutive vectors x_i is defined as

(2.1) $$x_{i+1} = V_i^{-1}H_i x_i \;,$$

where the matrices V_i and H_i are defined as,

(2.2) $$V_i = (D_i - \omega L_i)D_i^{-1}(D_i - \omega U_i)$$

and

(2.3) $$H_i = \omega^2 L_i D_i^{-1} U_i + (1-\omega)(L_i + U_i) \;,$$

with L_i and U_i being the lower and upper triangular parts of the matrix $(A-\mu_i B)$ respectively, and D_i being a diagonal matrix containing the diagonal entries of the matrix $(A-\mu_i B)$. Further, it can be easily seen that the relation (1.3) is satisfied for the above splitting.

The above splitting produces a convergent method (Ruhe [6],[7]) and sufficient conditions under which the μ_i, $i=1,2,\ldots$ forms a monotone sequence as given in the following theorem (Evans & Shanehchi, [3]).

Theorem 2.1:

If the starting vector x_1 in the iteration formula (2.1) is chosen such that the corresponding Rayleigh quotient μ_1 satisfies the inequality $\mu_1 < \min\{a_{i,i}/b_{i,i} \mid i=1,2,\ldots\}$, then the μ_i's will form a decreasing sequence. Alternatively, if this vector is selected such that μ_1 satisfies $\mu_1 > \max\{a_{i,i}/b_{i,i} \mid i=1,2,\ldots,n\}$, then the μ_i's will form an increasing sequence. In either case the method will converge in the following sense,

(2.4) $$\mu_i \to \bar{\mu} = \lambda \;,$$

where λ is an eigenvalue of the system (7) and

(2.5) $$r_i \to 0. \quad \square$$

Now the rate of convergence of the system (2.1) is determined by the convergence rate of the limiting iteration,

(2.6) $$x_{i+1} = V_\ell^{-1} H_\ell x_i \;,$$

where V_ℓ and H_ℓ denote the limits of the matrices V_i and H_i respectively.

In general, with the above splitting of the matrix C_i, the matrix H_i is no longer a semi-definite matrix, although the matrix V_i remains a positive (negative) definite matrix, provided that the conditions required for Theorem 2.1 are satisfied. We now attempt to select a suitable range for the preconditioning parameter ω such that for any value ω in that range, the matrix summation $V_i + H_i$ forms a definite matrix so that we will be able to apply Theorem 2.1 to prove the global convergence of the preconditioned method.

Lemma 2.1:

The matrix $V_i + H_i$, where V_i and H_i are defined in (2.2) and (2.3) is a positive definite matrix if ω is chosen to be in the range:

$$\frac{2-\sqrt{2}}{2} \leq \omega \leq \frac{2+\sqrt{2}}{2}, \tag{2.7}$$

and that the initial vector x_1 is chosen such that

$$\mu_1 = \frac{x_1^t A x_1}{x_1^t B x_1} < \min\left\{\frac{a_{i,i}}{b_{i,i}} \;\middle|\; i=1,2,\ldots,n\right\}. \tag{2.8}$$

Proof: The condition (2.8) ensures that:

$$d_{i,i} > 0, \quad i=1,2,\ldots,n,$$

therefore the matrix $D_i^{\frac{1}{2}}$ exists. However, the quadratic form of the matrix V_i+H_i for vector x_i can be expressed as follows:

$$x_i^t(V_i+H_i)x_i = x_i^t((D_i-\omega L_i)D_i^{-1}(D_i-\omega U_i)+\omega^2 L_i D_i^{-1}U_i+(1-\omega)(L_i+U_i))x_i$$

$$= x_i^t D_i x_i-(2\omega-1)x_i^t(L+U)x_i+2\omega^2 x_i^t LD_i^{-1}Ux_i \; . \tag{2.9}$$

With ω in the range (2.7), $(2\omega-1)^2 \leq 2\omega^2$ and since D_i is positive semi-definite it then follows that:

$$x_i^t(V_i+H_i)x_i \geq x_i^t D_i x_i-(2\omega-1)x_i^t(L_i+U_i)x_i+(2\omega-1)^2 x_i^t L_i D_i^{-1}U_i x_i$$

$$= x_i^t[(D_i-(2\omega-1)L_i)D_i^{-1}(D_i-(2\omega-1)U_i)]x_i$$

$$= x_i^t B_i^t B_i x_i > 0 ,$$

where $B_i = (D_i-(2\omega-1)L_i)D_i^{-\frac{1}{2}}$,

and the proof of the Lemma is complete. ⌑

To analyse the rate of convergence of the preconditioned method we again consider the limiting iteration formula (2.6) where the matrices V_ℓ and H_ℓ are defined as,

$$V_\ell = (D_\ell-\omega L_\ell)D_\ell^{-1}(D_\ell-\omega U_\ell) , \tag{2.10}$$

and

$$H_\ell = \omega^2 L_\ell D_\ell^{-1}U_\ell+(1-\omega)(L_\ell+U_\ell) . \tag{2.11}$$

The limit of the iteration matrix H_i can be expressed as,

$$\begin{aligned} H_\ell &= ((D_\ell-\omega L_\ell)D_\ell^{-1}(D_\ell-\omega U_\ell))^{-1}(\omega^2 L_\ell D_\ell^{-1}U_\ell+(1-\omega)(L_\ell+U_\ell)) \\ &= ((D_\ell-\omega L_\ell)D_\ell^{-1}(D_\ell-\omega U_\ell))^{-1}((D_\ell-\omega L_\ell)D_\ell^{-1}(D_\ell-\omega U_\ell)-D_\ell+L_\ell+U_\ell) \\ &= I-((D_\ell-\omega L_\ell)D_\ell^{-1}(D_\ell-\omega U_\ell))^{-1}C_\ell = I-B_\omega . \end{aligned} \tag{2.12}$$

The matrix B_ω is defined as the preconditioned matrix of the method. It can be seen that the matrix B_ω is similar to a positive semi-definite matrix $\bar{B}_\omega$ defined as

$$\bar{B}_\omega = P B_\omega P^{-1} , \tag{2.13}$$

where

$$\left.\begin{aligned} P &= D_\ell^{\frac{1}{2}}(D_\ell-\omega L_\ell), \text{ if } \lambda_1 \text{ is evaluated,} \\ P &= -(-D_\ell)^{\frac{1}{2}}(D_\ell-\omega L_\ell), \text{ if } \lambda_n \text{ is evaluated} \end{aligned}\right\} . \tag{2.14}$$

From the relation (2.12) we have that if γ_i and μ_i are the eigenvalues of the matrix H_ℓ and B_ω respectively, then they are real and are related through the relationship

$$(2.15) \qquad \gamma_i = 1-\mu_j, \quad \mu_j > 0, \; i,j=1,2,\ldots,n.$$

As mentioned earlier, the rate of convergence of this method is governed by the magnitude of the second eigenvalue of the matrix H_ℓ in absolute value since its largest eigenvalue converges to unity as the method converges to an eigenvalue of A.

Since it can be shown that the matrix C_ℓ is a semi-definite matrix, then if the matrix C_ℓ is consistently ordered (Evans & Missirlis, [2]) we have that the second smallest eigenvalue of the matrix B_ω i.e. μ_2 is bounded as follows,

$$(2.16) \qquad \mu_2 \leq \frac{1}{\omega(2-\omega)}, \quad \omega \in \left[\frac{2-\sqrt{2}}{2}, \frac{2+\sqrt{2}}{2}\right],$$

and

$$(2.17) \qquad \mu_2 \geq \begin{cases} \dfrac{1-\bar{p}}{1-\omega\bar{p}+\omega^2\bar{\beta}}, & \text{if } \bar{\beta}\geq 1/4 \text{ or if } \bar{\beta}\leq 1/4 \text{ and } \omega<\omega^*, \\[2ex] \dfrac{2}{2+\omega^2\bar{\beta}}, & \text{if } \bar{\beta}\leq 1/4 \text{ and } \omega\geq\omega^*, \end{cases}$$

where $\bar{p}$ denotes the second largest eigenvalue of the Jacobi iteration matrix of C_ℓ i.e.,

$$B = D_\ell^{-1}(L_\ell+U_\ell),$$

and

$$(2.18) \qquad \omega^* = \frac{2}{1+\sqrt{1-4\bar{\beta}}},$$

where

$$(2.19) \qquad \bar{\beta} = S(\tilde{L}_\ell\tilde{U}_\ell),$$

with $\tilde{L}_\ell$ and $\tilde{U}_\ell$ defined as,

$$(2.20) \qquad \tilde{L}_\ell = D_\ell^{-\frac{1}{2}}L_\ell D_\ell^{-\frac{1}{2}},$$

and

$$(2.21) \qquad \tilde{U}_\ell = D_\ell^{-\frac{1}{2}}U_\ell D_\ell^{-\frac{1}{2}}.$$

From the above bounds for μ_2 we can minimise the modulus of the second largest eigenvalue of H_ℓ denoted by $S_2(H_\ell)$ from which good estimates of the preconditioning parameter ω can be derived.

Thus, on equating (2.16) and (2.17) we have the quartic equation,

$$(2.22) \qquad y(\omega) = 2\bar{\beta}\omega^4-2(\bar{p}+2\bar{\beta})\omega^3+(1+5\bar{p}+\bar{\beta})\omega^2-(3\bar{p}+2)\omega+1 = 0,$$

and assuming that ω_2 is a root of this equation which lies in the interval $(1,1+\frac{\sqrt{2}}{2})$, then if $\bar{\beta}\geq 1/4$ we have the value of ω which minimises $S_2(H_\ell)$ defined as,

$$(2.23) \qquad \omega = \min\{\omega_2, \frac{\bar{p}}{2\bar{\beta}}\},$$

and the corresponding bounds can be seen to be,

$$(2.24) \qquad S_2(H_\ell) \leq \begin{cases} \dfrac{1}{\omega_2(2-\omega_2)}, & \text{if } \omega=\omega_2, \\[2ex] \dfrac{(4\bar{\beta}-\bar{p})\bar{p}}{(4\bar{\beta}-\bar{p}^2)}, & \text{if } \omega=\dfrac{\bar{p}}{2\bar{\beta}}. \end{cases}$$

However, if $\bar{\beta}<1/4$ then we have two separate cases based on different values of $\bar{\beta}$, which are obtained from confining ω^* within the given range.

Thus, if $\bar{\beta}\geq 0.2426$ then we take ω as defined in (2.18) and if $\bar{\beta}<0.2426$ then ω is chosen to be

$$(2.25) \qquad \omega = \min \{ \frac{\bar{p}}{2\bar{\beta}} , \omega^* \}.$$

Finally, the bounds for $S_2(H_\ell)$, when $\bar{\beta}<0.2426$ can be shown to be:

$$(2.26) \qquad S_2(H_\ell) \leqslant \begin{cases} \dfrac{4\bar{\beta}\bar{p}-\bar{p}^2}{4\bar{\beta}-\bar{p}^2} , & \text{if } \omega = \dfrac{\bar{p}}{2\bar{\beta}} , \\ \dfrac{\omega^*-1}{\omega^*} , & \text{if } \omega = \omega^* . \end{cases}$$

3. THE PRECONDITIONED CONJUGATE GRADIENT ALGORITHM

The determination of the eigenvalues and the corresponding eigenvectors of large order sparse symmetric matrices by the conjugate gradient algorithm [4] has recently been shown to be an efficient strategy for symmetric eigenvalue problems. Here we extend the practical algorithms devised by Ruhe & Wiberg [7] for the application of the C-G method in conjunction with the inverse iteration, to

$$(3.1) \qquad (\lambda_i B-A)x_{i+1} = x_i ,$$

where λ_i is a fairly good approximation to an eigenvalue.

Numerical experiments carried out on the (200×200) tridiagonal matrix A=(-1,2, -1), B=I derived from the well known central difference operator are shown in Table 1 and confirm that appreciable gains in efficiency can be achieved by the PCCG algorithm.

Precond. Parameter ω	Preconditioned C.G. Method			
	Eigenvalue		Eigenvector	
	No.of iter.	Time Units	No.of iter.	Time Units
0	100	364	190	794
1.0	40	251	75	492
1.1	38	238	67	441
1.2	38	238	65	429
1.3	38	238	65	378
1.4	38	238	57	321
1.5	32	201	48	321
1.6	27	170	39	263
1.7	21	133	29	199
1.8	16	103	22	154
1.9	11	71	17	122

TABLE 1

Finally, further numerical investigations are required in order to ascertain the pattern of behaviour of the optimum preconditioning parameter ω.

REFERENCES

1. D.J. EVANS, The use of preconditioning in iterative methods for solving linear equations with symmetric positive definite matrices, J.Inst.Math.Applics., 4 (1968), pp.295-314.

2. D.J. EVANS AND N.M. MISSIRLIS, On the preconditioned Jacobi method for solving large linear systems, Computing (1982), in press.

3. D.J. EVANS AND J. SHANEHCHI, Preconditioned iterative methods for the large sparse symmetric eigenvalue problem, Comp.Meth.Appl.Mech. & Eng. (1982), in press.

4. M.R. HESTENES AND E. STIEFEL, Methods of conjugate gradients for solving linear systems, Jour. of Res., N.B.S., 49 (1952), pp.409-436.

5. A. RUHE, SOR methods for the eigenvalue problem with large sparse matrices, Math. of Comp. 28 (1974), pp.697-710.

6. A. RUHE, Iterative eigenvalue algorithms based on convergent splittings, Jour. of Comp.Phys. 19 (1975), pp.110-120.

7. A. RUHE AND T. WIBERG, The method of conjugate gradients used in inverse iteration, B.I.T. 12 (1972), pp.543-554.

ON BOUNDS FOR SYMMETRIC EIGENVALUE PROBLEMS

Alan Jennings
Civil Engineering Department,
Queens University,
Belfast BT7 1NN, N. Ireland

Abstract

Bounds for the eigenvalues of $Ax = \lambda Bx$ are derived where A and B are symmetric and B is also positive definite with a matrix factor G such that $B = G^TG$. The method depends on it being possible to obtain a lower bound for the singular values of G by a method recently developed by the author.

1. Introduction

When Gerschgorin's theorem is used to determine the bounds for the eigenvalues of a symmetric matrix which is known to be positive definite the lower bound is often negative, thus revealing no useful information. However, in many cases where symmetric positive definite matrices arise, a factorization of the form

$$\bar{B} = F^TF \tag{1.1}$$

is also available or can easily be obtained. The author [1] has recently developed an alternative method of finding a lower bound for the eigenvalues of $\bar{B}$ by examining the structure of F to determine a lower bound for its singular values (which are related to the eigenvalues of $\bar{B}$). This method often, but not always, yields sharper lower bounds than does Gerschgorin's theorem.

The objective of this paper is to discuss extensions of this method which may be used to obtain bounds for the eigenvalues of symmetric generalised eigenvalue problems when one of the matrices is positive definite and there is a factorization of the form of equation (1.1) available.

2. Bounds for the Singular Values of a Matrix

The theorem defining a lower bound for the singular values of a matrix [1] is as follows:

Let F be of order mxn with singular values denoted by σ_i where $\sigma_1 \geq \sigma_2 \geq \dots \geq \sigma_n \geq 0$ and let $s = \{s_1\ s_2\ \dots\ s_n\}$ be a set of positive column scaling factors for F such that

$$\sum_{i=1}^{n} s_i^2 = 1 \tag{2.1}$$

then

$$\sigma_n^2 \geq \min_{j=1,n} \sum_{i=1}^{m} w_{ij}^2 \tag{2.2}$$

where

$$w_{ij} = \max(0,\ |f_{ij}|s_j - \sum_{k\neq j}|f_{ik}|s_k) \tag{2.3}$$

This theorem may be illustrated by the matrix

$$F = \begin{bmatrix} 0 & 3 & 1 \\ 1 & -1 & -1 \\ -2 & 1 & 0 \\ -1 & 1 & 2 \end{bmatrix} \tag{2.4}$$

With $s = \{\frac{1}{\sqrt{3}}\ \frac{1}{\sqrt{3}}\ \frac{1}{\sqrt{3}}\}$ the matrix of coefficients w_{ij} is

$$W = \begin{bmatrix} 0 & \frac{2}{\sqrt{3}} & 0 \\ 0 & 0 & 0 \\ \frac{1}{\sqrt{3}} & 0 & 0 \\ 0 & 0 & 0 \end{bmatrix} \tag{2.5}$$

In this case a bound of $\sigma_n \geq 0$ is obtained because the third column of W is null. However with s = {0.4990 0.4366 0.7485}

$$W = \begin{bmatrix} 0 & 0.5614 & 0 \\ 0 & 0 & 0 \\ 0.5614 & 0 & 0 \\ 0 & 0 & 0.5614 \end{bmatrix} \tag{2.6}$$

giving $\sigma_n \geq 0.5614$.

Because the singular values [2,3] of F are related to the eigenvalues of $\overline{B}$ according to $\sigma_i^2 = \overline{\lambda}_i$ this gives $\overline{\lambda}_n \geq 0.3152$ for the matrix

$$\bar{B} = \begin{bmatrix} 6 & -4 & -4 \\ -4 & 12 & 6 \\ -4 & 6 & 6 \end{bmatrix} \tag{2.7}$$

for which the Gerschgorin lower bound is negative. The column scaling factors giving this result are known to be the best possible on account of the following theorem [1]:

Let

$$t_j = \sum_{i=1}^{m} w_{ij}^2 \tag{2.8}$$

so that from equation (2.2)

$$\sigma_n^2 \geq \min_{j=1,n} (t_j) \tag{2.9}$$

then a necessary and sufficient condition for the bound to be the maximum possible according to the previous theorem is that

$$t_1 = t_2 = \ldots = t_n (=\text{opt}(t)) \tag{2.10}$$

Furthermore this condition is always attainable by some choice of s.

In many cases the basic procedure for obtaining bounds using the above theorem may be enhanced by the use of row partitioning of F, particularly where the row partitioning leads to block structures giving reducibility of partitions. These enhancements have been discussed by the author [1].

3. Linear Generalized Eigenvalue Problems

Consider the linear generalized eigenvalue problem

$$Ax = \lambda Bx \tag{3.1}$$

in which A and B are both symmetric matrices of order nxn and B is also positive definite. Let $\Lambda = \lceil \lambda_1 \ \lambda_2 \ \cdots \ \lambda_n \rfloor$ be a diagonal matrix of the eigenvalues ordered such that $\lambda_1 \geq \lambda_2 \geq \ \cdots \geq \lambda_n$. Without loss of generality an arbitrary diagonal elements may be introduced into the equations to give

(3.2) $$(DAD)(D^{-1}x) = \lambda(DBD)(D^{-1}x)$$

If $Q = [q_1 \ q_2 \ \cdots \ q_n]$ is the matrix of eigenvectors of equation (3.2), the full set of eigenvalue equations may be written in the form

(3.3) $$DADQ = DBDQ\Lambda$$

Any vector u satisfying the normality condition $u^Tu = 1$ may be expressed as a linear combination of the eigenvectors i.e.

(3.4) $$u = Qc$$

then if

(3.5) $$\eta = \frac{u^TDADu}{u^TDBDu}$$

it follows that

(3.6) $$c^TQ^TDADQc - \eta c^TQ^TDBDQc = 0$$

However it can be proved [4] that orthogonality conditions exist such that $Q^T(DBD)Q$ is a diagonal matrix, say K. Furthermore since B is positive definite, K is also, giving $k_i > 0$ for i=1,n. Hence substituting equation (3.3) into equation (3.6) gives

(3.7) $$c^TK\Lambda c - \eta c^TKc = 0$$

which can be expressed in the form

(3.8) $$\sum_{i=1}^{n} c_i^2 k_i(\lambda_i - \eta) = 0$$

yielding

(3.9) $$\lambda_n \leq \eta \leq \lambda_1$$

From equation (3.2) when u is the eigenvector corresponding to λ_1

(3.10) $$u^T DADu = \lambda_1 u^T DBDu$$

thus from equation (3.5) and (3.9)

(3.11) $$\lambda_1 = \max\left(\frac{u^T DADu}{u^T DBDu}\right)$$

The value of $u^T DADu$ lies within the spectrum of eigenvalues of DAD which is similar to $VDADV^{-1}$ where V is any non-singular matrix (only diagonal V matrices will be considered). Therefore, if the j-th Gerschgorin disc of matrix $VDADV^{-1}$ intercepts the real axis at g_j and $\bar{g}_j$ such that $g_j \geq \bar{g}_j$,

(3.12) $$u^T DADu \leq \max(g_j)$$

Also

(3.13) $$u^T DBDu \geq \min(t_j)$$

where t_j is in accordance with §2 where

(3.14) $$\bar{B} = DBD$$

If max (g_j) is positive it follows from equation (3.11) that

(3.15(a)) $$\lambda_1 \leq \frac{\max(g_j)}{\min(t_j)}$$

otherwise

(3.15(b)) $$\lambda_1 \leq 0$$

Using similar arguments relating to the smallest eigenvalue λ_n rather than the largest λ_1 it follows that

(3.16) $$\lambda_n = \min\left(\frac{u^T DADu}{u^T DBDu}\right)$$

and, if $\min(\bar{g}_j)$ is negative

(3.17(a)) $$\lambda_n \geq \frac{\min(\bar{g}_j)}{\min(t_j)}$$

otherwise

(3.17(b)) $\lambda_n \geq 0$

Equations (3.15) and (3.17) give meaningful bounds on the eigenvalues provided that $\min(t_j) > 0$.

4. Scaling Factors

Whereas §3 shows that bounds for the eigenvalues of symmetric linear generalised eigenvalue problems can be obtained provided that $\min(t_j) > 0$, scope is available for optimization of the bounds since there is the possibility of varying the column scaling factors s_j applied to F, the coefficients of D and the similarity transformation matrix V. It is already known that, if the values of d_i and v_i are held constant, the largest value of mint(t_j) is achieved when the values of s_j are chosen so that equation (2.10) is satisfied. It is therefore necessary to investigate the effect of varying D and V.

The factorization (1.1) of $\overline{B}$ corresponds to the factorization

(4.1) $$B = G^TG$$

where G and F are related by

(4.2) $$F = GD$$

Hence equation (2.3) gives

(4.3) $$w_{ij} = \max(0, |g_{ij}|d_js_j - \sum_{k\neq j}|g_{ik}|d_ks_k)$$

If the values of w_{ij} were originally computed with D=I and with column scaling factors s_j^*, all these coefficients will be unchanged provided that

(4.4) $$d_js_j = s_j^* \quad (k=1,n)$$

which may be written in matrix form as DS = S*.

Since $\Sigma s_j^2 = 1$, it is necessary that

(4.5) $$\sum_{j=1}^{n} \left(\frac{s_j^*}{d_j}\right)^2 = 1$$

Therefore if the coefficients of D satisfy equation (4.5), the column scaling factors s_j may be chosen such that $\min(t_j)$ remains the same as for the initial analysis with D=I. It is therefore possible to use such adjustments together with a suitable choice for V to optimize the Gerschgorin bounds of $VDADV^{-1}$.

5. Optimization of the Bounds

In equation (3.12) $\max(g_j)$ is the Gerschgorin bound for the maximum eigenvalue of

$$\bar{A} = VDADV^{-1} \tag{5.1}$$

Choosing $V = \bar{V}S^{-1}$ and restricting $\bar{V}$ to a diagonal matrix, it may be shown that

$$\bar{A} = S^{-2}(\bar{V}S^*AS^*\bar{V}^{-1}) \tag{5.2}$$

which means that the rows of $\bar{A}$ are proportional to the rows of $\bar{V}S^*AS^*\bar{V}^{-1}$. If $h_j (j=1,n)$ are the larger intercepts of the Gerschgorin row discs of $\bar{V}S^*AS^*\bar{V}^{-1}$ with the real axis

$$g_j = h_j/s_j^2 \tag{5.3}$$

If $g_i > \min(g_j)$ and the values of h_j are non-negative it is possible to increase s_i and decrease other values of s_j to ensure conformity with equation (4.5) in such a way that $\min(g_j)$ will be increased. Hence if $\bar{V}$ is kept constant the maximum value attainable by $\min(g_j)$ will occur when

$$g_1 = g_2 = \ldots = g_n (=g \text{ say}) \tag{5.4}$$

From equations (4.5), (5.3) and (5.4) it follows that

$$g = \sum_{j=1}^{n} h_j \tag{5.5}$$

and

$$s_j = (h_j/g)^{\frac{1}{2}} \tag{5.6}$$

Alternatively, if any of the values of h_j are negative, an arbitrarily small corresponding coefficient s_j may be employed. Hence in the evaluation of equation (5.5) any negative values of h_j could be replaced by zero and the solution may still be stated as conforming to equation (5.4).

The above discussion shows that any finite bound obtained from equation (3.15) which does not satisfy equation (5.4) cannot be the optimum bound, and it was previously proved that any bound which does not satisfy equation (2.10) cannot be the optimum either. Hence it follows that the optimum bound must be one satisfying both equations (2.10) and (5.4) simultaneously. If the column scaling parameters s_j^* are obtained first to satisfy equation (2.10) and yield opt(t), the global optimum bound can be obtained by choosing $\bar{V}$ such that g is minimised according to equation

(5.5). The values of S must conform to equation (5.6) but need not be computed.

The minimum eigenvalue bound obtained from equation (3.17) can similarly be analysed. If $\bar{h}_j$ are the smaller intercepts of the Gerschgorin row discs of $\bar{V}S^*AS^*\bar{V}^{-1}$ the real axis and are non-positive, the optimum value, $\bar{g}$, of $\min(\bar{g}_j)$ is obtained when

$$(5.7) \qquad \bar{g} = \sum_{j=1}^{n} \bar{h}_j$$

and

$$(5.8) \qquad s_j = (\bar{h}_j/\bar{g})^{\frac{1}{2}}$$

with $\bar{V}$ being chosen to minimise $\bar{g}$.

6. On the Use of Diagonal Similarity Transformation Matrices

Consider the case where all the values of h_j are non-negative. Since

$$(6.1) \qquad h_i = a_{ii}(s_i^*)^2 + \sum_{j\neq i} |a_{ij}| s_i^* s_j^* \bar{v}_i/\bar{v}_j$$

allowing for symmetry in A, equation (5.5) gives

$$(6.2) \qquad g = \sum_{i=1}^{n} a_{ii}(s_i^*)^2 + \sum_{i=1}^{n} \sum_{j=1}^{i-1} |a_{ij}| s_i^* s_j^* \left(\frac{\bar{v}_i}{\bar{v}_j} + \frac{\bar{v}_j}{\bar{v}_i}\right)$$

On the other hand if any of the values of h_j are negative g will have a value greater than that given by equation (6.2). It may be shown that g attains its minimum value when $\bar{v}_i = 1(i=1,\ldots,n)$. Hence, if $\bar{V} = I$ gives all non-negative values for h_j, the resulting value of g will be the optimum. Similarly it may be shown that, if $\bar{V} = I$ gives all non-positive value for $\bar{h}_j$, the resulting values of $\bar{g}$ will be the optimum.

If, on the other hand, $\bar{V} = I$ gives some of the values of h_j negative, it is possible to use $\bar{V}$ to adjust them so that they are all either non-negative or non-positive. This will give an improved, although not necessarily an optimum bound. For instance, if

$$(6.3) \qquad A = \begin{bmatrix} -8 & 1 & 1 \\ 1 & 0 & 1 \\ 1 & 0 & 0 \end{bmatrix}$$

and $s_1^* = s_2^* = s_3^* = 1/\sqrt{3}$, then $\overline{V}$ gives the values of h_j as -3, 0.6667 and 0.6667 and g = 1.333. Using $\overline{V} = \lceil 4 \quad 1 \quad 1 \rfloor$ gives the values of h_j as 0, 0.4167 and 0.4167 and hence g = 0.833. Alternatively if

$$A = \begin{bmatrix} -8 & 2 & 3 \\ 2 & -6 & 1 \\ 3 & 1 & -2 \end{bmatrix} \tag{6.4}$$

and $s_1^* = s_2^* = s_3^* = 1/\sqrt{3}$, then $\overline{V} = I$ gives the values of h_j as -1, -1 and 0.6667. Using $\overline{V} = \lceil 1 \quad 1 \quad 0.5 \rfloor$ gives h_j as 0, -0.6667 and 0 proving that $\lambda_1 \leq 0$.

7. Some Examples

Consider the equations $Ax = \lambda Bx$ with

$$A = \begin{bmatrix} 6 & 2 & -8 & 5 \\ 2 & 4 & -1 & -10 \\ -8 & -1 & 14 & -10 \\ 5 & -10 & -10 & 40 \end{bmatrix} \tag{7.1(a)}$$

and

$$B = \begin{bmatrix} 1 & 1 & -2 & 0 \\ 1 & 2 & -2 & -2 \\ -2 & -2 & 5 & -1 \\ 0 & -2 & -1 & 6 \end{bmatrix} = \begin{bmatrix} 1 & 0 & 0 & 0 \\ 1 & -1 & 0 & 0 \\ -2 & 0 & 1 & 0 \\ 0 & 2 & -1 & 1 \end{bmatrix} \begin{bmatrix} 1 & 1 & -2 & 0 \\ 0 & -1 & 0 & 2 \\ 0 & 0 & 1 & -1 \\ 0 & 0 & 0 & 1 \end{bmatrix} \tag{7.1(b)}$$

Because F is triangular the non-zero elements of W must occur on the leading diagonal and the optimum for min(t) occurs when $s^* = \{0.9058 \quad 0.3397 \quad 0.2265 \quad 0.1132\}$ with opt(t) = 0.0128. However the values of h_j obtained from the rows of S^*AS^* are 7.69, 1.54, 2.69 and 1.67 giving, from equation (5.5), g = 13.59. Hence

$$\lambda_1 \leq \frac{13.59}{0.0128} = 1062$$

With $\overline{V} = I$ the values of $\overline{h}_j$ are 2.15, -0.61, -1.26 and -0.64 equation (5.7) gives $\overline{g} = -2.51$. However, since $\overline{h}_1$ is positive it is possible to take advantage of a variation in $\overline{V}$. Thus, if $\overline{V} = \lceil 1.778, 1, 1, 1 \rfloor$, then the values of $\overline{h}_j$ are 0, -0.346, -0.538 and -0.417 giving $\overline{g} = 1.30$. Hence

$$\lambda_n \geq -\frac{1.30}{0.0128} = -102$$

(The eigenvalues are 15, 5, 5 and -1). Whereas the bounds in this case are very slack indeed, at least bounds are obtained.

Allen and Bulson[5] derive equations from the buckling of a set of three rods hinged together to form a cantilever column with the hinge rotations restrained by springs giving

$$A = \begin{bmatrix} 6 & -4 & 1 \\ -4 & 6 & -2 \\ 1 & -2 & 2 \end{bmatrix} \tag{7.2(a)}$$

and

$$B = \begin{bmatrix} 2 & -1 & 0 \\ -1 & 2 & -1 \\ 0 & -1 & 1 \end{bmatrix} = \begin{bmatrix} 1 & -1 & \\ & 1 & -1 \\ & & 1 \end{bmatrix} \begin{bmatrix} 1 & & \\ -1 & 1 & \\ & -1 & 1 \end{bmatrix} \tag{7.2(b)}$$

The non-zero elements of W can only occur on the leading diagonal and the optimum for min(t) occurs when s = {0.2673 0.5345 0.8018} giving opt(t) = 0.07145. The values of h_j obtained from the rows of S*AS* are 1.215, 3.143 and 2.357 giving, from equation (5.5), g = 6.714. Thus

$$\lambda_1 \leq \frac{6.714}{0.07145} = 94$$

(The largest eigenvalue is given as 5.170).

Allen and Bulson also derive equations for the buckling of a non-uniform cantilever strut divided into unequal segments giving

$$A = \begin{bmatrix} 1 & 0 & -0.6 & 0.1 \\ 0 & 0.6667 & -0.1 & 0.0667 \\ -0.6 & -0.1 & 0.6 & -0.1 \\ 0.1 & 0.0667 & -0.1 & 0.2667 \end{bmatrix} \tag{7.3(a)}$$

and

$$B = \begin{bmatrix} 1.0444 & -0.0667 & -0.6 & 0.6 \\ -0.0667 & 2.1333 & -0.6 & 0.4 \\ -0.6 & -0.6 & 0.6 & -0.6 \\ 0.6 & 0.4 & -0.6 & 0.8 \end{bmatrix} \tag{7.3(b)}$$

for which it is possible to derive the matrix factor

$$F = \begin{bmatrix} -0.6667 & 1 & 0 & 0 \\ 0 & 0.57735 & 0 & 0 \\ 0.7746 & 0.7746 & -0.7746 & 0.7746 \\ 0 & 0.4472 & 0 & 0.4472 \end{bmatrix} \tag{7.4}$$

The non-zero elements of W lie on the diagonal with the optimum column scaling factors given by $s^* = \{0.3255 \quad 0.1376 \quad 0.8808 \quad 0.3152\}$ such that opt(t) = 0.00630. The values of h_j obtained from the rows of A are 0.288, 0.028, 0.786 and 0.068 giving g = 1.170 and

$$\lambda_1 \leq \frac{1.170}{0.00630} = 186$$

(The value quoted for λ_1 is 11.73).

In both of the buckling problems λ is inversely proportion to the applied load and hence the eigenvalue upper bound gives a lower bound to the buckling load. It is a lower bound rather than an upper bound which is most useful, but it would be very much more useful if the bound was tight rather than slack.

REFERENCES

1. ALAN JENNINGS, Bounds for the singular values of a matrix, IMAJNA, to be published.

2. A.S. HOUSEHOLDER, The Theory of Matrices in Numerical Analysis, Dover, New York, 1964.

3. J.H. WILKINSON, The Algebraic Eigenvalue Problem, Oxford, Clarendon Press, 1965.

4. ALAN JENNINGS, Matrix Computation for Engineering and Scientists, Wiley, London, 1977.

5. H.G. ALLEN, and P.S. BULSON, Background to Buckling, McGraw-Hill, London, 1980.

SECTION C

GENERALIZED SINGULAR VALUES AND DATA ANALYSIS

A METHOD FOR COMPUTING THE GENERALIZED SINGULAR VALUE DECOMPOSITION

G W Stewart*
Dept of Computer Science
University of Maryland
Collage Park MD 20742
U S A

1. Introduction

The object of this paper is to describe an algorithm for computing the generalized singular value decomposition of a partitioned matrix

$$X = \begin{bmatrix} X_1 \\ X_2 \end{bmatrix}.$$

Only the simplest case where X is of full rank and X_1 and X_2 are square will be treated, and we shall only sketch algorithm. Extensions and a detailed analysis will appear later.

The generalized singular value decomposition was introduced by Van Loan [7,8]. Here a variant due to Paige and Saunders [4] will be treated. They have shown that there are orthogonal matrices U_1 and U_2 and a non-singular matrix V such that

$$(1.1) \qquad \begin{bmatrix} U_1^T & 0 \\ 0 & U_2^T \end{bmatrix} \begin{bmatrix} X_1 \\ X_2 \end{bmatrix} = \begin{bmatrix} D_1 \\ D_2 \end{bmatrix} V,$$

where D_1 and D_2 are diagonal. For applications of this decomposition see [8].

Since U_1 and U_2 are orthogonal, it follows from (1.1) that

$$(1.2) \qquad V^{-T} X_i^T X_i V^{-1} = D_i^2 \qquad (i=1,2).$$

*University of Maryland and National Bureau of Standards. This work was supported in part by the Air Force Office of Sponsored Research under Contract No.

Since D_i^2 is diagonal, (1.2) implies that the columns of V^{-1} are the eigenvectors of the generalized eigenvalue problem

$$(1.3) \qquad (X_1^T X_1)z = \lambda(X_2^T X_2)z.$$

This suggests that the generalized singular value decomposition be calculated by first solving the eigenvalue problem for the columns of V^{-1} and the matrices D_i, and then computing the U_i from the formula

$$(1.4) \qquad U_i = X_i V^{-1} D_i^{-1} \qquad (i=1,2).$$

However this procedure is objectionable on three counts. First, the passage from X_i to $X_i^T X_i$ involves a loss of information about X_i. Second, there are no unconditionally stable algorithms for solving the generalized eigenvalue problem (1.3). Finally, the formula (1.4) is impossible if D_i is singular, and when some diagonal element of D_i is small, the columns of U_i can deviate from orthogonality owing to rounding error. This latter phenomenon has been described by Golub and Kahan [3] in connection with the ordinary singular value decomposition.

The algorithm proposed here circumvents these difficulties by working directly with X. In the next section we show how the problem of computing the decomposition (1.1) can be reduced to that of computing the generalized singular value decomposition of a matrix with orthonormal columns. In §3, the existence of this latter decomposition, which is important in its own right, is established. The proof is constructive and suggests an algorithm, which, however, is unstable. In §4 we show how the algorithm can be made stable by a reorthogonalization technique. The paper concludes with a numerical example.

2. The generalized singular value decomposition

In this section we shall establish the existence of the generalized singular value decomposition under the assumption that X is of full column rank and X_1 and X_2 are square. Let

(2.1) $$X = QR$$

be the QR factorization of X, i.e. let X be factored into the product of a matrix Q with orthonormal columns and an upper triangular matrix R. Because X has full column rank, R is nonsingular. We now employ the following lemma.

Lemma 2.1. Let Q be a matrix with orthonormal columns and suppose Q is partitioned in the form

$$Q = \begin{bmatrix} Q_1 \\ Q_2 \end{bmatrix},$$

where Q_1 and Q_2 are square. Then there are orthogonal matrices U_1, U_2, and W such that

(2.2) $$\begin{bmatrix} U_1^T & 0 \\ 0 & U_2^T \end{bmatrix} \begin{bmatrix} Q_1 \\ Q_2 \end{bmatrix} W = \begin{bmatrix} C \\ S \end{bmatrix},$$

where C and S are nonnegative diagonal matrices satisfying

$$C^2 + S^2 = I.$$

This lemma will be proved in §3. Its application to the generalized singular value decomposition is immediate; for it follows from (2.1) and (2.2) that

(2.3) $$\begin{bmatrix} U_1^T & 0 \\ 0 & U_2^T \end{bmatrix} \begin{bmatrix} X_1 \\ X_2 \end{bmatrix} = \begin{bmatrix} U_1^T & 0 \\ 0 & U_2^T \end{bmatrix} \begin{bmatrix} Q_1 \\ Q_2 \end{bmatrix} WW^T R = \begin{bmatrix} C \\ S \end{bmatrix} W^T R.$$

Hence if we set

$$V = W^T R,$$

$D_1 = C$, and $D_2 = S$, equation (2.3) becomes identical with (1.1).

Since there are stable algorithms for computing the QR decomposition (e.g. see [2]), we have reduced the problem of computing the generalized singular value decomposition to that of computing the decomposition in Lemma 2.1, which will be the subject of the rest of the paper.

3. Existence of the CS decomposition

We shall call the decomposition of Lemma 2.1 the CS decomposition of Q . The name derives from the fact that the diagonal elements γ_i and τ_i of C and S satisfy $\gamma_i^2+\tau_i^2=1$. Hence for some angle θ_i we can write $\gamma_i=\cos\theta_i$ and $\tau_i=\sin\theta_i$, and thus the letters C and S stand for cosine and sine. The decomposition and its extensions is important in the analysis of the relation between two subspaces of R^n [1,4].

To establish the existence of the CS decomposition, we begin by observing that from (2.2)

$$U_1^T Q_1 W = C, \tag{3.1}$$

which is the singular value decomposition of Q_1. Thus, U_1, W, and C are determined by Q_1, and it remains only to determine U_2.

Now the matrix

$$\begin{bmatrix} U_1^T & 0 \\ 0 & I \end{bmatrix} \begin{bmatrix} Q_1 \\ Q_2 \end{bmatrix} W = \begin{bmatrix} C \\ Q_2 W \end{bmatrix}$$

has orthonormal columns. Hence, with

$$\bar{Q}_2 \triangleq Q_2 W,$$

we have $C^2+\bar{Q}_2^T\bar{Q}_2 = I$ or

$$\bar{Q}_2^T\bar{Q}_2=I-C^2.$$

Since I and C^2 are diagonal, so is $\bar{Q}_2^T\bar{Q}_2$, which means that the nonzero columns of $\bar{Q}_2$ are orthogonal to one another. If all the columns of $\bar{Q}_2$ are nonzero, we may set

$$S^2=\bar{Q}_2^T\bar{Q}_2 \tag{3.2}$$

and

$$U_2=\bar{Q}_2 S^{-1}, \tag{3.3}$$

in which case $U_2^T U_2 = I$ and $U_2^T \bar{Q}_2 = S$, which establishes the decomposition. If $\bar{Q}_2$ has zero columns, we may normalize the nonzero columns as above, and replace the zero columns with an orthonormal basis for the orthogonal complement of the column space of $\bar{Q}_2$. It is easily verified that the U_2 so defined is orthogonal, and $S \triangleq U_2^T \bar{Q}_2$ is diagonal, which again establishes the decomposition.

The above proof is, in principle, an algorithm for computing the CS decomposition. There are standard methods [3], for computing the singular value decomposition (3.1). The normalization defined by (3.2) and (3.3) is trivial to compute. In the event that $\bar{Q}_2$ has zero columns, the required orthonormal basis can be found from the QR decomposition of $\bar{Q}_2$.

However, in the presence of rounding error, the process can fail completely. The problem occurs when some column $\bar{q}_i^{(2)}$ of $\bar{Q}_2$ is small, and therefore consists largely of rounding error. The corresponding column of U_2 is $u_i^{(2)} = \bar{q}_i^{(2)}/||\bar{q}_i^{(2)}||$, and the division by the small number $||\bar{q}_i^{(2)}||$ magnifies the rounding error and causes $u_i^{(2)}$ to fail to be orthogonal to the other vectors.

A numerical example may make this point clear. Exhibit 3.1 contains a partitioned matrix Q whose columns are orthogonal to working accuracy (in this and all subsequent examples, the computations are done in 32-bit, hexadecimal, floating-point arithmetic). Exhibit 3.2 shows the matrix $\bar{Q}_2$. The first two columns of $\bar{Q}_2$, which correspond to the small singular values of Q_2, are small, and it might be expected that they fail to be orthogonal to the others. Exhibit 3.2 also shows the upper half of the cross-product matrix

$$(3.4) \qquad A = \bar{Q}_2^T \bar{Q}_2 .$$

The angle that the i-th column of $\bar{Q}_2$ makes with the j-th is given by

$$(3.5) \qquad \theta_{ij} = \cos^{-1}\left[\frac{a_{ij}}{\sqrt{a_{ii}a_{jj}}}\right].$$

Exhibit 3.1

A Partioned Orthonormal Matrix

Q1

2.1597E-01	1.5968E-02	-4.2403E-01	-3.8387E-01	-3.8411E-01
-1.5968E-02	1.8403E-01	-3.7597E-01	-4.1613E-01	-4.1589E-01
-4.2403E-01	3.7597E-01	3.5968E-02	-2.3872E-02	-2.4112E-02
3.8387E-01	-4.1613E-01	2.3872E-02	-1.5888E-02	-1.6048E-02
-3.8411E-01	4.1589E-01	-2.4112E-02	1.6048E-02	1.6008E-02

Q2

2.1597E-01	1.5968E-02	-4.2403E-01	-3.8387E-01	-3.8411E-01
-1.5968E-02	1.8403E-01	-3.7597E-01	-4.1613E-01	-4.1589E-01
-4.2403E-01	3.7597E-01	3.5968E-02	-2.3872E-02	-2.4112E-02
3.8387E-01	-4.1613E-01	2.3873E-02	-1.5888E-02	-1.6048E-02
-3.8411E-01	4.1589E-01	-2.4112E-02	1.6048E-02	1.6008E-02

SINGULAR VALUES

Q1	Q2
1.0000	$1.0000 . 10^{-6}$
1.0000	$2.0000 . 10^{-4}$
0.1000	0.9950
$4.0000 . 10^{-4}$	1.0000
$2.0000 . 10^{-4}$	1.0000

Exhibit 3.2

Q2*W

4.3340E-05	1.1241E-04	3.9800E-01	-3.9984E-01	-4.0016E-01
-2.8156E-05	-7.4327E-05	3.9800E-01	-3.9984E-01	-4.0016E-01
-2.8968E-05	-7.4387E-05	-5.9699E-01	-3.9984E-01	-4.0016E-01
-2.9087E-05	-7.4327E-05	3.9800E-01	6.0016E-01	-3.9976E-01
-2.9624E-05	-7.4506E-05	3.9800E-01	-4.0024E-01	5.9984E-01

THE CROSS-PRODUCT MATRIX

5.2338E-09	1.3488E-08	-2.9623E-08	-8.9195E-08	-6.2591E-07
	3.4769E-08	3.2916E-07	-2.7106E-07	-4.5015E-07
		9.9000E-01	5.9605E-07	3.5763E-07
			1.0000E+00	-2.3842E-07
				1.0000E+00

The angles in degrees between the first column and the others are

0.93 89.98 89.99 89.50

Thus the first column is almost parallel to the second and fails to be orthogonal to the others.

4. Reorthogonalization

From the foregoing, it is evident that we must further transform the matrix Q in order to restore the orthogonality in $\bar{Q}_2$ that has been lost to rounding error. From the nature of the decomposition we seek, it follows that the transformations must take the form

$$\begin{bmatrix} K^T & 0 \\ 0 & L^T \end{bmatrix} \begin{bmatrix} U_1^T & 0 \\ 0 & I \end{bmatrix} \begin{bmatrix} Q_1 \\ \bar{Q}_2 \end{bmatrix} WJ = \begin{bmatrix} K^T CJ \\ L^T \bar{Q}_2 J \end{bmatrix} ,$$

where J, K, and L are orthogonal. Since L cannot affect inner products among columns of $\bar{Q}_2$, we may take $L = I$. Thus we must find orthogonal matrices J and K such that

1. The columns of $\bar{Q}_2 J$ may be normalized to give an orthogonal U_2,

2. $K^T CJ$ is diagonal.

Turning first to the determination of J, we note that replacing $\bar{Q}_2$ by $\bar{Q}_2 J$ replaces the cross-product matrix A by $J^T AJ$ [cf. (3.4)]. In view of (3.5), we wish to make the off-diagonal elements of $J^T AJ$ as small as possible. In other words, we wish to diagonalize A by an orthogonal similarity transformation. We shall use Jacobi's method to do this,since this algorithm works well on matrices whose elements vary widely in size. Rutishauser [5] has published an elegant implementation of this method, and we refer the reader to his paper for details. What follows here is a brief description.

In a typical step of the method, an element a_{ij} is selected to be anihilated. This is done by means of a plane rotation J_{ij} in the (i,j)-plane; i.e. a matrix of the form

$$J_{ij} = \begin{bmatrix} I_{i-1} & 0 & 0 & 0 & 0 \\ 0 & c & 0 & s & 0 \\ 0 & 0 & I_{j-i-1} & 0 & 0 \\ 0 & -s & 0 & c & 0 \\ 0 & 0 & 0 & 0 & I_{p-j} \end{bmatrix}$$

where $c^2+s^2=1$. It is always possible to determine c and s with $|s|\leq|c|$ so that the (i,j)-element of $J_{ij}^T A J_{ij}=0$.

The introduction of zeros at a later stage will usually destroy zeros introduced earlier; however, if the order in which the elements of A are anihilated is suitably chosen, the net effect will be to drive all the elements to zero - ultimately quadratically. Whenever an off-diagonal element satisfies

$$|a_{ij}|\leq \ \varepsilon_M\sqrt{a_{ii}a_{jj}},$$

where ε_M is the rounding unit for the arithmetic being used, it is set to zero and no rotation is performed. As the rotations are generated, they are accumulated in $\bar{Q}_2$ and V.

It is important to remember that the object of the rotations is to orthogonalize the columns of $\bar{Q}_2$. If ever the elements of A become inaccurate owing to rounding error, the process may converge to a product J of the J_{ij}'s that fails to orthogonalize the columns of $\bar{Q}_2$. We therefore take the precaution of saving the initial diagonal elements $\alpha_i=a_{ii}$ of A. If ever an element a_{ii} falls below some tolerance, say

(4.1) $\quad a_{ii}\leq 0.1\alpha_i$,

the i-th row and column of A is recomputed from the current $\bar{Q}_2$ and α_i is reset.

We turn now to the determination of an orthogonal matrix K such that $K^T CJ$ is diagonal. The suprising fact is that <u>provided certain unnecessary rotations are not performed</u>, <u>we may take</u> K=J. To see this, note that performing a Jacobi rotation in the (i,j)-plane affects only the subproblem

$$\begin{bmatrix} \gamma_i & 0 \\ 0 & \gamma_j \\ \bar{q}_i^{(2)} & \bar{q}_j^{(2)} \end{bmatrix} .$$

Since the columns of Q are assumed to be orthogonal, we must have

$$\gamma_i^2 + ||q_i^{(2)}||^2 \equiv \gamma_i^2 + a_{ii} = 1 + \varepsilon_i,$$

$$\gamma_j^2 + ||\bar{q}_j^{(2)}||^2 \equiv \gamma_j^2 + a_{jj} = 1 + \varepsilon_j,$$

and

$$\bar{q}_i^{(2)T} \bar{q}_j^{(2)} \equiv a_{ij} = \varepsilon_{ij}.$$

for some ε_i, ε_j and ε_{ij} that are small compared to one. The basic result is summarized in the following theorem, which is given without proof.

<u>Theorem</u> 4.1 Let

$$|\varepsilon_i|, \ |\varepsilon_j|, \ |\varepsilon_{ij}| \leq 1$$

and

$$\tau = \gamma_i + \gamma_j.$$

If $u_i = \bar{q}_i^{(2)} / ||\bar{q}_i^{(2)}||$ and $u_j = \bar{q}_j^{(2)} / || \bar{q}_i^{(2)} ||$, then approximately

(4.2) $$|u_i^T u_j| \leq \frac{\varepsilon}{\sqrt{1-\tau^2}}$$

Moreover, if δ denotes the largest element of $|J_{ij}^T C J_{ij} - C|$, then

(4.3) $$\delta \leq \frac{\sqrt{2} \cdot \varepsilon}{\tau}.$$

The inequality (4.3) says that if τ is not small we shall make no great error in approximating $J_{ij}^T CJ_{ij}$ by C itself. On the other hand if τ is small, the inequality (4.2) says that there was no need to orthogonalize $\bar{q}_i^{(2)}$ and $\bar{q}_j^{(2)}$ in the first place, since the vectors u_i and u_j obtained by normalizing them are as orthogonal as the columns Q. We therefore recommend that when

(4.4) $\tau \leq 0.7,$

the rotation be skipped. Otherwise the rotation is accumulated into U_1, as well as $\bar{Q}_2$ and V. The number 0.7 in (4.4) makes the denominators in (4.2) and (4.3) approximately equal.

5. A numerical example

We continue with the example of §3. Exhibit 5.1 gives the cosines and sines of one Jacobi sweep across the matrix, along with the final cross-product matrix. The first rotation J_{12} differs from the identity; however, $\gamma_1 = \gamma_2 = 1$, and hence $J_{12}^T C J_{12}$ is effectively C.

Note that the rotation results in cancellation in a_{11}, which must then be recomputed [cf. (4.1)] . As an experiment, this recomputation was surpressed in a later run, with the result that the final U_2 deviated considerably from orthogonality.

Note that no rotation is made in the (5,4)-plane, since it is forbidden by (4.4) ($\tau=6\cdot10^{-4}$). Again we experimented by forcing this rotation. The result was that U_1CW^T failed to reproduce Q_2 to full accuracy.

Exhibit 5.2 shows the second pass over A. Except for J_{12}, this amounts to a polishing up of the orthogonality of $\bar{Q}_2$. To test the final results we computed $\mu(U_2^T U_2 - I)$, $\mu(Q_1 - U_1 CW^T)$, and $\mu(Q_2 - U_2 SW^T)$, where $\mu(A) = \max |a_{ij}|$. These numbers are

$$\mu(I-U_2^T U_2) = 9.5.10^{-7}$$
$$\mu(Q_1-U_1 CW^T) = 3.4.10^{-7}$$

and

$$\mu(Q_2-U_2 SW^T) = 1.9-10^{-6}$$

This is a perfectly satisfactory result at the level of precision used in the calculation.

Exhibit 5.1

First Jacobi Sweep

$$\begin{array}{cc} i & j \\ \left[\begin{array}{cc} a_{ii} & a_{ij} \\ & a_{jj} \\ C & S \end{array}\right] \end{array}$$

1 2	
5.2338E-09	1.3488E-08
	3.4769E-08
9.3230E-01	3.6169E-01
1 3	
5.2338E-09	-1.4667E-07
	9.9000E-01
1.0000E+00	-1.4815E-07
1 4	
5.23338E-09	1.4883E-08
	1.0000E+00
1.0000E+00	1.4883E-08
1 5	
5.2338E-09	-4.2072E-07
	1.0000E+00
1.0000E+00	-4.2072E-07

2 3	
3.4769E-08	2.9617E-07
	9.9000E-01
1.0000E+00	2.9916E-07
2 4	
3.4769E-08	-2.8497E-07
	1.0000E+00
1.0000E+00	-2.8497E-07
2 5	
3.4769E-08	-6.4606E-07
	1.0000E+00
1.0000E+00	-6.4606E-07

NEW CROSS-PRODUCT MATRIX

7.7971E-13	-2.2797E-13	-2.7740E-11	2.4102E-11	1.2847E-11
	4.0001E-08	4.0090E-13	-1.5403E-13	
		9.9000E-01		
			1.0000E+00	
				1.0000E+00

Exhibit 5.2

Second Jacobi Sweep

$$\begin{matrix} i & j \\ \begin{bmatrix} a_{ii} & a_{ij} \\ & a_{jj} \end{bmatrix} \\ C & S \end{matrix}$$

1 2		2 3	
7.7971E-13	-2.2797E-13	4.0001E-08	4.0106E-13
	4.0001E-08		9.9000E-01
1.0000E+00	-5.6992E-06	1.0000E+00	4.0511E-13
1 3		2 4	
7.7971E-13	-2.7740E-11	4.0001E-08	-1.5417E-13
	9.9000E-01		1.0000E+00
1.0000+00	-2.8020E-11	1.0000E+00	-1.5417E-13
1 4		2 5	
7.7971E-13	2.4102E-11	4.0001E-08	-7.3216E-17
	1.0000E+00		1.0000E+00
1.0000E+00	2.4102E-11	1.0000E+00	-7.3215E-17
1 5			
7.7971E-13	1.2847E-11		
	1.0000E+00		
1.0000E+00	1.2847E-11		

FINAL CROSS-PRODUCT MATRIX

7.7971E-13	1.4954E-23	2.0907E-32	-3.9799E-33	-1.0949E-39
	4.0001E-08	-1.0415E-34	2.2669E-38	
		9.9000E-01		
			1.0000E+00	
				1.0000E+00

REFERENCES

1. CHANDLER DAVIS AND W. M. KAHAN, The rotation of eigenvectors by a perturbation.III, SIAM J. Numer. Anal, 7 (1970), pp.1-46.

2. J. J. DONGARRA, C. B. MOLER, J. R. BUNCH, AND G. W. STEWART, LINPACK User's Guide, Society for Industrial and Applied Mathematics, Philadelphia, 1979.

3. G. GOLUB AND W. KAHAN, Calculating the singular values and pseudo-inverse of a matrix, SIAM J. Numer. Anal., 2 (1965), pp.205-224

4. C. C. PAIGE AND M. A. SAUNDERS, Towards a generalized singular value decomposition, SIAM J. Numer. Anal., 18 (1981), pp. 398-405.

5. H. RUTISHAUSER, The Jacobi method for real symmetric matrices, Numer. Math. 9 (1966), pp. 1-10.

6. G. W. STEWART, On the perturbation of pseudo-inverses, projections and linear least squares problems, SIAM Rev., 19 (1977), pp. 634-662.

7. CHARLES F. VAN LOAN, A general matrix eigenvalue algorithm, SIAM J Numer. Anal., 12 (1975), pp 819-834.

8. CHARLES F. VAN LOAN, Generalizing the singular value decomposition, SIAM J. Numer. Anal., 13 (1976), pp. 76-83.

PERTURBATION ANALYSIS FOR THE GENERALIZED EIGENVALUE AND THE GENERALIZED SINGULAR VALUE PROBLEM*

Ji-guang Sun
Computing Center, Academia Sinica
Peking, China
Visiting the University of Bielefeld
Federal Republic of Germany

Abstract

The author has obtained some results in his recent work, which generalize some classical perturbation theorems for the standard eigenvalue problem $Ax=\lambda x$ to regular matrix pencils, and give a positive answer for an open question proposed by G. W. Stewart. A perturbation analysis for the generalized singular value decomposition suggested by Van Loan, C. C.Paige and M. A. Saunders has also been carried out.

1. Introduction

Let A and B be complex matrices of order n. The perturbation analysis of the generalized eigenvalue problem

$$\beta Ax = \alpha Bx$$

is important not only to the theory of matrices, but also to numerical analysis. G. W. Stewart [18]-[22] and C.R. Crawford[2] have made important contributions in this area. During the last years the author has been engaged in this subject and has obtained some results, which generalize some classical perturbation theorems for the standard eigenvalue problem

* This work was supported by the Alexander von Humboldt Foundation in Federal Republic of Germany.

$Ax=\lambda x$ (e.g. Bauer-Fike theorem, Hoffman-Wielandt theorem, Weyl-Lidskii theorem, Davis-Kahan $\sin\theta$ theorem, and so on. ref. [1],[3], [6]-[8], [10], [12], [33]), and give a positive answer for an open question proposed by G. W. Stewart [22]. A perturbation analysis for the generalized singular value decomposition suggested by Van Loan [31], C. C. Paige and M. A. Saunders [16] has also been carried out.

The following results about the perturbation theorems of eigenvalues are well known to numerical analysts. Suppose that A and C are nxn matrices with eigenvalues $\lambda(A)=\{\lambda_i\}$ and $\lambda(C)=\{\mu_i\}$ resp. We have

Bauer-Fike theorem. [1]. If A is diagonalizable, i.e. there is a non-singular nxn matrix Q such that

$$Q^{-1}AQ = \operatorname{diag}(\lambda_1,\ldots,\lambda_n),$$

then the spectral variation of C with respect to A

$$S_A(C) \equiv \max_j \{\min_i |\lambda_i-\mu_j|\} \le \|Q\|_2 \, \|Q^{-1}\|_2 \, \|A-C\|_2. \tag{1.1}$$

Here $\|\cdot\|_2$ denotes the spectral norm.

Hoffman-Wielandt theorem [7]. If A and C are both normal, then the eigenvalue distance of A and C

$$d(A,C) \equiv \min_\pi \sqrt{\sum_{i=1}^{n} |\lambda_i-\mu_{\pi(i)}|^2} \le \|A-C\|_F \tag{1.2}$$

Here $\|\cdot\|_F$ denotes the Frobenius norm and the minimum is taken with respect to all permutations π of the set $\{1,\ldots,n\}$.

Weyl-Lidskii theorem [10],[33] If A and C are both Hermitian, then the eigenvalue variation of A and C

$$v(A,C) \equiv \min_\pi \{\max_i |\lambda_i-\mu_{\pi(i)}|\} \le \|A-C\|_2. \tag{1.3}$$

Here the minimum is the same as above in (1.2).

From the point of view of geometry, in the inequalities (1.1)-(1.3) the perturbation from points $\{\lambda_i\}$ to $\{\mu_i\}$ in the complex plane is bounded by a certain distance (i.e. a certain Euclidean metric) between the points A and C in the complex n^2-dimensional space $\mathbb{C}^{n\times n}$. We seek the perturbation bounds of generalized eigenvalues exactly from this point of view.

Besides we can state Davis-Kahan sin θ theorems for eigenspaces of a Hermitian matrix as follows.

Davis-Kahan sin θ theorem [3]. Suppose that A and $\tilde{A}=A+E$ are nxn Hermitian matrices, $Q=(\underset{\ell}{Q_1},\underset{n-\ell}{Q_2})$ and $\tilde{Q}=(\underset{\ell}{\tilde{Q}_1},\underset{n-\ell}{\tilde{Q}_2})$ are nxn unitary matrices such that

$$Q^HAQ=\begin{pmatrix} A_1 & 0 \\ 0 & A_2 \end{pmatrix}\begin{matrix}\ell \\ n-\ell\end{matrix}, \quad \tilde{Q}^H\tilde{A}\tilde{Q}=\begin{pmatrix} \tilde{A}_1 & 0 \\ 0 & \tilde{A}_2 \end{pmatrix}\begin{matrix}\ell \\ n-\ell\end{matrix}$$
$$\qquad\quad \ell \quad n-\ell \qquad\qquad\qquad\quad \ell \quad n-\ell$$

Assume that there is an interval $[\beta,\alpha]$ and a number $\delta>0$ such that

$$\lambda(A_1)\subset[\beta,\alpha] \quad \text{and} \quad \lambda(A_2)\subset(-\infty,\beta-\delta]\cup[\alpha+\delta,+\infty)$$

$$(\text{or } \lambda(A_2)\subset[\beta,\alpha] \quad \text{and} \quad \lambda(A_1)\subset(-\infty,\beta-\delta]\cup[\alpha+\delta,+\infty)). \text{ Then for}$$

every unitary invariant matrix norm

(1.4) $$\|\sin\Theta\| \leq \|EQ_1\|/\delta,$$

where

$$\Theta \equiv \arccos(Q_1^H\tilde{Q}_1\tilde{Q}_1^HQ_1)^{\frac{1}{2}} \geq 0.$$

Davis-Kahan generalized sin Θ theorem [3]. Suppose A, $\tilde{A}=A+E$, Q, $\tilde{Q}$, A_1 and $\tilde{A}_2$ are the same as above. If

$$\delta \equiv \min_{i,j}\{|\lambda_i-\tilde{\lambda}_j|: \lambda_i\in\lambda(A_1), \tilde{\lambda}_j\in\lambda(\tilde{A}_2)\}>0,$$

then

(1.5) $$\|\sin\Theta\|_F \leq \|EQ_1\|_F/\delta.$$

The above classical theorems are very important. But what are the corresponding results for the generalized eigenvalue problem?

The purpose of the present lecture is to explain the major idea of treating the perturbation analysis of the generalized eigenvalue problem, and to present some main results in the author's paper [4] (with L Elsner) and [23] - [29].

2. Definitions and basic results

The symbol $\mathbb{C}^{m\times n}$ denotes the set of complex mxn matrices, $\mathbb{C}^n = \mathbb{C}^{n\times 1}$ and $\mathbb{C} = \mathbb{C}^1$. $\bar{A}$ and A^T are for conjugate and transpose of A resp., and $A^H=\bar{A}^T$. $A > 0$ denotes that A is a positive definite matrix. The column space of $Z \in \mathbb{C}^{n\times \ell}$ $(0<\ell<n)$ is denoted by $R(Z)$. For $x \in \mathbb{C}^n$, $||x||$ is the Euclidean vector norm.

Definition 2.1 A matrix pair (A,B), A and $B \in \mathbb{C}^{n\times n}$, is called a regular pair, if $\det(A+\lambda B) \not\equiv 0$ [5].

Definition 2.2 Let (A,B) be a regular pair. A non-zero vector $x\in\mathbb{C}$ is an eigenvector of (A,B) corresponding to the generalized eigenvalue (α,β), α and $\beta \in\mathbb{C}$, if

$$(\alpha,\beta) \neq (0,0) \quad \text{and} \quad \beta Ax = \alpha Bx.$$

The number pair (α,β) is the homogeneous coordinate of a point in the complex projective plane. If $\beta\neq 0$, then $\lambda = \frac{\alpha}{\beta}$ is a finite generalized eigenvalue of (A,B). $\lambda(A,B)$ denotes the set of all generalized eigenvalues of (A,B).

Definition 2.3 A regular pair (A,B) is called a diagonalizable pair, if there exists a basis of $\mathbb{C}^n$ formed from eigenvectors of (A,B).

Definition 2.4 A regular pair (A,B) is called a normal pair, if there exists an orthonormal basis of $\mathbb{C}^n$ formed from eigenvectors of (A,B).

Definition 2.5 Let Hermitian matrices A and $B \in \mathbb{C}^{n\times n}$. (A,B) is called a definite pair , if

$$c(A,B) \equiv \min_{||x||=1} \{|\, x^H(A+iB)x\,|\} > 0.$$

The symbols R(n), Dg(n), N(n) and D(n) denote the set of nxn regular pairs, diagonalizable pairs, normal pairs and definite pairs resp.

From the definitions we can prove the following theorems [4], [24].

Theorem 2.1 Suppose that $(A,B)\in R(n)$. Then $(A,B)\in Dg(n)$ if there exist nonsingular S and $Q \in \mathbb{C}^{n\times n}$ such that

$$S^HAQ=\mathrm{diag}(\alpha_1,\dots,\alpha_n),\quad S^HBQ=\mathrm{diag}(\beta_1,\dots,\beta_n).$$

Theorem 2.2 Suppose that $(A,B)\in R(n)$. Then $(A,B)\in N(n)$ if there exist a nonsingular S and a unitary $U\in \mathbb{C}^{n\times n}$ such that

$$S^H AU = \operatorname{diag}(\alpha_1,\dots,\alpha_n),\ S^H BU=\operatorname{diag}(\beta_1,\dots,\beta_n).$$

The following result is well known [22], [30].

Theorem 2.3 Let $(A,B)\in D(n)$. Then there exists a nonsingular $Q\in\mathbb{C}^{n\times n}$ such that

$$Q^H AQ=\operatorname{diag}(\alpha_1,\dots,\alpha_n),\quad Q^H BQ=\operatorname{diag}(\beta_1,\dots,\beta_n).$$

Definition 2.6 A subspace $\mathcal{X}$ in $\mathbb{C}^n$ is called a generalized invariant subspace for (A,B) if

$$\dim(A\mathcal{X}+B\mathcal{X}) \leq \dim(\mathcal{X}).$$

Particularly, if $(A,B)\subset Dg(n)$ then the generalized invariant subspace $\mathcal{X}$ is called an eigenspace for (A,B). We have proved that [29], if $(A,B)\in Dg(n)$, then $\mathcal{X}$ is an eigenspace for (A,B) iff $\mathcal{X}$ is spanned by a set of eigenvectors of (A,B). Moreover, we may adopt the following decompositions in order to study perturbation bounds of any eigenspace for $(A,B)\in Dg(n)$:

$$S^H AQ = \begin{pmatrix} A_1 & 0 \\ 0 & A_2 \end{pmatrix}, \quad S^H BQ = \begin{pmatrix} B_1 & 0 \\ 0 & B_2 \end{pmatrix}.$$

Here $S=(S_1,S_2)$, $Q=(Q_1,Q_2)$, $S_1^H S_1=Q_1^H Q_1=I^{(\ell)}$, $S_2^H S_2=Q_2^H Q_2=I^{(n-\ell)}$, $\mathcal{X}=R(Q_1)$ is an ℓ-dimensional eigenspace for (A,B), $(A_1,B_1)\in Dg(\ell)$ and $(A_2,B_2)\in N(n-\ell)$; if $(A,B)\in D(n)$, then $S=Q$, $(A_1,B_1)\in D(\ell)$ and $(A_2,B_2)\in D(n-\ell)$.

Let $H, K\in\mathbb{C}^{n\times n}$ and $H, K\geq 0$. It is easy to see that if $(\lambda,\mu)\in\lambda(H,K)$, then $(|\lambda|, |\mu|)\subset\lambda(H,K)$. Therefore we suggest the following generalized singular value of the matrix pair {A,B}, if

$$(\alpha,\beta) = (\lambda^{\frac12},\mu^{\frac12}),\ \text{provided}\ (\lambda,\mu)\in\lambda(A^HA,\ B^HB)\ \text{and}\ \lambda,\mu\geq 0.$$

The set of generalized singular values of {A,B} is denoted by $\sigma\{A,B\}$.

Definition 2.8 Let $A\in\mathbb{C}^{m\times n}$ and $B\in\mathbb{C}^{p\times n}$. A matrix pair {A,B} is called an (m,p,n) - Grassmann pair, if rank $(A^H, B^H) = n$.

The symbol G(m,p,n) denotes the set of (m,p,n) - Grassmann pairs.

Van Loan [31], Paige and Saunders [16] have suggested forms of the generalized singular value decomposition (GSVD) of two matrices having the same number of columns. But we can also deduce the GSVD for every $\{A,B\} \in G(m,p,n)$ with ease from the theorem 2.3, since $(A^H A, B^H B) \in D(n)$.

Theorem 2.4 (GSVD) For given $\{A,B\} \subset G(m,p,n)$, there exist unitary $U \in \mathbb{C}^{m\times m}$, unitary $V \in \mathbb{C}^{p\times p}$ and nonsingular $Q \in \mathbb{C}^{n\times n}$ giving

$$U^H A Q = \Sigma_a, \quad V^H B Q = \Sigma_b,$$

$$\underset{m\times n}{\Sigma_a} = \begin{pmatrix} \Lambda & \\ & O_a \end{pmatrix} \begin{matrix} r+s \\ m-r-s \end{matrix} \qquad \underset{p\times n}{\Sigma_b} = \begin{pmatrix} O_b & \\ & \Omega \end{pmatrix} \begin{matrix} p+r-n \\ n-r \end{matrix}$$
$$\qquad\quad r+s \quad n-r-s \qquad\qquad\qquad\qquad r \quad n-r$$

where O_a and O_b are null matrices, and

$$\Lambda = \mathrm{diag}(\alpha_1, \ldots, \alpha_{r+s}), \quad \Omega = \mathrm{diag}(\beta_{r+1}, \ldots, \beta_n)$$

with

$$1 = \alpha_1 = \ldots = \alpha_r > \alpha_{r+1} \geq \ldots \geq \alpha_{r+s} > \alpha_{r+s+1} = \ldots = \alpha_n = 0,$$

$$0 = \beta_1 = \ldots = \beta_r < \beta_{r+1} \leq \ldots \leq \beta_{r+s} < \beta_{r+s+1} = \ldots = \beta_n = 1$$

and

$$\alpha_i^2 + \beta_i^2 = 1 \quad \text{for } 1 \leq i \leq n.$$

Obviously $\sigma\{A,B\} = \{(\alpha_i, \beta_i)\}_{i=1}^n$.

The above GSVD motivates the following definition.

Definition 2.9 Suppose that $\{A,B\} \in G(m,p,n)$ and $\max\{n-p,0\} < \ell < \min\{m,n\}$. Let $X_1 \subset \mathbb{C}^m$, X_3 and $X_4 \subset \mathbb{C}^n$ be subspaces of dimension ℓ, $X_2 \subset \mathbb{C}^p$ be a subspace of dimension ℓ if $p \geq n$ and of dimension $p+\ell-n$ if $p<n$. Then χ_1, χ_2, χ_3 and χ_4 form a set of generalized singular subspaces (GSSS) for $\{A,B\}$, if

$$A\chi_3\subset\chi_1,\ B\chi_3\subset\chi_2,\ A^H\chi_1\subset\chi_4,\ B^H\chi_2\subset\chi_4.$$

The author [27] has proved that we may adopt the following decompositions in order to study perturbation bounds of any GSSS for $\{A,B\}\in G(m,p,n)$:

$$(2.1)\quad \begin{aligned} &A(\underset{\ell}{Q_1},\underset{n-\ell}{Q_2})=(\underset{\ell}{U_1},\underset{m-\ell}{U_2})\begin{pmatrix}A_1 & 0\\ 0 & A_2\end{pmatrix},\quad B(Q_1,Q_2)=(\underset{k}{V_1},\underset{p-k}{V_2})\begin{pmatrix}B_1 & 0\\ 0 & B_2\end{pmatrix},\\ &A^H(U_1,U_2)=(\underset{\ell}{P_1},\underset{n-\ell}{P_2})\begin{pmatrix}A_1'^H & 0\\ 0 & A_2'^H\end{pmatrix},B^H(V_1,V_2)=(P_1,P_2)\begin{pmatrix}B_1'^H & 0\\ 0 & B_2'^H\end{pmatrix}\end{aligned}$$

Here $U=(U_1,U_2)$ and $V=(V_1,V_2)$ are unitary matrices,

$$Q_1^HQ_1=P_1^HP_1=I^{(\ell)},\quad Q_2^HQ_2=P_2^HP_2=I^{(n-\ell)},$$

$$\{A_1,B_1\}\text{ and }\{A_1',B_1'\}\in G(\ell,k,\ell),$$

$$\{A_2,B_2\}\text{ and }\{A_2',B_2'\}\in G(m-\ell,p-k,n-\ell),$$

$$(2.2)\quad k=\begin{cases}\ell & \text{if } p\geq n\\ p+\ell-n & \text{if } p<n\end{cases};$$

and

$$(2.3)\quad \chi_1=R(U_1),\chi_2=R(V_1),\ \chi_3=R(Q_1),\ \chi_4=R(P_1)$$

form a set of GSSS for $\{A,B\}$.

We shall use the same notation for the perturbed $\{\tilde{A},\tilde{B}\}\subset G(m,p,n)$, except that all quantities will be marked with tildes.

3. Some metrics on the Grassmann manifold

Let

$$\mathbb{C}(p,p+q)=\{Z\in\mathbb{C}^{p\times(p+q)}:\ \text{rank}(Z)=p\}.$$

We use Z to represent a point of $\mathbb{C}(p,p+q)$. Points of $\mathbb{C}(p,p+q)$ are divided into equivalence classes as follows: two points Z_1 and Z_2 are said to belong to the same equivalence class (symbolically $Z_1 \sim Z_2$), if there exists a nonsingular $T \in \mathbb{C}^{p \times p}$ such that $Z_1 = TZ_2$. We consider every equivalence class of $\mathbb{C}(p,p+q)$ as a point and consequently obtain a complex projective space, i.e. the Grassmann manifold $G_{p,p+q}$ [11],[13]. We usuall use a matrix $Z \in \mathbb{C}^{p \times (p+q)}$ whose rank is p, or use a subspace $\mathcal{X} = R(Z^T)$, to represent a point of $G_{p,p+q}$.

It is easy to see that for every $(A,B) \in R(n)$ we have rank $(A,B)=n$, and (TA,TB) for a nonsingular $T \in \mathbb{C}^{n \times n}$ has the same eigenvalues and eigenvectors as those of (A,B). Hence a major idea is that we can consider $Z=(A,B) \in R(n)$ as a point on $G_{n,2n}$. Similarly, for $\{A,B\} \in G(m,p,n)$ we can consider $\begin{pmatrix} A \\ B \end{pmatrix}^T$ as a point on $G_{n,m+p}$. Specially we consider an eigenvalue (α,β) (similarly, a singular value (α,β)) as a point on the complex projective plane $G_{1,2}$. On the other hand we know that every subspace $\mathcal{X}=R(Z_1)$, $Z_1 \in \mathbb{C}^{n \times \ell}$ with $0 < \ell < n$, is a point on $G_{\ell,n}$. It is therefore necessary to investigate metrics on the Grassmann manifold in order to study the perturbation theory for the generalized eigenvalue problem.

A real-valued function $d(.,.)$ defined on $G_{p,p+q}$ is called a unitary-invariant metric (UIM) if it satisfies the following conditions:

(i) $d(Z,W) \geq 0$, $d(Z,W)=0$ iff $Z \sim W$;

(ii) $d(Z,W)=d(W,Z)$;

(iii) $d(Z,W) \leq d(Z,Y)+d(Y,W)$;

(iv) $d(ZU,WU)=d(Z,W)$.

Here Z,W and Y are any points on $G_{p,p+q}$, U is any $(p+q)\times(p+q)$ unitary matrix.

Now we cite some examples [11], [19], [29].

<u>Examples</u> For any points Z and W on $G_{p,p+q}$, let $X_1=(ZZ^H)^{-\frac{1}{2}}Z$, $Y_1=(WW^H)^{-\frac{1}{2}}W$, $\mathcal{X}=R(X_1^T)$, $\mathcal{Y}=R(Y_1^T)$, P_x and P_y be the orthogonal projectors onto $\mathcal{X}$ and $\mathcal{Y}$ resp., $\|\cdot\|$ be any unitary-invariant norm (UIN) on $\mathbb{C}^{(p+q)\times(p+q)}$. Then

$$\ell(Z,W)=\arccos(\det X_1Y_1^HY_1X_1^H)^{\frac{1}{2}},$$
$$d_\ell(Z,W)= \sin \ell(Z,W)$$

and

$$||P_X - P_Y||$$

are UIM on $G_{p,p+q}$.

It is well known that every UIN is obtained as a symmetric gauge function of the singular values, and the converse also holds (ref [15], [17]). From this we can prove the following conclusion [29].

<u>Theorem 3.1</u> Let

$$\Theta(X,Y) = \arccos (X_1 Y_1^H Y_1 X_1^H)^{\frac{1}{2}} \geq 0$$

and $||\cdot||_{(p)}$ be an UIN on $\mathbb{C}^{p\times p}$. Then there exists an UIN $||\cdot||_{(p+q)}$ on $\mathbb{C}^{(p+q)\times(p+q)}$ such that

$$||P_X - P_Y||_{(p+q)} = ||\sin \Theta(X,Y)||_{(p)}.$$

Conversely, let $||\cdot||_{(p+q)}$ be an UIN on $\mathbb{C}^{(p+q)\times(p+q)}$, then there exists an UIN $||\cdot||_{(p)}$ on $\mathbb{C}^{p\times p}$ such that

$$||\sin\Theta(X,Y)||_{(p)} = ||P_X - P_Y||_{(p+q)}.$$

Here X and Y are any points on $G_{p,p+q}$.

Hence, for any UIN $||\cdot||$ on $\mathbb{C}^{p\times p}$

$$||\sin\Theta(X,Y)||$$

is an UIM on $G_{p,p+q}$.

Particularly we write

$$||\sin \Theta(X,Y)||_2 = d_2(Z,W)$$

and

$$||\sin \Theta(X,Y)||_F = d_F(Z,W).$$

The following relations are valid:

$$d_2(Z,W) = ||P_X - P_Y||_2, \quad d_F(Z,W) = \frac{1}{\sqrt{2}} ||P_X - P_Y||_F$$

and

$$d_2(Z,W) \leq d_\ell(Z,W) \leq d_F(Z,W).$$

If $p=q=1$, then $d_2=d_\ell=d_F$ is the ordinary chordal metric

$$\rho((\alpha,\beta),(\gamma,\delta)) = \frac{|\alpha\delta-\beta\gamma|}{\sqrt{(|\alpha|^2+|\beta|^2)(|\gamma|^2+|\delta|^2)}}, \quad (\alpha,\beta) \text{ and } (\gamma,\delta) \in G_{1,2}.$$

4. Perturbation theorems of generalized eigenvalues

Suppose that $Z=(A,B)$ and $W=(C,D) \in R(n)$, $\lambda(A,B) = \{(\alpha_i,\beta_i)\}$ and $\lambda(C,D) = \{(\gamma_i,\delta_i)\}$. In order to characterize the perturbation of $\{(\alpha_i,\beta_i)\}$ to $\{(\gamma_i,\delta_i)\}$, we introduce the following quantities:

(a). The generalized spectral variation of W with respect to Z

$$S_Z(W) = \max_j \{\min_i \rho((\alpha_i,\beta_i),(\gamma_j,\delta_j))\},$$

(b). The generalized eigenvalue distance of Z and W

$$d(Z,W) = \min_\pi \sqrt{\sum_{i=1}^{n} \rho^2((\alpha_i,\beta_i),(\gamma_{\pi(i)},\delta_{\pi(i)}))}$$

and

(c). The generalized eigenvalue variation of Z and W

$$v(Z,W) = \min_\pi \{\max_i \rho((\alpha_i,\beta_i),(\gamma_{\pi(i)},\delta_{\pi(i)}))\}.$$

Here π runs through all permutations of $\{1,\ldots,n\}$.

A part of the results in author's recent articles [4], [24] and [25] will be stated as follows.

<u>Theorem 4.1</u> (The Bauer-Fike type theorem). Let $Z=(A,B) \in D_g(n)$ with decompositions

$$S^H AQ = \text{diag}(\alpha_i) \equiv \Lambda, \quad S^H BQ = \text{diag}(\beta_i) \equiv \Omega \tag{4.1}$$

and $W=(C,D) \in R(n)$. Then

(4.2) $$s_Z(W) \leq ||Q||_2 \, ||Q^{-1}||_2 \, d_2(Z,W).$$

Outline of proof.

We may assume that $ZZ^H=I$, $WW^H=I$. Let P_x and P_y be the orthogonal projectors onto $X=R(Z^T)$ and $Y=R(W^T)$ resp., (γ,δ) be an eigenvalue of (C,D) such that $|\gamma|^2+|\delta|^2=1$, and x with $||x||=1$ corresponding eigenvector: Then from $\delta Cx=\gamma Dx$ we obtain

$$\delta Ax-\gamma Bx = Z(P_x-P_y)\begin{pmatrix}\delta x\\ -\gamma x\end{pmatrix}$$

and thus

(4.3) $$||\delta Ax-\gamma Bx|| \leq d_2(Z,W)$$

On the other hand from (4.1) one has

$$(S^HS)^{-1} \geq ||Q^{-1}||_2^2 \, (\Lambda\Lambda^H+\Omega\Omega^H)^{-1},$$

and so

$$||\delta Ax-\gamma Bx|| \geq ||Q^{-1}||_2^{-1} \, ||Q||_2^{-1} \min_i \rho((\alpha_i,\beta_i),(\gamma,\delta)).$$

This together with (4.3) gives (4.2). □

Corollary 4.1 Let $Z=(A,B)\in N(n)$ and $W=(C,D)\in R(n)$. Then

$$s_Z(W) \leq d_2(Z,W).$$

Corollary 4.2 Let $Z=(A,B)\in D(n)$ and $W=(C,D)\in R(n)$. Then

$$s_Z(W) \leq \frac{||Z||_2}{c(A,B)} d_2(Z,W).$$

Theorem 4.2 (The Hoffman-Wielandt type theorem). Let $Z=(A,B)$ and $W=(C,D)\in N(n)$. Then

(4.4) $$d(Z,W) \leq d_F(Z,W).$$

Outline of proof

Substituting $Z=(A,B)$ and $W=(C,D)$ in $d_F(Z,W)$, and utilizing theorem 2.2, we get

(4.5) $$d_F^2(Z,W) = n-g(V),$$

where $V=(v_{ij})$ is an nxn unitary matrix, and

$$g(V) = \sum_{i,j=1}^{n} \theta_{ij} \, |v_{ji}|^2,$$

$$\theta_{ij} = \frac{|\alpha_i\bar{\gamma}_j+\beta_i\,\bar{\delta}_i|^2}{(|\alpha_i|^2+|\beta_i|^2)(|\gamma_j|^2+|\delta_j|^2}, \quad (\alpha_i,\beta_i)\in\lambda(A,B),$$

$$(\gamma_j,\delta_j)\in\lambda(C,D),\ 1\leq i,j\leq n.$$

Defining $W=(w_{ij})$ by $w_{ij}=|v_{ji}|^2$, W is a bistochastic matrix and we have

$$g(V) = f(W)$$

with a linear function f. As the bistochastic matrices form a convex polyhedron the vertices of which are the permutation matrices there exists a permutation matrix P such that $f(W) \leq f(P)$. This together with (4.5) gives (4.4). □

Theorem 4.3 (The Weyl-Lidskii type theorem). Let $Z=(A,B)\in D(n)$, $\tilde{A}=A+E$ and $\tilde{B} = B+F$ be Hermitian matrices. Assume that

$$\delta((A,B),(\tilde{A},\tilde{B})) \equiv \max_{||x||=1} \left\{\sqrt{\frac{(x^HEx)^2+(x^HFx)^2}{(x^HAx)^2+(x^HBx)^2}}\right\} < 1.$$

Then $\tilde{Z}=(\tilde{A},\tilde{B})\in D(n)$ too, and

(4.6) $$v(Z,\tilde{Z})\leq d_s(Z,\tilde{Z})$$

where

$$d_s(Z,\tilde{Z}) = \max_{||x||=1} \left\{ \rho((x^HAx,x^HBx),\ (x^H\tilde{A}x,x^H\tilde{B}x))\right\}$$

is a projective metric in D(n).

Outline of proof

Associating with each eigenvalue an angle θ in an appropriate way and ordering the eigenangles θ_i $[\tilde{\theta}_i]$ of the unperturbed[perturbed]problem. The proof of (4.6) is based on a minmax characterization of eigenangles (ref. [22]).

Let $Z=(A,B)$ and $W=(C,D)\in D(n)$. Z and W are equivalent (symbolically $Z\sim W$) if there exists a real number r such that $(C,D)=(rA,rB)$, or $(A,C)=(rB,rD)$, or $(B,D)=(rA,rC)$. We have proved that if Z, $\tilde{Z}$ and $W\in D(n)$, then

(i) $d_s(Z,\tilde{Z})\geq 0$, $d_s(Z,\tilde{Z})=0$ iff $Z\sim\tilde{Z}$;

(ii) $d_s(Z,\tilde{Z})=d_s(\tilde{Z},Z)$;

(iii) $d_s(Z,\tilde{Z})\leq d_s(Z,W)+d_s(W,\tilde{Z})$.

Hence $d_s(Z,\tilde{Z})$ is a projective metric in $D(n)$. □

Remark 4.1

By a limiting procedure from the theorems 4.1, 4.2 and 4.3 we can deduce Bauer-Fike theorem, Hoffman-Wielandt theorem and Weyl-Lidskii theorem resp. (i.e. (4.2) ⇒ (1.1), (4.4) ⇒ (1.2), (4.6) ⇒ (1.3)).

Remark 4.2

The theorem 4.3 is an improvement of a Stewart's inequality

$$v(Z,\tilde{Z}) \leq \sqrt{||E||_2^2+||F||_2^2}/c(A,B)$$

because of $d_s(Z,\tilde{Z}) \leq \delta((A,B),(\tilde{A},\tilde{B})) \leq \sqrt{||E||_2^2+||F||_2^2}/c(A,B)$.

As an example let $Z=(A,B)\in D(n)$ and $\tilde{Z}=(\tilde{A},\tilde{B})=(1+r)(A,B)$, here r satisfies $r\sqrt{||A||_2^2+||B||_2^2}/c(A,B) < 1$ but ≈ 1. Obviously $(\tilde{A},\tilde{B})\in D(n)$ too, and $\lambda(A,B)=\lambda(\tilde{A},\tilde{B})$. We have

$$0=v(Z,\tilde{Z})=d_s(Z,\tilde{Z}) < \sqrt{||E||_2^2+||F||_2^2}/c(A,B) \approx 1.$$

5. Perturbation theorems of eigenspaces

G. W. Stewart[22] has obtained perturbation bounds for eigenspace of a definite pair on the Frobenius norm under some conditions and pointed out: " For the Hermitian eigenvalue problem, Davis and Kahan have been able to obtain bounds on the spectral norm by imposing additional restrictions on the location of the eigenvalues. Whether such bound can be obtained for the definite generalized eigenvalue problem is an open question" (1979). The author has proved the following theorem (not only on the spectral norm) which gives a positive answer for this question (see [26]).

Theorem 5.1 (The sin Θ theorem for definite pair). Let (A,B) and $(\tilde{A},\tilde{B})=(A+E,B+F)\in D(n)$. Assume that

(i) (A,B) and $(\tilde{A},\tilde{B})$ have decompositions

$$(5.1)\qquad Q^H AQ = \begin{pmatrix} A_1 & 0 \\ 0 & A_2 \end{pmatrix}, \quad Q^H BQ = \begin{pmatrix} B_1 & 0 \\ 0 & B_2 \end{pmatrix},$$

$$(5.2)\qquad \tilde{Q}^H \tilde{A}\tilde{Q} = \begin{pmatrix} \tilde{A}_1 & 0 \\ 0 & \tilde{A}_2 \end{pmatrix}, \quad Q^H BQ = \begin{pmatrix} \tilde{B}_1 & 0 \\ 0 & \tilde{B}_2 \end{pmatrix}$$

Here $Q=(Q_1,Q_2)$, $\tilde{Q}=(\tilde{Q}_1,\tilde{Q}_2)$, $Q_1^H Q_1=\tilde{Q}_1^H\tilde{Q}_1=I^{(\ell)}$, $Q_2^H Q_2=\tilde{Q}_2^H\tilde{Q}_2=I^{(n-\ell)}$, (A_1,B_1) and $(\tilde{A}_1,\tilde{B}_1)\in D(\ell)$, (A_2,B_2) and $(\tilde{A}_2,\tilde{B}_2)\in D(n-\ell)$.

(ii) there exist $\alpha \geq 0$, $\delta > 0$ satisfying $\alpha+\delta \leq 1$, and γ a real number, such that

$$(5.3)\qquad \lambda(A_1,B_1) \subset \{(\alpha,\beta)\in G_{1,2} : \rho((\gamma,1),(\alpha,\beta)) \leq \alpha\} \equiv \Delta_1$$

and

$$(5.4)\qquad \lambda(\tilde{A}_2,\tilde{B}_2) \subset \{(\alpha,\beta)\in G_{1,2} : \rho((\gamma,1),(\alpha,\beta)) \geq \alpha+\delta\} \equiv \Delta_2$$

or $\lambda(\tilde{A}_2,\tilde{B}_2)\subset\Delta_1$ and $\lambda(A_1,B_1)\subset\Delta_2$).

Then for every UIN $\|\cdot\|$ and for $\chi=R(Q_1)$, $\tilde{\chi}=R(\tilde{Q}_1)$, we have

$$(5.5)\qquad \|\sin\Theta(\chi,\tilde{\chi})\| \leq \frac{p(\alpha,\delta;\gamma)\sqrt{\|A^2+B^2\|_2}}{c(A,B)\,c(\tilde{A},\tilde{B})} \cdot \frac{\sqrt{\|EQ_1\|^2+\|FQ_1\|^2}}{\delta},$$

where

$$p(\alpha,\delta;\gamma) = \frac{q(\gamma)\,[(\alpha+\delta)\sqrt{1-\alpha^2}+\alpha\sqrt{1-(\alpha+\delta)^2}]}{2\alpha+\delta}$$

and

$$q(\gamma) = \begin{cases} \sqrt{2} & \text{if } \gamma \neq 0 \\ 1 & \text{if } \gamma = 0 \end{cases}$$

<u>Outline of proof</u>

1°. The perturbation equations.

Setting

$$P' = Q^{-H} = (\underset{\ell}{P_1'}, \underset{n-\ell}{P_2'}) , \qquad P = P' \begin{pmatrix} (P_1'^H P_1')^{-\frac{1}{2}} & 0 \\ 0 & (P_2'^H P_2')^{-\frac{1}{2}} \end{pmatrix}$$

and

$$(P_j'^H P_j')^{\frac{1}{2}} (A_j, B_j) = (A_j', B_j') \qquad j=1,2;$$

then (5.1) may be written as

$$(5.6) \qquad (AQ, BQ) = P \left(\begin{pmatrix} A_1' & 0 \\ 0 & A_2' \end{pmatrix}, \begin{pmatrix} B_1' & 0 \\ 0 & B_2' \end{pmatrix} \right) ,$$

where $Q=(Q_1,Q_2)$ and $P=(P_1,P_2)$ satisfy

$$Q_1^H Q_1 = P_1^H P_1 = I^{(\ell)}, \quad Q_2^H Q_2 = P_2^H P_2 = I^{(n-\ell)}, \quad Q_1^H P_2 = 0, \quad Q_2^H P_1 = 0.$$

Similarly we obtain

$$(5.7) \qquad (\tilde{A}\tilde{Q}, \tilde{B}\tilde{Q}) = \tilde{P} \left(\begin{pmatrix} \tilde{A}_1' & 0 \\ 0 & \tilde{A}_2' \end{pmatrix}, \begin{pmatrix} \tilde{B}_1' & 0 \\ 0 & \tilde{B}_2' \end{pmatrix} \right) ,$$

where $\tilde{Q}=(\tilde{Q}_1,\tilde{Q}_2)$ and $\tilde{P}=(\tilde{P}_1,\tilde{P}_2)$ satisfy

$$\tilde{Q}_1^H \tilde{Q}_1 = \tilde{P}_1^H \tilde{P}_1 = I^{(\ell)}, \tilde{Q}_2^H \tilde{Q}_2 = \tilde{P}_2^H \tilde{P}_2 = I^{(n-\ell)}, \quad \tilde{Q}_1^H \tilde{P}_2 = 0, \quad \tilde{Q}_2^H \tilde{P}_1 = 0.$$

Defining the residuals

$$(5.8) \qquad R_A = \tilde{A} Q_1 - P_1 A_1' \ (= E Q_1), \quad R_B = \tilde{B} Q_1 - P_1 B_1' \ (= F Q_1)$$

and utilizing (5.2), (5.6)-(5.8), we get

$$(5.9)\qquad \begin{cases} R_A^H\tilde{Q}_2 = Q_1^H\tilde{P}_2\hat{A}_2^H - A_1'^H P_1^H\tilde{Q}_2 \\ R_B^H\tilde{Q}_2 = Q_1^H\tilde{P}_2\hat{B}_2^H - B_1'^H P_1^H\tilde{Q}_2 \end{cases}$$

where $\hat{A}_2 = \tilde{A}_2(\tilde{P}_2'^H\tilde{P}_2')^{\frac{1}{2}}$, $\hat{B}_2 = \tilde{B}_2(\tilde{P}_2'^H\tilde{P}_2')^{\frac{1}{2}}$.

Let

$$P_1^H\tilde{Q}_2 = X,\quad Q_1^H\tilde{P}_2 = Y,\quad -R_A^H\tilde{Q}_2 = C,\ -R_B^H\tilde{Q}_2 = D,$$

then equations (5.9) may be written as

$$(5.10)\qquad A_1'^H X - Y\hat{A}_2^H = C,\ B_1'^H X - Y\hat{B}_2^H = D,$$

where $A_1', B_1' \in \mathbb{C}^{\ell\times\ell}$, $\hat{A}_2, \hat{B}_2 \in \mathbb{C}^{(n-\ell)\times(n-\ell)}$, and the unknowns X, $Y \in \mathbb{C}^{\ell\times(n-\ell)}$.

2^0. Observe that

$$||Y|| = ||\sin\Theta(\mathcal{X},\tilde{\mathcal{X}})||$$

for any unitary-invariant matrix norm $||\cdot||$, therefore our goal is to estimate the $||Y||$ from the above.

After the simplification of the equations (5.10) in terms of the theorem 2.3, utilizing the conditions (5.3) and (5.4) we obtain the inequality (5.5). □

Under weaker conditions we have proved the following theorem.

<u>Theorem 5.2</u> (The generalized sin θ theorem for definite pair). Let (A,B) and $(\tilde{A},\tilde{B}) = (A+E, B+F) \in D(n)$. Assume that

(i) the same assumption as (i) in theorem 5.1;

(ii) $$\delta \equiv \min_{i,j}\left\{ p((\alpha_i,\beta_i),(\tilde{\alpha}_j,\tilde{\beta}_j)) : \begin{array}{l} (\alpha_i,\beta_i) \in \lambda(A_1,B_1) \\ (\tilde{\alpha}_j,\tilde{\beta}_j) \in \lambda(\tilde{A}_2,\tilde{B}_2) \end{array}\right\} > 0.$$

Then

$$(5.11)\qquad ||\sin\Theta(\mathcal{X},\tilde{\mathcal{X}})||_F \le \frac{\sqrt{||A^2+B^2||_2}}{c(A,B)\,c(\tilde{A},\tilde{B})}\ \frac{\sqrt{||EQ_1||_F^2 + ||FQ_1||_F^2}}{\delta}$$

Remark 5.1

By a limiting procedure from the theorems 5.1 and 5.2 we can deduce Davis-Kahan sin Θ theorem and the generalized sin Θ theorem resp. (i.e. (5.5) $\Rightarrow$ (1.4), (5.11) $\Rightarrow$ (1.5)).

Remark 5.2

It is worth noting that the upper bound in (5.11) is independent of the dimension ℓ of the eigenspace $X = R(Q_1)$, but the upper bound obtained by G.W. Stewart [22] contains a factor $\sqrt{\ell}$.

6. Perturbation analysis for the generalized singular value problem

Theorem 6.1 (The Hoffman-Wielandt type theorem) [27]. Let $\{A,B\}$ and $\{\tilde{A},\tilde{B}\} \in C(m,p,n)$ with the decompositions represented in the theorem 2.4 (CSVD), where we use the same notation for $\{\tilde{A},\tilde{B}\}$ except that all quantities are marked with tildes. Set

$$Z = \begin{pmatrix} A \\ B \end{pmatrix}^T, \quad \tilde{Z} = \begin{pmatrix} \tilde{A} \\ \tilde{B} \end{pmatrix}^T, \quad \rho_{ii} = \rho((\alpha_i,\beta_i),(\tilde{\alpha}_i,\tilde{\beta}_i)), \quad i=1,\dots,n.$$

Then

$$\prod_{i=1}^{n} (1-\rho_{i,i}^2) \geq 1-d_\ell^2(Z,\tilde{Z}) \tag{6.1}$$

Outline of proof

Substituting $Z = \begin{pmatrix} A \\ B \end{pmatrix}^T$ and $\tilde{Z} = \begin{pmatrix} \tilde{A} \\ \tilde{B} \end{pmatrix}^T$ in $d_\ell(Z,\tilde{Z})$, utilizing the theorem 2.4 and the following facts:

(1). Let

$$\Gamma = \operatorname{diag}(\gamma_1,\dots,\gamma_n), \quad \Delta = \operatorname{diag}(\delta_1,\dots,\delta_n)$$

with

$$0 \leq \gamma_1 \leq \dots \leq \gamma_n \leq 1, \qquad 0 \leq \delta_1 \leq \dots \leq \delta_n \leq 1.$$

Then

$$\sup_{UU^H=I,\,VV^H=I} |\det(I+\Gamma U\Delta V)| = \prod_{i=1}^{n} (1+\gamma_i\delta_i). \quad [11]$$

(2). Let $Z \in \mathbb{C}^{n\times n}$. Then any analytical function of several complex variables $f(Z)$ on the domain $\{Z: ZZ^H \leq I\}$ attains its maximum modulus on the characteristic manifold $\{Z \in \mathbb{C}^{n\times n}: ZZ^H = I\}$. [9]

From this we obtain the inequality (6.1). □

Corollary 6.1 Assume the hypotheses of the theorem 6.1. Then

$$(6.2) \qquad \sum_{i=1}^{n} \rho_{ii}^2 \leq n\left(1 - \sqrt[n]{1 - d_\ell^2(Z,\tilde{Z})}\right).$$

Theorem 6.2 (The Weyl-Lidskii type theorem)[28]. Let $\{A,B\} \in G(m,p,n)$ and $\{\tilde{A},\tilde{B}\} = \{A+E, B+F\}$. Assume that

$$\tau(\{A,B\},\{\tilde{A},\tilde{B}\}) \equiv \max_{||x||=1} \left\{ \sqrt{\frac{||Ex||^2 + ||Fx||^2}{||Ax||^2 + ||Bx||^2}} \right\} < 1.$$

Then $\{\tilde{A},\tilde{B}\} \in G(m,p,n)$ too, and

$$(6.3) \qquad v(\{A,B\},\{\tilde{A},\tilde{B}\}) \leq d_r(\{A,B\},\{A,B\}).$$

Here

$$v(\{A,B\},\{\tilde{A},\tilde{B}\}) = \min_{\pi} \{\max \rho((\alpha_i,\beta_i),(\tilde{\alpha}_{\pi(i)},\tilde{\beta}_{\pi(i)}))\},$$

$\{(\alpha_i,\beta_i)\} = \sigma\{A,B\}$, $\{(\tilde{\alpha}_i,\tilde{\beta}_i)\} = \sigma\{\tilde{A},\tilde{B}\}$, π runs through all permutations of $\{1,\ldots,n\}$, and

$$d_r(\{A,B\},\{\tilde{A},\tilde{B}\}) = \max_{||x||=1} \{\rho((||Ax||, ||Bx||),(||\tilde{A}x||, ||\tilde{B}x||))\}$$

Outline of proof

Associating with each singular value an angle θ in an appropriate way and ordering the singular angles θ_i $[\tilde{\theta}_i]$ of the unperturbed[perturbed] problem. The proof of (6.3) is based on a minmax characterization of singular angles. □

Observe that

$$d_r(\{A,B\},\{\tilde{A},\tilde{B}\}) \leq \tau(\{A,B\},\{\tilde{A},\tilde{B}\}) \leq \frac{\sigma_{max}\begin{pmatrix}E\\F\end{pmatrix}}{\sigma_{min}\begin{pmatrix}A\\B\end{pmatrix}}$$

here $\sigma_{max}(\cdot)$ and $\sigma_{min}(\cdot)$ denote maximum and minimum singular value of a matrix resp. Therefrom we get

Corollary 6.2 Let $\{A,B\} \in G(m,p,n)$ and $\{\tilde{A},\tilde{B}\} = \{A+E,B+F\}$. Assume that

$$\sigma_{max}\begin{pmatrix}E\\F\end{pmatrix} < \sigma_{min}\begin{pmatrix}A\\B\end{pmatrix}.$$

Then $\{\tilde{A},\tilde{B}\} \in G(m,p,n)$ too, and

$$v(\{A,B\},\{\tilde{A},\tilde{B}\}) \le \sigma_{max}\begin{pmatrix}E\\F\end{pmatrix} \Big/ \sigma_{min}\begin{pmatrix}A\\B\end{pmatrix}. \tag{6.4}$$

Remark 6.1 From the theorem 6.1 (or corollary 6.1) and the theorem 6.2 (or corollary 6.2) we can deduce the following classical results for the ordinary singular value problem (i.e. (6.1)⇒ (6.2)⇒ (6.5) , (6.3) ⇒(6.4) ⇒(6.6)):

Suppose that A and $\tilde{A} \in \mathbb{C}^{m\times n}$ with singular values $\sigma_1 \ge \ldots \ge \sigma_n \ge 0$ and $\tilde{\sigma}_1 \ge \ldots \ge \tilde{\sigma}_n \ge 0$ resp. Then [14]

$$\sqrt{\sum_{i=1}^{n} |\sigma_i - \tilde{\sigma}_i|^2} \le ||A-\tilde{A}||_F \tag{6.5}$$

and

$$|\sigma_i - \sigma_i| \le ||A-\tilde{A}||_2, \quad i = 1,\ldots,n. \tag{6.6}$$

Theorem 6.3 (The sinθ theorem for the GSSS) [27]. Let $\{A,B\}$ and $\{\tilde{A},\tilde{B}\} = \{A+E, B+F\} \in G(m,p,n)$ with the decompositions (2.1). Assume that there are $\alpha \ge 0$ and $\delta > 0$ satisfying $\alpha+\delta \le 1$ such that

$$\rho((0,1),(\alpha_i,\beta_i)) \ge \alpha+\delta \qquad \forall(\alpha_i,\beta_i) \in \sigma\{A_1,B_1\} \tag{6.7}$$

and

$$\rho((0,1),(\tilde{\alpha}_j,\tilde{\beta}_j)) \le \alpha \qquad \forall(\tilde{\alpha}_j,\tilde{\beta}_j) \in \sigma\{\tilde{A}_2,\tilde{B}_2\} \tag{6.8}$$

Then for every unitary-invariant matrix norm $||\cdot||$, we have

$$\left.\begin{aligned} ||\sin\Theta(\chi_1,\tilde{\chi}_1)|| &\le \frac{\tau_1}{\delta}\left[\frac{\gamma_a+\omega_1\gamma_b}{\sigma_m} + \frac{\omega_1\omega_2(\omega_3\gamma_a'+\gamma_b')}{\tilde{\sigma}_m}\right], \\ ||\sin\Theta(\chi_2,\tilde{\chi}_2) &\le \frac{\tau_2}{\delta}\left[\frac{\omega_3\omega_4(\gamma_a+\omega_1\gamma_b)}{\sigma_m} + \frac{\omega_3\gamma_a'+\gamma_b'}{\tilde{\sigma}_m}\right], \end{aligned}\right\} \tag{6.9}$$

$$\|\sin\Theta(X_3,\tilde{X}_3)\| \leq \frac{\tau_2}{\delta}\cdot\frac{\sigma_M}{\tilde{\sigma}_m}\left[\frac{\omega_2\omega_3\omega_4\gamma_a+\gamma_b}{\sigma_m}\quad\frac{\omega_2(\omega_3\gamma_a'+\gamma_b')}{\tilde{\sigma}_m}\right]$$

and

$$\|\sin\Theta(X_4,\tilde{X}_4)\| \leq \frac{\tau_1}{\delta}\cdot\frac{\tilde{\sigma}_M}{\sigma_m}\left[\frac{\omega_4(\gamma_a+\omega_1\gamma_b)}{\sigma_m}\quad\frac{\gamma_a'+\omega_1\omega_2\omega_4\gamma_b'}{\tilde{\sigma}_m}\right]$$

Here X_i and $\tilde{X}_i$ $(1\leq i\leq 4)$ are represented by (2.3), and

$$\tau_1=\frac{(\alpha+\delta)(1-\alpha^2)}{2\alpha+\delta},\quad \tau_2=\frac{(\alpha+\delta)^2\sqrt{1-\alpha^2}}{2\alpha+\delta},$$

$$\omega_1=\frac{\alpha}{\sqrt{1-\alpha^2}},\quad \omega_2=\sqrt{\frac{1-(\alpha+\delta)^2}{1-\alpha^2}},\quad \omega_3=\frac{\sqrt{1-(\alpha+\delta)^2}}{\alpha\ \delta},\quad \omega_4=\frac{\alpha}{\alpha+\delta},$$

$$\gamma_a=\|EQ_1\|,\ \gamma_b=\|FQ_1\|,\ \gamma_a'=\|E^HU_1\|,\ \gamma_b'=\|F^HV_1\|,$$

$$\sigma_M=\sigma_{max}\begin{pmatrix}A\\B\end{pmatrix},\ \sigma_m=\sigma_{min}\begin{pmatrix}A\\B\end{pmatrix},\ \tilde{\sigma}_M=\sigma_{max}\begin{pmatrix}\tilde{A}\\\tilde{B}\end{pmatrix},\ \tilde{\sigma}_m=\sigma_{min}\begin{pmatrix}\tilde{A}\\\tilde{B}\end{pmatrix}$$

Outline of proof

1°. The perturbation equations.
By (2.1) we define the residuals

(6.10) $$R_A=\tilde{A}Q_1-U_1A_1,\ R_B=\tilde{B}Q_1-V_1B_1,\ R_{A^H}=\tilde{A}^HU_1-P_1A_1'^H,\ R_{B^H}=\tilde{B}^HV_1-P_1B_1'^H.$$

Evidently

$$R_A=EQ_1,\ R_B=FQ_1,\ R_{A^H}=E^HU_1,\ R_{B^H}=F^HV_1.$$

Writing

$$\tilde{U}_2^HU_1=X_1,\ \tilde{V}_2^HV_1=X_2,\ \tilde{P}_2^HQ_1=X_3,\ \tilde{Q}_2^HP_1=X_4,$$

$$\tilde{U}_2^HR_A=C_1,\ \tilde{V}_2^HR_B=C_2,\ \tilde{Q}_2^HR_{A^H}=C_3,\ \tilde{Q}_2^HR_{B^H}=C_4;$$

then from (6.10) we obtain the perturbation equations

$$(6.11)\quad \left.\begin{array}{l} \tilde{A}_2'X_3 - X_1A_1 = C_1 \ , \quad \tilde{B}_2'X_3 - X_2B_1 = C_2 \\ \tilde{A}_2^H X_1 - X_4 A_1'^H = C_3 \ , \quad \tilde{B}_2^H X_2 - X_4 B_1'^H = C_4 \end{array}\right\}$$

Where the unknowns $X_1 \in \mathbb{C}^{(m-\ell)\times\ell}$, $X_2 \in \mathbb{C}^{(p-k)\times k}$, X_3 and $X_4 \in \mathbb{C}^{(n-\ell)\times\ell}$, k is represented by (2.2).

2°. Observe that

$$||X_j|| = ||\sin\Theta(X_j,\tilde{X}_j)|| \ , \ j=1,2,3,4$$

for any unitary-invariant matrix norm $||\cdot||$. Therefore our goal is to estimate the $||X_j||$ (j=1,2,3,4) from the above.

After the simplification of the equations (6.11) in terms of the theorem 2.4, utilizing the conditions (6.7) and (6.8) we obtain the inequalities (6.9). □

For the ordinary singular value problem, G.W. Stewart[19] has suggested the following definition.

Definition 6.1 Let $A \in \mathbb{C}^{m\times n}$, $X \subset \mathbb{C}^n$ and $Y \in \mathbb{C}^m$ be subspaces of dimension $\ell < \min\{m,n\}$. Then X and Y form a pair of singular subspaces (SSS) of A if $A X \subset Y$ and $A^H Y \subset X$.

It is easy to see that we may adopt the following decomposition in order to study perturbation bounds of any SSS for $A \in \mathbb{C}^{m\times n}$:

$$(6.12)\qquad \underset{\ell \quad n-\ell}{A(V_1,V_2)} = \underset{\ell \quad m-\ell}{(U_1,U_2)} \underset{\ell \quad n-\ell}{\begin{pmatrix} A_1 & 0 \\ 0 & A_2 \end{pmatrix}} \begin{matrix} \ell \\ m-\ell \end{matrix}$$

Here (U_1,U_2) and (V_1,V_2) are unitary matrices. $X=R(V_1)$ and $Y=R(U_1)$ form a pair of SSS for A. We shall use the same notation for $\tilde{A} \in \mathbb{C}^{m\times n}$, except that all quantities will be marked with tildes.

From the theorem 6.3 we can deduce the following Wedin's result [32].

Theorem 6.4 (The $\sin\theta$ theorem for the SVD). Let A and $\tilde{A}=A+E \in \mathbb{C}^{m\times n}$ with the decompositions (6.12). Assume that there are $\alpha \geq 0$ and $\delta > 0$ such that

$$\sigma_i \geq \alpha+\delta \qquad \forall\sigma_i \in \sigma(A_1), \quad \tilde{\sigma}_j \leq \alpha \qquad \forall\tilde{\sigma}_j \in \lambda(\tilde{A}_2).$$

Here $\sigma(\cdot)$ denotes the set of all singular values of a matrix. Then for $\mathcal{X}=R(V_1)$, $\mathcal{Y}=R(U_1)$, $\tilde{\mathcal{X}}=R(\tilde{V}_1)$, $\tilde{\mathcal{Y}}=R(\tilde{U}_1)$ and for every unitary-invariant matrix norm $||\cdot||$, we have

$$||\sin\Theta(\mathcal{X},\tilde{\mathcal{X}})|| \leq \frac{\alpha+\delta}{(2\alpha+\delta)\delta}\left(\frac{\alpha}{\alpha+\delta}||EV_1|| + ||E^HU_1||\right)$$

and

$$||\sin\Theta(\mathcal{Y},\tilde{\mathcal{Y}})|| \leq \frac{\alpha+\delta}{(2\alpha+\delta)\delta}\left(||EV_1|| + \frac{\alpha}{\alpha+\delta}||E^HU_1||\right).$$

Acknowledgement

I would like to thank professor F. L. Bauer for his encouragement and professor L. Elsner for his support. I also thank the assistance of the Alexander von Humboldt Foundation in FRG.

References

1. F.L. BAUER AND C.T. FIKE, Norms and exclusion theorem, Numer. Math. 2 (1960), 137-141.

2. C.R. CRAWFORD, A stable generalized eigenvalue problem, SIAM J. Numer. Anal. 8 (1976), 854-860.

3. C. DAVIS AND W. KAHAN, The rotation of eigenvectors by a perturbation. III, SIAM J. Numer. Anal. 7 (1970), 1-46.

4. L. ELSNER AND J.G. SUN, Perturbation theorems for the generalized eigenvalue problem, submitted to Linear Algebra and Appl.

5. F.R. GANTMACHER, The Theory of Matrices, trans. K. A. Hirsch, Chelsea, 1959.

6. P. HENRICI, Bounds for iterates, inverses, spectral variation and fields of values of non-normal matrices, Numer. Math. 4 (1962), 24-39

7. A. J. HOFFMAN AND H. W. WIELANDT, The variation of the spectrum of a normal matrix, Duke Math. Journal 20 (1953), 37-39.

8. A. S. HOUSEHOLDER, The Theory of Matrices in Numerical Analysis, Blaisedell, New York, 1964.

9. L. K. HUA, Harmonic Analysis of Functions of Several Complex Variables in the Classical Domains, Amer. Math. Soc. Providence, Rhode Island, 1963.

10. T. KATO, Perturbation Theory for Linear Operators, Springer Verlag, New York, 1966.

11. Q. K. LU, The elliptic geometry of extended spaces, Acta Math. Sinica, 13 (1963), 49-62; translated as Chinese Math. 4 (1963), 54-69.

12. M. MARCUS AND H. MINC, A Survey of Matrix Theory and Matrix Inequalities, Allyn and Bacon, Boston, 1964.

13. Y. MATUSHIMA, Differentiable Manifolds, New York, 1972.

14. L. MIRSKY, Symmetric gauge functions and unitarily invariant norms, Quart, J. Math. Oxford, 11 (1960), 50-59.

15. J. VON NEUMANN, Some matrix-inequalities and metrization of matrix-space, Bull. Inst. Math. Mēcan. Univ. Kouybycheff Tomsk, 1(1935-37), 286-300.

16. C. C. PAIGE AND M. A. SAUNDERS, Towards a generalized singular value decomposition, SIAM J. Numer. Anal. 18(1981), 398-405.

17. R. SCHATTEN, Norm Ideals of Completely Continuous Operators, Springer, Berlin, 1960.

18. G. W. STEWART, On the sensitivity of the eigenvalue problem $Ax=\lambda Bx$, SIAM J. Numer. Anal. 9(1972), 669-686.

19. G. W. STEWART, Error and perturbation bounds for subspaces associated with certain eigenvalue problems, SIAM Rev. 15 (1973), 727-769.

20. G. W. STEWART, Gerschgorin theory for the generalized eigenvalue problem $Ax=\lambda Bx$, Math. Comp. 29 (1975), 600-606.

21. G. W. STEWART, Perturbation theory for the generalized eigenvalue problem, Recent Advances in Numerical Analysis, (proc. Sympos., Math. Res. Center, Univ. Wisconsin, Madison, Wis., 1978), pp. 193-206.

22. G. W. STEWART, Perturbation bounds for the definite generalized eigenvalue problem, Linear Algebra and Appl. 23 (1979), 69-83.

23. J. G. SUN, Invariant subspaces and generalized invariant subspaces (I), (II), Math. Numer. Sinica 2 (1980), 1-13, 113-123.

24. J. G. SUN, The perturbation bounds of generalized eigenvalues of a class of matrix-pairs, Math. Numer. Sinica 4 (1982), 23-29.

25. J. G. SUN, A note on Stewart's theorem for definite matrix pairs, submitted to Linear Algebra and Appl.

26. J. G. SUN, The perturbation bounds for eigenspaces of a definite matrix pair, I. The $\sin\theta$ theorems, II. The $\sin 2\theta$ theorems, sub-

mitted to Numer. Math.

27. J. G. SUN, Perturbation analysis for the generalized singular value problem, submitted to SIAM J. Numer. Anal.

28. J. G. SUN, On the perturbation of generalized singular values, to appear in Math. Numer. Sinica.

29. J. G. SUN, Some metrics on a Grassmann manifold and perturbation estimates for eigenspaces (I), (II), submitted to Acta Math. Sinica.

30. F. UHLIG, A recurring theorem about pairs of quadratic forms and extensions: A survey, Linear Algebra and Appl. 25(1979),219-237.

31. CHARLES F. VAN LOAN, Generalizing the singular value decomposition, SIAM J. Numer. Anal. 13 (1976), 76-83

32. P.-Å. WEDIN, Perturbation bounds in connection with singular value decomposition, BIT, 12(1972), 99-111

33. J.H. WILKINSON, The Algebraic Eigenvalue Problem, Clarendon Press, Oxford, 1965.

A Generalized SVD Analysis of Some Weighting Methods for Equality Constrained Least Squares

Charles Van Loan
Department of Computer Science
Cornell University
Ithaca, New York, 14853, USA

Abstract

The method of weighting is a useful way to solve least squares problems that have linear equality constraints. New error bounds for the method are derived using the generalized singular value decomposition. The analysis clarifies when the weighting approach is succesful and suggests modifications when it is not.

1. Introduction

The problem we consider is how to find a vector $x \in R^n$ that solves the equality constrained problem

$$\text{(LSE)} \qquad \min_{Bx=d} \|Ax - b\|_2$$

where $A \in R^{m\times n}$ $(m\geq n)$, $b \in R^m$, $B \in R^{p\times n}$ $(p\leq n)$ and $d \in R^p$. We will assume that $\text{rank}(B) = p$ and that the nullspaces of the two matrices satisfy $N(A) \cap N(B) = \{0\}$. These conditions ensure that (LSE) has a unique solution which we designate by x_{LSE}.

Important settings where this problem arises include constrained surface fitting, penalty function methods in nonlinear optimization, and geodetic least squares adjustment.

Several methods for solving the LSE problem are discussed in Lawson and Hanson [7, Chapters 20-22]. In one approach Q-R factorizations are used to compute the projections of x_{LSE} onto $N(B)^{\perp}$ and $N(B)$:

(1.1)

$$\text{(a)}\quad B^T = [Q_1, Q_2] \begin{bmatrix} R_B \\ 0 \end{bmatrix} \begin{matrix} p \\ n-p \end{matrix} \qquad \text{(Q-R)}$$

$$\text{(b)}\quad R_B^T y_1 = d \;;\; x_1 := Q_1 y_1$$

$$\text{(c)}\quad AQ_2 = [U_1, U_2] \begin{bmatrix} R_A \\ 0 \end{bmatrix} \begin{matrix} n-p \\ m-n+p \end{matrix} \qquad \text{(Q-R)}$$

$$\text{(d)}\quad R_A y_2 = U_1^T (b - Ax_1) \;;\; x_{LSE} := x_1 + Q_2 y_2$$

This algorithm is easy to implement using the LINPACK routines. (It is a MATLAB "5-liner".)

Unfortunately, (1.1) is not a viable method for solving the large sparse LSE problem because the matrix AQ_2 will generally be dense. In this context the method of weighting is of interest. The idea behind this approach is simply to compute the solution $x(\mu)$ to the unconstrained problem

$$(1.2)\qquad \min_{x \in R^n} \left\| \begin{pmatrix} \mu B \\ A \end{pmatrix} x - \begin{pmatrix} \mu d \\ b \end{pmatrix} \right\|_2$$

for a large value of $\mu \in R$. It is widely known that $x(\mu) \to x_{LSE}$ as $\mu \to \infty$. Thus, existing software for sparse LS problems can "in principal' be used to generate an approximation to x_{LSE} of arbitrary quality.

However, several issues associated with the method of weighting demand our attention. At what rate do the quantities $x(\mu)$ and $d - Bx(\mu)$ converge? Are there practical ways to estimate the accuracy of $x(\mu)$? How can we cope with the numerical problems that can be expected to arise when μ is extremely large? We are prompted to discuss these issues because of the increasing importance of the sparse LSE problem. But our analytic and algorithmic developments will also be of interest to solvers of small LSE problems since the simplicity of the weighting method makes it extremely attractive and popular.

Our discussion is structured as follows. First, we analyze the properties of x_{LSE} and $x(\mu)$ using the generalized singular value decomposition. The limitations of the theory are then made obvious by reviewing the numerical difficulties associated with large μ. Next, we propose two techniques that can be used both to improve $x(\mu)$ and to estimate its error. One technique involves extrapolation and the other iterative improvement. We conclude with some remarks about the practical implementation of our ideas.

2. A GSVD Analysis of $x(\mu)$

The generalized singular value decomposition (GSVD) of A and B is useful as a tool for analyzing the method of weighting. This decomposition is as follows:

Theorem 2.1

If $A \in R^{m\times n}$ $(m \geq n)$ and $B \in R^{p\times n}$ $(p \leq n)$ satisfy $N(A) \cap N(B) = \{0\}$ then there exist

$$U = [u_1, \ldots, u_m] \in R^{m\times m} \qquad \text{(orthogonal)}$$
$$V = [v_1, \ldots, v_p] \in R^{p\times p} \qquad \text{(orthogonal)}$$
$$X = [x_1, \ldots, x_n] \in R^{n\times n} \qquad \text{(nonsingular)}$$

such that

$$U^T AX = D_A = \operatorname{diag}(\alpha_1, \ldots, \alpha_n)$$

and

$$V^T BX = D_B = \operatorname{diag}(\beta_1, \ldots, \beta_p).$$

Without loss of generality we may assume

$$||X||_2 = 1 \quad \text{and} \quad ||X^{-1}||_2 = \sigma_1/\sigma_n$$

where σ_1 and σ_n are the largest and smallest singular values of $\begin{bmatrix} A \\ B \end{bmatrix}$.

Proof

Let

$$\begin{bmatrix} A \\ B \end{bmatrix} = \begin{bmatrix} Q_1 \\ Q_2 \end{bmatrix} \operatorname{diag}(\sigma_i) Z^T$$

be the SVD of $\begin{bmatrix} A \\ B \end{bmatrix}$ with $Q_1^T Q_1 + Q_2^T Q_2 = I_n$, $\sigma_1 \geq \ldots \geq \sigma_n \geq 0$, and $Z^T Z = I_n$.

Let

$$\begin{bmatrix} Q_1 \\ Q_2 \end{bmatrix} = \begin{bmatrix} U & 0 \\ 0 & V \end{bmatrix} \begin{bmatrix} C \\ S \end{bmatrix} W^T$$

be the C-S decomposition of $\begin{bmatrix} Q_1 \\ Q_2 \end{bmatrix}$ where $U \in R^{m \times m}$, $V \in R^{p \times p}$, and $W \in R^{n \times n}$ are orthogonal and

$$C = \operatorname{diag}(c_1, \ldots, c_n) \in R^{m \times n} \qquad c_i \geq 0$$

and

$$S = \operatorname{diag}(s_1, \ldots, s_p) \in R^{p \times n} \qquad 0 \leq s_1 \leq \ldots \leq s_p$$

satisfy $C^T C + S^T S = I_n$. This decomposition is discussed in Stewart [9] who also presents an effective algorithm for computing it in [10].

The theorem follows by setting $D_A = \sigma_n C$, $D_B = \sigma_n S$, and $X^{-1} = \frac{1}{\sigma_n} W^T \operatorname{diag}(\sigma_i) Z^T$. Note that $\sigma_n > 0$ because $N(A) \cap N(B) = N\left(\begin{bmatrix} A \\ B \end{bmatrix}\right) = \{0\}$ ∎.

A number of elementary consequences of the GSVD are repeatedly used in the sequel. These are summarized in the following result:

Corollary 2.2

Suppose the GSVD is computed as indicated by the proof of Theorem 2.1. If rank(B) = p and

$$\mu_i = \alpha_i / \beta_i \qquad i = 1, \ldots, p$$

then

(a) $\alpha_i^2 + \beta_i^2 = \sigma_n^2 \qquad i = 1,\ldots,p$

(b) $\alpha_1 \geq \ldots \geq \alpha_q > \alpha_{q+1} = \ldots = \alpha_p = 0$ where $q = \dim[N(A)^\perp \cap N(B)^\perp]$

(c) $\alpha_{p+1} = \ldots = \alpha_n = \sigma_n$

(d) $0 < \beta_1 \leq \ldots \leq \beta_p$

(e) $\mu_1 \geq \ldots \geq \mu_p \geq 0$

(f) $Ax_i = \alpha_i u_i \qquad i = 1,\ldots,n$

(g) $Bx_i = \beta_i v_i \qquad i = 1,\ldots,p.$

Proof

Contentions (a) and (c) follow from $D_A^T D_A + D_B^T D_B = \sigma_n^2 I_n$ while (d) and (e) are true because $s_1 \leq \ldots \leq s_p$ and $p = \text{rank}(B)$. The equations $AX=UD_A$ and $BX = VD_B$ establish (f) and (g). Finally, if $\alpha_q > \alpha_{q+1} = \ldots = \alpha_p = 0$ then it follows from $A^T U = X^{-T} D_A^T$ and $B^T V = X^{-T} D_B^T$ that the first q columns of X^{-T} span the subspace $\text{Ran}(A^T) \cap \text{Ran}(B^T) = N(A)^\perp \cap N(B)^\perp$. This proves (b). □

The μ_i are called the generalized singular values of (A,B). We mention that a more general version of the GSVD is given in [11].

The GSVD can be used to diagonalize the LSE problem. In particular, by setting

$$(2.1) \qquad \begin{aligned} \tilde{b} &= U^T b = (u_1^T b,\ldots,u_m^T b)^T \\ \tilde{d} &= V^T d = (v_1^T d,\ldots,v_p^T d)^T \\ y &= X^{-1} x = (y_1,\ldots,y_n)^T \end{aligned}$$

we obtain

$$(\text{LSE}') \qquad \min_{D_B y = \tilde{d}} \| D_A y - \tilde{b} \|_2 \quad .$$

It is not hard to show that

$$y_{LSE} = \left(\frac{v_1^T d}{\beta_1}, \ldots, \frac{v_p^T d}{\beta_p}, \frac{u_{p+1}^T b}{\alpha_{p+1}}, \ldots, \frac{u_n^T b}{\alpha_n} \right)^T$$

solves this problem. Since $\alpha_{p+1} = \ldots = \alpha_n = \sigma_n$ we have

$$(2.2) \qquad x_{LSE} = Xy_{LSE} = \sum_{i=1}^{p} \frac{v_i^T d}{\beta_i} x_i + \frac{1}{\sigma_n} \sum_{i=p+1}^{n} (u_i^T b)\, x_i \, .$$

If we define

$$(2.4) \qquad \rho_i = u_i^T b - \mu_i v_i^T d \qquad i=1,\ldots,q$$

and

$$(2.5) \qquad r_{LS} = \sum_{i=q+1}^{p} (u_i^T b) u_i + \sum_{i=n+1}^{m} (u_i^T b) u_i$$

then it is easy to show that the constrained minimum residual is given by

$$(2.6) \qquad r_{LSE} = b - Ax_{LSE} = r_{LS} + \sum_{i=1}^{q} \rho_i u_i \, .$$

Note that $r_{LS} = b - Ax_{LS}$ where

$$x_{LS} = \sum_{i=1}^{q} \frac{u_i^T b}{\alpha_i} x_i + \frac{1}{\sigma_n} \sum_{i=p+1}^{n} (u_i^T b) x_i$$

is the solution of minimum 2-norm to the unconstrained LS problem $\min \|Ax-b\|_2$. Observe that

$$(2.7) \qquad \Delta^2 \equiv \|r_{LSE}\|_2^2 - \|r_{LS}\|_2^2 = \sum_{i=1}^{q} \rho_i^2$$

measures how much the residual increases as a result of the constraints.

It is clear from (2.2) that x_{LSE} is sensitive to perturbation whenever σ_n or $\beta_1 = \sigma_n/\sqrt{1+\mu_1^2}$ is small. Because it "displays" these critical quantities, it is advisable to solve ill-conditioned LSE problems via the GSVD. The sensitivity of the LSE problem is investigated in [5] and [12].

The GSVD is also convenient for analyzing $x(\mu)$. Since this vector solves (1.2) it must satisfy the normal equations

$$(2.8) \qquad (A^T A + \mu^2 B^T B) x(\mu) = A^T b + \mu^2 B^T d.$$

Under (2.1) this transforms to $(D_A^T D_A + \mu^2 D_B^T D_B) y(\mu) = D_A^T \tilde{b} + \mu^2 D_B^T \tilde{d}$ where $x(\mu) = Xy(\mu)$. It is easy to deduce the solution to this diagonal system and that

(2.9) $$x(\mu) = \sum_{i=1}^{p} \frac{\alpha_i u_i^T b + \mu^2 \beta_i v_i^T d}{\alpha_i^2 + \mu^2 \beta_i^2} x_i + \frac{1}{\sigma_n} \sum_{i=p+1}^{n} (u_i^T b) x_i .$$

By subtracting (2.2) from this equation we find

(2.10) $$e(\mu) \equiv x(\mu) - x_{LSE} = \sum_{i=1}^{q} \left(\frac{\mu_i}{\mu_i^2 + \mu^2} \right) * \left(\frac{\rho_i}{\beta_i} \right) x_i .$$

Note that the error is confined to span $\{x_1, \ldots, x_q\}$ and that it tends to zero as $\mu \to \infty$. An alternative proof of the latter fact may be found in [7, Chapter 22].

In Section 4 we establish the following inequalities:

$$\| x(\mu) - x_{LSE} \|_2 \leq \frac{\Delta}{2\mu\sigma_n} \sqrt{1+\mu_1^2}$$

$$\| d - Bx(\mu) \|_2 \leq \frac{\Delta}{2\mu}$$

$$0 \leq \| r_{LSE} \|_2 - \| b - Ax(\mu) \|_2 \leq \frac{\Delta \mu_1}{2\mu}$$

These results suggest that a large weight might be required if either μ_1 is large or σ_n is small.

3. Difficulties Associated With a Large Weight

It is widely appreciated that care must be exercised when Householder matrices are used to compute the Q-R factorization of a matrix whose rows vary greatly in norm, e.g., the matrix in (1.2) . Powell and Reid [8] examined this problem in conjunction with the Businger-Golub algorithm [4] and advise incorporation of row interchanges, much as in Gaussian elimination. Specifically, they recommend that the k-th column be searched and its largest entry pivoted to the (k,k) position before the k-th Householder matrix is applied.

Note that near-domination of the pivot elements will result if the Businger - Golub algorithm is applied to our heavily weighted matrix in (1.2) but <u>not</u> if we apply it to the mathematically equivalent problem

$$(3.1) \qquad \min \left\| \begin{pmatrix} A \\ \mu B \end{pmatrix} x - \begin{pmatrix} b \\ \mu d \end{pmatrix} \right\|_2 \qquad \mu >> 1.$$

To appreciate the difference between the "B-over-A" and the "A-over-B" formulations consider the LSE problem

$$A = \begin{bmatrix} 1 & 2 \\ 3 & 4 \end{bmatrix} , \quad b = \begin{bmatrix} 1 \\ 1 \end{bmatrix} , \quad B = (1,-1) , \quad d = (2) .$$

This problem is well-conditioned and has exact solution $x_{LSE} = \frac{1}{29}(39, -19)^T$. In the following table we record the error $\| x(\mu) - x_{LSE} \|_2$ for both approaches as a function of μ:

Method \ μ	10^1	10^3	10^5	10^7	10^9	10^{11}	10^{13}	10^{15}	10^{17}
B-over-A	10^{-3}	10^{-7}	10^{-11}	10^{-15}	10^{-17}	10^{-17}	10^{-17}	10^{-17}	10^{-17}
A-over-B	10^{-3}	10^{-7}	10^{-11}	10^{-11}	10^{-10}	10^{-7}	10^{-6}	10^{-4}	10^{-2}

These computations were performed using VAX double precision arithmetic (macheps $\approx 10^{-17}$) in the MATLAB environment.

The divergence of the two methods in the vicinity of $\mu = (\text{macheps})^{-1/2}$ is fairly typical, even for ill-conditioned examples. However, for ill-conditioned LSE problems, the "optimum" weight for the B-over-A approach is usually several orders of magnitude greater than $(\text{machep})^{-1/2}$. Thus it is preferable (although not always critical) to solve (1.2) instead of (3.1). In the remainder of this paper, we will always use the B-over-A formulation (1.2).
However, it is not always easy to arrange the rows of the unconstrained problem so that the constraint equations come first. The minimization of fill-in may dictate some other row ordering in the case when A and B are large and sparse.

Column ordering is also important in the method of weighting. At the suggestion of the referee, we considered the example

$$A = \begin{bmatrix} 1 & 1 & 1 \\ 1 & 3 & 1 \\ 1 & -1 & 1 \\ 1 & 1 & 1 \end{bmatrix} \qquad b = \begin{bmatrix} 1 \\ 2 \\ 3 \\ 4 \end{bmatrix}$$

$$B = \begin{bmatrix} 1 & 1 & 1 \\ 1 & 1 & -1 \end{bmatrix} \qquad d = \begin{bmatrix} 7 \\ 4 \end{bmatrix}$$

This example is well conditioned and has exact solution $x_{LSE} = \frac{1}{8}(46,-2,12)^T$. Let $R(\mu) = (r_{ij}(\mu))$ be the 3-by-3 upper triangular matrix obtained by computing the Q-R decomposition of $\begin{pmatrix}\mu B\\ A\end{pmatrix}$. It is clear that $r_{11}(\mu)/r_{22}(\mu)$ becomes very large as $\mu \to \infty$. This occurs because the first two columns of B are dependent. On the other hand, if we apply Q-R with column pivoting to $\begin{pmatrix}\mu B\\ A\end{pmatrix}$ then $r_{11}(\mu)$ and $r_{22}(\mu)$ have the same order of magnitude as $\mu \to \infty$. The decision to column pivot is important as the following table of relative errors in $x(\mu)$ indicates:

Column Pivoting \ μ	10^1	10^3	10^5	10^7	10^9	10^{11}	10^{13}	10^{15}
No	10^{-2}	10^{-7}	10^{-11}	10^{-11}	10^{-9}	10^{-7}	10^{-5}	10^{-3}
Yes	10^{-2}	10^{-7}	10^{-10}	10^{-14}	10^{-16}	10^{-16}	10^{-16}	10^{-16}

We infer that the weighting method can be unstable if (1.2) is solved using the Q-R decomposition without pivoting.

4. An Extrapolation Procedure

As we have seen, the importance of row and column ordering in (1.2) increases with increasing μ. Hence, it is potentially interesting to see how the method might be used with "safe" weights since the ordering of rows and columns is often inconvenient.

Suppose $x(\mu)$ and $x(\gamma\mu)$ have been computed for some $\mu > 0$ and $\gamma > 1$. Using (2.10) it can be shown that if

$$(4.1)\qquad x^{(1)}(\gamma\mu) = x(\gamma\mu) + \frac{1}{\gamma^2-1}\left[x(\gamma\mu) - x(\mu)\right]$$

then

$$x^{(1)}(\gamma\mu) = x_{LSE} + \sum_{i=1}^{q} \frac{\mu_i^2}{\mu_i^2+\mu^2} \cdot \frac{\mu_i}{\mu_i^2+(\gamma\mu)^2} \cdot \frac{\rho_i}{\beta_i}\, x_i$$

Thus, the error expansion for $x^{(1)}(\gamma\mu)$ is the same as that for $x(\gamma\mu)$ except the coeffient of x_i is diminished by the factor $\mu_i^2/(\mu_i^2+\mu^2)$.

The calculation (4.1) is merely the first extrapolation in the following Richardson scheme:

(4.2)

For $j=0,1,\ldots$
- $x^{(0)}(\gamma^j\mu) = x(\gamma^j\mu)$
- For $k=1,2,\ldots,j$
  - $$x^{(k)}(\gamma^j\mu) = x^{(k-1)}(\gamma^j\mu) + \frac{1}{\gamma^{2k}-1} * \left[x^{(k-1)}(\gamma^j\mu) - x^{(k-1)}(\gamma^{j-1}\mu)\right]$$

A messy but straightforward induction argument based on (2.10) can be used to show that

$$(4.3)\qquad e^{(k)}(\gamma^j\mu) \equiv x^{(k)}(\gamma^j\mu) - x_{LSE} = \sum_{i=1}^{q} \varepsilon_i^{(k)}(j)\, x_i$$

where

$$(4.4)\qquad \varepsilon_i^{(k)}(j) = \frac{\rho_i}{\beta_i} * \frac{\mu_i}{\mu_i^2+(\mu\gamma)^2} * \prod_{\ell=j-k}^{j-1} \left(\frac{\mu_i^2}{\mu_i^2+\mu^2\gamma^{2\ell}}\right)$$

Other quantities of interest associated with $x^{(k)}(\gamma^j\mu)$ include the B-residual

$$(4.5)\qquad s^{(k)}(\gamma^j\mu) = d - Bx^{(k)}(\gamma^j\mu) = -\sum_{i=1}^{q} \beta_i \varepsilon_i^{(k)}(j) v_i$$

and the A-residual

$$(4.6)\qquad r^{(k)}(\gamma^j\mu) = b - Ax^{(k)}(\gamma^j\mu) = r_{LSE} - \sum_{i=1}^{q} \alpha_i \varepsilon_i^{(k)}(j) u_i$$

Using these formulae we have

Theorem 4.1

If $\theta = \mu_1^2/(\mu_1^2 + \mu^2\gamma^{2(j-k)})$ and $x^{(k)}(\gamma^j\mu)$ is generated via (4.2), then

(a) $$||x^{(k)}(\gamma^j\mu) - x_{LSE}||_2 \le \frac{\Delta\sqrt{1+\mu_1^2}}{2\mu\gamma^j\sigma_n}\theta^k$$

(b) $$||d - Bx^{(k)}(\gamma^j\mu)||_2 \le \frac{\Delta}{2\mu\gamma^j}\theta^k$$

(c) $$0 \le ||r_{LSE}||_2 - ||b - Ax^{(k)}(\gamma^j\mu)||_2 \le \frac{\mu_1\Delta}{2\mu\gamma^j}\theta^k.$$

Proof

Using (4.4) and the inequality

$$\frac{\mu_i}{\mu_1^2 + \mu^2\gamma^{2j}} \le \frac{1}{2\mu\gamma^j}$$

it follows that $|\varepsilon_i^{(k)}(j)| \le \frac{|\rho_i|}{\beta_i}\frac{\theta^k}{2\mu\gamma^j}$. Since $||X||_2 = 1$

we have by taking norms in (4.3) and using (2.7) that

$$||e^{(k)}(\gamma^j\mu)||_2 \le \frac{\Delta}{2\mu\beta_1\gamma^j}\theta^k.$$

Inequality (a) follows since $\beta_1^{-2} = (1+\mu_1^2)/\sigma_n^2$. To prove (b) and the upper bound in (c), take norms in (4.5) and (4.6) respectively and use the orthonormality of $\{u_1\}$ and $\{v_i\}$.

To establish the lower bound in (c), note from (4.6) and (2.6) that

$$r^{(k)}(\gamma^j\mu) = r_{LS} + \sum_{i=1}^{q} (\rho_i - \alpha_i \varepsilon_i^{(k)}(j))\, u_i .$$

Since $r_{LS} \in \text{span}\,\{u_1, \ldots, u_q\}^{\perp}$ it follows that

$$\|r^{(k)}(\gamma^j\mu)\|_2^2 = \|r_{LS}\|_2^2 + \sum_{i=1}^{q} (\rho_i - \alpha_i \varepsilon_i^{(k)}(j))^2 .$$

The desired bound follows from (2.7) and the easily verified inequality $\rho_i^2 \leq (\rho_i - \alpha_i \varepsilon_i^{(k)}(j))^2$. (Note that $\|r^{(k)}(\gamma^j\mu)\|_2$ increases monotonically to $\|r_{LSE}\|_2$ as μ increases.) □

The main observation to make from the theorem is that the error, A-residual, and B-residual improve by a factor θ with each extrapolation. For example, in a small problem with $\mu_1 \approx 10^3$, we set $\mu = 100$, $\gamma = 2$ and found the following relative errors in $x^{(k)}(\gamma^j\mu)$:

$\gamma^j\mu$	k=0	k=1	k=2	k=3	k=4
100	10^{0}				
200	10^{0}	10^{-1}			
400	10^{-1}	10^{-2}	10^{-2}		
800	10^{-2}	10^{-3}	10^{-4}	10^{-4}	
1600	10^{-2}	10^{-4}	10^{-5}	10^{-6}	10^{-6}
3200	10^{-3}	10^{-5}	10^{-7}	10^{-8}	10^{-8}

In another example ($\mu_1 \cong 10^5$) we found the relative errors dimishing as follows

$\gamma^j\mu$	k=0	k=1	k=2	k=3
10^4	10^{0}			
10^6	10^{-1}	10^{-1}		
10^8	10^{-5}	10^{-6}	10^{-6}	
10^{10}	10^{-9}	10^{-11}	10^{-11}	10^{-11}

The corresponding entries for the relative B-residual $\|d-Bx^{(k)}(\gamma^j\mu)\| / \|B\|_2 * \|x_{LSE}\|_2$ were smaller by a factor of 10^4.

Based on numerous small examples run with MATLAB we conclude

(a) The extrapolation cannot "take hold" until $\gamma^j \approx \mu_1$.

(b) In ill-conditioned examples, there is no significant reduction in the error beyond the second or third extrapolation.

(c) There is no point in extrapolating if $\gamma^j\mu > (\text{macheps})^{-1/2}$ when the unconstrained LS problem is in A-over-B form.

(d) Once the extrapolates "settle down", $||x^{(k)}(\gamma^j\mu) - x^{(k-1)}(\gamma^j\mu)||_2$ gives good estimates of the error in $x^{(k-1)}(\gamma^{j-1}\mu)$.

(e) Most time is spent in computing the Q-R decompositions for each $x(\gamma^j\mu)$. The extrapolates are "free".

5. The Iterative Improvement of $x(\mu)$

A routine Lagrange multiplier argument can be used to show that x_{LSE} and $r_{LSE} = b - Ax_{LSE}$ satisfy

$$\begin{bmatrix} b \\ d \\ 0 \end{bmatrix} = \begin{bmatrix} I & 0 & A \\ 0 & 0 & B \\ A^T & B^T & 0 \end{bmatrix} \begin{bmatrix} r_{LSE} \\ \lambda_{LSE} \\ x_{LSE} \end{bmatrix} \tag{5.1}$$

Here, λ_{LSE} is the corresponding Lagrange multiplier.

The unconstrained LS problem (1.2) can also be posed as a linear equation problem:

$$\begin{bmatrix} b \\ d \\ 0 \end{bmatrix} = \begin{bmatrix} I & 0 & A \\ 0 & \mu^{-2}I & B \\ A^T & B^T & 0 \end{bmatrix} \begin{bmatrix} b - Ax(\mu) \\ \mu^2(d-Bx(\mu)) \\ x(\mu) \end{bmatrix}$$

(The last row in this equation is just the normal equation system (2.8).) This suggests the following iteration for improving $x(\mu)$:

$$x := x(\mu) \;;\quad \lambda := \mu^2(d - Bx(\mu)) \;;\quad r := b - Ax(\mu)$$

Repeat:

$$(5.3)\qquad \begin{bmatrix}\delta_1\\ \delta_2\\ \delta_3\end{bmatrix} = \begin{bmatrix} b\\ d\\ 0\end{bmatrix} - \begin{bmatrix} I & 0 & A\\ 0 & 0 & B\\ A^T & B^T & 0\end{bmatrix}\begin{bmatrix} r\\ \lambda\\ x\end{bmatrix}$$

$$\begin{bmatrix} I & 0 & A\\ 0 & \mu^{-2}I & B\\ A^T & B^T & 0\end{bmatrix}\begin{bmatrix}\Delta r\\ \Delta\lambda\\ \Delta x\end{bmatrix} = \begin{bmatrix}\delta_1\\ \delta_2\\ \delta_3\end{bmatrix}$$

$$x := x + \Delta x \;;\; \lambda := \lambda + \Delta\lambda \quad;\; r := r + \Delta r$$

In this scheme, if the matrix that defines the corrections Δr, $\Delta\lambda$, and Δx is replaced by the matrix of (5.1), then we obtain the iterative improvement scheme of Bjork and Golub [2]. These authors were interested in how to improve the computed solution $\hat{x}_{LSE}$ obtained via (1.1). We, however, are only interested in getting a solution approximately as good as $\hat{x}_{LSE}$. Consequently, we need not resort to multiple precision computation of the residuals δ_i.

The iteration (5.3) undergoes considerable simplification when we observe that δ_1 and δ_3 are always zero. (Use induction.) In light of this observation, it follows that Δx satisfies

$$\left\| \begin{pmatrix}\mu B\\ A\end{pmatrix}\Delta x - \begin{pmatrix}\mu\delta_2\\ 0\end{pmatrix}\right\|_2 = \min$$

where $\delta_2 = d - Bx$. Thus, (5.3) collapses to

$$x(\mu, 0) := x(\mu)$$

For $k = 0, 1, \ldots$

$$\delta^{(k)} := d - Bx(\mu, k)$$

(5.4) Compute the solution $\Delta x^{(k)}$ to

$$\min_z \left\| \begin{pmatrix}\mu B\\ A\end{pmatrix} z - \begin{pmatrix}\mu\delta^{(k)}\\ 0\end{pmatrix}\right\|_2$$

$$x(\mu, k+1) = x(\mu, k) + \Delta x^{(k)}$$

Note that only one Q-R factorization, that of $\begin{pmatrix}\mu B\\ A\end{pmatrix}$, is needed to execute (5.4).

The convergence of the method is established in the following result:

Theorem 5.1

In the notation of Section 2, the vectors $x(\mu,k)$ generated by (5.4) satisfy

$$x(\mu,k) = x_{LSE} + e(\mu,k)$$

where

$$e(\mu,k) = \sum_{i=1}^{q} \frac{\rho_i}{\beta_i} \frac{\mu_i}{\mu_i^2+\mu^2} \theta_i^k x_i$$

with $\theta_i = \mu_i^2/(\mu_i^2 + \mu^2)$.

Proof

We use induction observing from (2.10) that the theorem is true for k=0. Since $\delta^{(k)} = -Be(\mu,k)$ we have by induction that

$$\delta^{(k)} = -\sum_{i=1}^{q} \rho_i \frac{\mu_i}{\mu_i^2+\mu^2} \theta_i^k v_i.$$

Now $(A^TA+\mu^2B^TB)^{-1}B^T = X(D_A^TD_A + \mu^2D_B^TD_B)^{-1}D_B^TV^T$ and so

$$\Delta x^{(k)} = \mu^2(A^TA+\mu^2B^TB)^{-1}B^T\delta^{(k)}$$

$$= -\mu^2 \sum_{i=1}^{q} \frac{\rho_i}{\beta_i} \frac{\mu_i}{\mu_i^2+\mu^2} \theta_i^{k+1} x_i .$$

The theorem follows directly from the equation $e(\mu,k+1) = e(\mu,k)+\Delta x^{(k)}$. □

Arguments very similar to those used in the proof of Theorem 4.1 can be used to establish the following inequalities

$$||x(\mu,k) - x_{LSE}||_2 \leq \frac{\Delta\sqrt{1+\mu_1^2}}{2\mu\sigma_n} \theta_1^k$$

$$||d - Bx(\mu,k)||_2 \leq \frac{\Delta}{2\mu} \theta_1^k$$

$$0 \leq ||r_{LSE}||_2 - ||b - Ax(\mu,k)||_2 \leq \frac{\Delta\mu_1}{2\mu} \theta_1^k.$$

These bounds closely correspond to those for the extrapolation method.

To examine the effectiveness of the iteration, we applied it to a problem in which μ_1=5000. The relative error of $x(\mu,k)$ is tabulated in the following table:

μ	k=0	k=1	k=2	k=3	k=4
10^3	10^0	10^0	10^0	10^0	10^0
10^4	10^{-1}	10^{-2}	10^{-3}	10^{-3}	10^{-4}
10^5	10^{-3}	10^{-6}	10^{-8}	10^{-11}	10^{-14}
10^8	10^{-9}	10^{-14}	10^{-14}	10^{-14}	10^{-14}

Note that the iteration cannot substantially improve the accuracy of $x(\mu,0)$ unless $\mu_1^2/(\mu_1^2+\mu^2)$ is somewhat less than unity.

If the A-over-B formulation to compute $x(\mu,0)$ is used, then improvements in the iterates should not be expected if $\mu > (\text{macheps})^{-\frac{1}{2}}$.

In practice, $||\Delta x^{(k)}||_2$ is a very good estimate of the error in $x(\mu,k)$ once convergence begins to set in.

6. Conclusions

Much more work is needed before the ideas in this paper can take on a practical form. Some topics for future research include the following.

(i) A closer examination of ill-conditioning is needed. Does the weighting method merely fail to compute the "unstable" components of x_{LSE}?

(ii) It is not necessary to re-compute the Q-R factorization of $\begin{pmatrix}\mu B\\ A\end{pmatrix}$ for each different μ when extrapolating. Instead, the Q-R factorization of A can be updated [1]. How could this idea be implemented with the George-Heath algorithm [6]?

(iii) What are good values for γ and μ when extrapolating? How many extrapolation steps can we "afford"?

(iv) Are there ways to implement (5.4) without having to store the complete Q-R factorization of $\begin{pmatrix}\mu B\\ A\end{pmatrix}$?

It is our intention to continue investigating these questions.

Acknowledgements

I would like to thank Gene Golub, Per-Åke Wedin, and the referee for their constructive comments. In addition, the support of the National Science Foundation and the Swedish Natural Science Research Council is gratefully acknowledged. This paper was produced while I was a visitor at the University of Umeå.

REFERENCES

1. Å. BJÖRK (1981), "A general updating algorithm for constrained linear least squares problems," Report LiTH-MAT-R-81-18, Department of Mathematics, University of Linkoping, Sweden.

2. Å. BJÖRK AND G.H. GOLUB (1967), "Iterative refinement of linear least squares solutions by Householder transformation," BIT, 7, 327-337.

3. Å. BJÖRK AND I.S. DUFF (1980), "A direct method for the solution of sparse linear least squares problems," Lin. Alg.&Applic., 34, 43-67.

4. P. BUSINGER AND G.H. GOLUB (1965), "Linear least squares solutions by Householder transformations," Numer. Math. 7, 169.

5. L. ELDEN (1980), "Perturbation theory for the least squares problem with linear equality constraints", SIAM J. Numer. Anal., 17, 338-350.

6. A. GEORGE AND M. HEATH (1980), "Solution of sparse linear least squares problems using Givens rotations," Lin. Alg.& Applic. 34, 69-84.

7. C.L LAWSON AND R.J. HANSON (1974), "Solving least squares problems, Prentice-Hall, Englewood Cliffs NJ.

8. M.J.D. POWELL AND J.K. REID, (1969), "On applying Householder's method to linear least squares problems," Proc. IFIP Congress, 1968.

9. G.W. STEWART (1977)," On the perturbation of pseudo-inverses, projections and linear least squares problems," SIAM Review, 19, 634-662.

10. G.W. STEWART (1982), " A Method for Computing the Generalized Singular value decomposition," this volume.

11. C.VAN LOAN (1976), "Generalizing the singular value decomposition," SIAM J. Numer. Anal., 13, 76-83.

12. P-Å WEDIN (1979), "Notes on the constrained least squares problem. A new approach based on generalized inverses," Report UMINF 75.79, Institute of Information Processing, University of Umeå, Sweden.

13. C.B. MOLER, (1980), "MATLAB- An Interactive matrix Laboratory," Dept of Computer Science, University of New Mexico, Albuquerque, New Mexico.

On Angles between Subspaces of a Finite Dimensional Inner Product Space

Per Åke Wedin

Institute of Information Processing, Dept of Numerical Analysis
University of Umeå, S-901 87 UMEÅ, SWEDEN

Abstract In this paper the basic results on angles between subspaces of C^n are presented in a way that differs from the abstract mathematical narrative of Kato [10] et al. First the intuitively clear properties of the angle between one dimensional subspaces of R^2 are stated using orthogonal projections. Then, using the singular value decomposition, a very simple representation is derived for pairs of orthogonal projections in C^n. Through this representation the angles between two subspaces of C^n are related to the principal angles between certain invariant two dimensional subspaces. It is seen how the properties of angles between general subspaces can be derived from the corresponding simple properties of angles between two dimensional subspaces. The results are used to estimate angles between certain subspaces. These estimates are not new. But the point made here is that when the relevant perturbation identities are known it is easy to use an angle function to get perturbation bounds. Finally in appendix 1 a new perturbation bound is given for pairs of oblique projections.

1. Introduction to pairs of projections in R^2

Let two one dimensional subspaces M_1 and M_2 in C^n be spanned by the vectors u and v. The angle Θ between M_1 and M_2 is defined as

$$\Theta = \measuredangle (M_1,M_2) = \cos^{-1} \frac{|u^H v|}{||u||\cdot||v||} \qquad (1.1)$$

It follows from this definition that $0 \leq \theta \leq \pi/2$. The norm used here and elsewhere without a subscript is the 2-norm defined by $||u|| = (u^H u)^{1/2}$. Let us now specialize to the case when M_1 and M_2 are one dimensional subspaces of R^2. Choose the vectors u and v that span M_1 and M_2 such that $u^H u = v^H v = 1$ and $u^H v \geq 0$. Let w be a unit vector orthogonal to u. Choose the sign of w such that

$$v = u \cos\theta + w \sin\theta , \qquad 0 \leq \theta \leq \pi/2 \tag{1.2}$$

Since $u^H v$ equals $\cos\theta$ the vector w can be computed from formula (1.2) when $\theta \neq 0$. See figure 1.

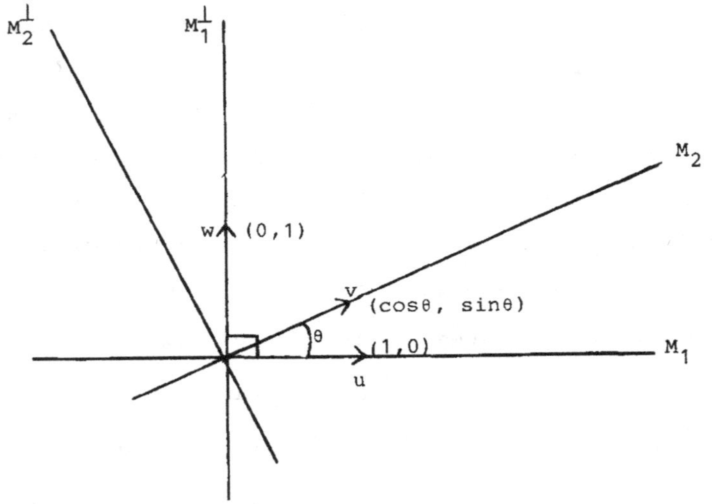

Figure 1

With respect to the coordinate system defined by the orthogonal basis (u,w):

$$u = \begin{pmatrix} 1 \\ 0 \end{pmatrix} \in M_1 \quad \text{and} \quad v = \begin{pmatrix} \cos\theta \\ \sin\theta \end{pmatrix} \in M_2 \tag{1.3}$$

Now let P_{M_1} and P_{M_2} be the orthogonal projections onto M_1 and M_2 respectively.

These projections are given by the following matrices

$$P_{M_1} = \begin{pmatrix}1\\0\end{pmatrix}(1,0)\ , \qquad P_{M_2} = \begin{pmatrix}\cos\theta\\ \sin\theta\end{pmatrix}(\cos\theta,\sin\theta) \tag{1.4}$$

with respect to the basis (u,w).

Hence

$$P_{M_1}P_{M_2} = \begin{pmatrix}1\\0\end{pmatrix}\underbrace{(1,0)\begin{pmatrix}\cos\theta\\ \sin\theta\end{pmatrix}}_{\cos\theta}(\cos\theta,\sin\theta) =$$

$$= \begin{pmatrix}1\\0\end{pmatrix}\underbrace{\cos\theta}_{\substack{\text{singular}\\ \text{value of } P_{M_1}P_{M_2}}}(\cos\theta,\sin\theta) \tag{1.5}$$

$$(I-P_{M_1})P_{M_2} = \begin{pmatrix}0\\1\end{pmatrix}\underbrace{\sin\theta}_{\substack{\text{singular value}\\ \text{of } (I-P_{M_1})P_{M_2}}}(\cos\theta,\sin\theta). \tag{1.6}$$

and

$$P_{M_1}(I-P_{M_2}) = \begin{pmatrix}1\\0\end{pmatrix}(-\sin\theta)(-\sin\theta,\cos\theta) \tag{1.7}$$

The important identity

$$P_{M_2}-P_{M_1} = (I-P_{M_1})P_{M_2} - P_{M_1}(I-P_{M_2}) \tag{1.8}$$

makes it possible to derive the singular value decomposition of $P_{M_2} - P_{M_1}$ from the formulas (1.6) and (1.7):

$$P_{M_2}-P_{M_1} = \begin{pmatrix}0\\1\end{pmatrix}\sin\theta\,(\cos\theta,\sin\theta) - \begin{pmatrix}1\\0\end{pmatrix}(-\sin\theta)(-\sin\theta,\cos\theta) =$$

$$= \sin\theta\begin{pmatrix}-\sin\theta & \cos\theta\\ \cos\theta & \sin\theta\end{pmatrix} \tag{1.9}$$

Hence we get

$$||P_{M_2}-P_{M_1}|| = ||(I-P_{M_1})P_{M_2}|| = ||P_{M_1}(I-P_{M_2})|| = \sin\theta. \quad (1.10)$$

from (1.6), (1.7) and (1.9). Similarly it follows from (1.5) that

$$||P_{M_1}P_{M_2}|| = \cos\theta$$

Let us also introduce the oblique projection S of R^2 onto M_1 along $M_2^{\perp}$, the orthogonal complement of M_2.

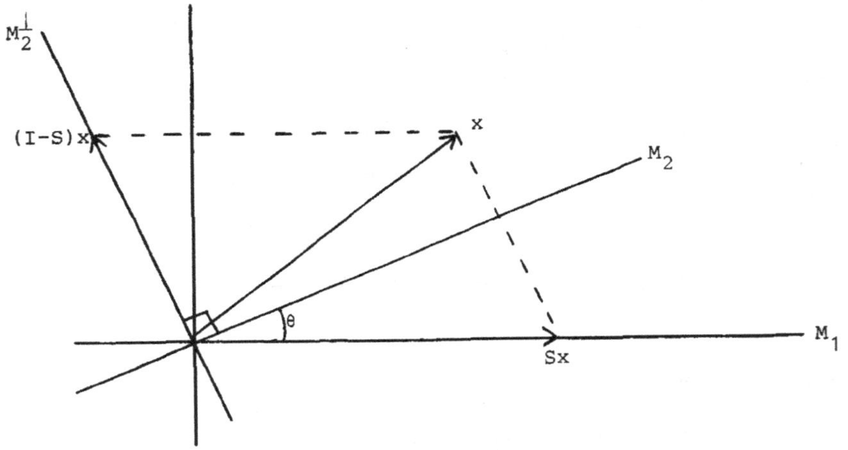

Figure 2

By definition the range of S is spanned by u and the nullspace of S is spanned by a vector orthogonal to v. It is easily checked that S can be written $u(v^Hu)^{-1}v^H$. Hence with respect to the basis {u,w} we get

$$S = \begin{pmatrix}1\\0\end{pmatrix} \frac{1}{\cos\theta} (\cos\theta, \sin\theta) \quad (1.11)$$

Compare this representation of S to the product

$$P_{M_2}P_{M_1} = \begin{pmatrix}\cos\theta\\\sin\theta\end{pmatrix} \cos\theta \; (1,0) \quad (1.12)$$

From (1.11) and (1.12) we conclude that

$$S = (P_{M_2} P_{M_1})^+ \tag{1.13}$$

i.e. S is the pseudo-inverse of $P_{M_2} P_{M_1}$. The identity (1.13) also holds in the general case when M_1 and $M_2^\perp$ are complimentary subspaces of C^n i.e.

$$\dim(M_1) = \dim(M_2),\ M_1^\perp \cap M_2 = \{0\},\ M_1 \cap M_2^\perp = \{0\} \tag{1.14}$$

The identity (1.13) is important because it is the key to a perturbation theory for oblique projections. Through this identity fairly wellknown perturbation identities for orthogonal projections and pseudo-inverses can be combined to give a perturbation identity for oblique projections. The resulting identity is surprisingly simple. For details see appendix 1.

2. The principal angles between two subspaces of C^n

The following theorem gives a nice representation of a pair of orthogonal projections in C^n.

Theorem Let M_1 and M_2 be two subspaces of C^n. Assume that

$$\left.\begin{array}{l} \operatorname{rank}(P_{M_1} P_{M_2}) = k+r \quad \text{and} \\ P_{M_1} P_{M_2} \text{ has } k \text{ singular values equal to } 1 \end{array}\right\} \tag{2.1}$$

Then there exists an orthogonal basis of C^n such that with respect to this basis P_{M_1} and P_{M_2} are represented by P_1 and P_2 respectively where

$$P_1 = \overbrace{\begin{pmatrix} \overset{k}{I} & \overbrace{\underset{Q}{\,}}^{2r} & \overbrace{\bigcirc}^{n-(k+2r)} \\ & \ddots & \\ & & Q \\ \bigcirc & & & D_1 \end{pmatrix}} \quad ; \quad P_2 = \begin{pmatrix} \overset{k}{I} & \overbrace{E(\theta_{k+1})}^{2r} & \bigcirc \\ & \ddots & \\ & & E(\theta_{k+r}) \\ \bigcirc & & & D_2 \end{pmatrix} \qquad (2.2)$$

and

$$\left. \begin{aligned} Q &= \begin{pmatrix} 1 \\ 0 \end{pmatrix} (1,0) = \begin{pmatrix} 1 & 0 \\ 0 & 0 \end{pmatrix} \\ E(\theta) &= \begin{pmatrix} \cos\theta \\ \sin\theta \end{pmatrix} (\cos\theta, \ \sin\theta) \end{aligned} \right\} \qquad (2.3)$$

D_1 and D_2 are diagonal matrices with only the numbers 0 and 1 in the diagonal and with

$$D_1 D_2 = 0. \qquad (2.4)$$

The numbers

$$0 < \theta_{k+1} \le \cdots \le \theta_{k+r} < \pi/2 \qquad (2.5)$$

are called the _acute principal_ angles between the subspaces M_1 and M_2. Usually we take

$$\theta_1 = \theta_2 = \cdots = \theta_k = 0. \qquad (2.6)$$

Note. The identity (2.3) says that it is possibly to find a basis of C^n such that with respect to this basis P_{M_1} becomes a diagonal matrix and P_{M_2} a blockdiagonal matrix with 2x2 blocks. The 2x2 blocks Q and $E(\theta_k)$ of this representation correspond to the representation given by (1.4) for two orthogonal projections in R^2. Hence the representation (2.2) can be used to prove identities corresponding to (1.8) and (1.10) for subspaces of C^n.

The following is another formulation of the theorem. Assume that the projections P_{M_i}, $i = 1,2$ are given in matrix form. Then there exists an orthogonal nxn matrix Z such that

$$P_{M_i} = ZP_iZ^H , \quad i = 1,2$$

with the matrices P_i defined by (2.2).

The proof of the theorem will be constructive. We shall show how the matrices P_i, $i = 1,2$ and Z can be computed from the singular value decomposition of $P_{M_1}P_{M_2}$. An algorithm for computing the principle angles and the SVD of $P_{M_1}P_{M_2}$ was first given by Björck, Golub -73[4].

Assume that M_1 and M_2 are spanned by the linearly independent column vectors of A_1 and A_2 respectively. Make the QR-decomposition of these matrices i.e. take

$$A_i = Q_iR_i \qquad i = 1,2$$

where Q_i has orthonormal columns. We get

$$P_{M_i} = Q_iQ_i^H , \quad i = 1,2 \tag{2.7}$$

and

$$P_{M_1}P_{M_2} = Q_1(Q_1^HQ_2)Q_2^H \tag{2.8}$$

Take the SVD of $Q_1^HQ_2$ to get

$$Q_1^HQ_2 = X\Sigma Y^H. \tag{2.9}$$

Define

$$U = Q_1X \quad \text{and} \quad V = Q_2Y. \tag{2.10}$$

U and V have orthonormal columns and hence $P_{M_1}P_{M_2}$ has the SVD:

$$P_{M_1}P_{M_2} = U\Sigma V^H \tag{2.11}$$

From assumption (2.1) about the rank of $P_{M_1}P_{M_2}$ it follows that Σ is a $(k+r) \times (k+r)$-matrix. Take

$$\Sigma = \text{diag}(\sigma_1,\ldots,\sigma_{k+r}).$$

Since $\sigma_{max} = ||P_{M_1}P_{M_2}|| \leq ||P_{M_1}|| \, ||P_{M_2}|| = 1$ all singular values are less than or equal to 1. It was assumed in (2.1) that k singular values are equal to 1. Hence the singular values can be ordered like this

$$1 = \sigma_1 = \ldots = \sigma_k > \sigma_{k+1} \geq \ldots \geq \sigma_{k+r} > 0$$

Let σ_i be a singular value with the corresponding singular vectors u_i and v_i. From the definition of the SVD and since $P_{M_i}^H = P_{M_i}$ it follows that

$$P_{M_1}P_{M_2}v_i = u_i\,\sigma_i$$
$$(P_{M_1}P_{M_2})^H u_i = P_{M_2}P_{M_1}u_i = v_i\,\sigma_i.$$

Hence $u_i \in M_1$, $v_i \in M_2$ and

$$\begin{cases} P_{M_1}v_i = u_i\,\sigma_i & (v_i \text{ is projected onto } u_i \text{ by } P_{M_1}) \\ \\ P_{M_2}u_i = v_i\,\sigma_i & (u_i \text{ is projected back onto } v_i \text{ by } P_{M_2}) \end{cases} \tag{2.12}$$

Take

$$\theta_i = \cos^{-1}(\sigma_i) \tag{2.13}$$

and note that the principle angles θ_i defined in this way satisfy (2.5) or (2.6). Then the identity (2.12) can be illustrated by the following figure. Note that $u_i = v_i$ when $\sigma_i = 1$.

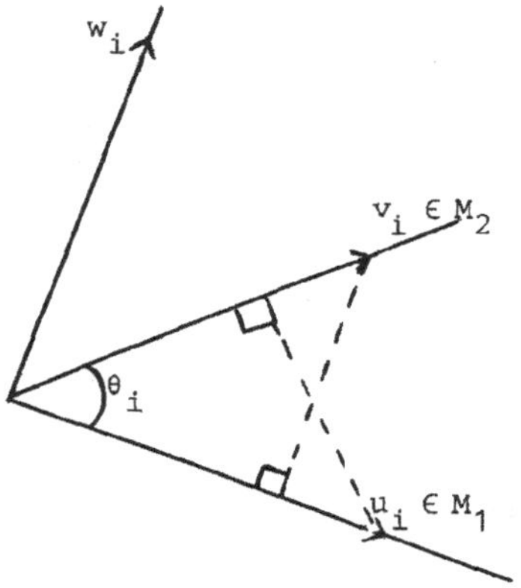

Figure 3

From the assumption (2.1) it is seen that there are r <u>2-dimensional</u> subspaces spanned by the pairs of vectors (u_i, v_i), $i = k+1,\dots,k+r$. Such a subspace has two important properties:

i) it is invariant under both P_{M_1} and P_{M_2} i.e.
if x belongs to the subspace then $P_{M_i}x$, $i=1,2$ does so, too.

ii) it is orthogonal to all other subspaces of this kind.

The first statement is a consequence of (2.12). Statement ii) says that subspaces spanned by pairs of vectors u_i, v_i are mutually orthogonal.

That follows from the identity

$$v_i^H u_j = (P_{M_2} v_i)^H u_j = v_i^H P_{M_2} u_j \underset{\substack{\uparrow \\ (2.12)}}{=} \sigma_j v_i^H v_j = \begin{cases} 0 & \text{if } i \neq j \\ \sigma_j = \cos\theta_j & \text{if } i=j \end{cases} \qquad (2.14)$$

As in the twodimensional case the vector w_i defined by

$$v_i = u_i \underbrace{\cos\theta_i}_{=\sigma_i} + w_i \sin\theta_i \quad , \ i = k+1,\dots,k+r$$

becomes a unit vector orthogonal to u_i. The identity (2.14)also implies that the vectors

$$u_1, u_2, \dots, u_k \quad \text{and} \quad u_{k+1}, w_{k+1}, u_{k+2}, w_{k+2}, \dots, u_{k+r}, w_{k+r}$$

are orthogonal and span the same subspace that is spanned by the columns of U and V.
Take

$$Z_1 = (u_1, \dots, u_k, \ u_{k+1}, \ w_{k+1}, \dots, u_{k+r}, \ w_{k+r})$$

Obviously Z_1 is an $n \times (k+2r)$ matrix with orthonormal columns that satisfy

$$\left.\begin{aligned} P_{M_1} Z_1 &= Z_1 \begin{pmatrix} I & & & 0 \\ & Q & & \\ & & \ddots & \\ 0 & & & Q \end{pmatrix} \\ \text{and} \qquad & \\ P_{M_2} Z_1 &= Z_1 \begin{pmatrix} I & & & 0 \\ & E(\theta_{k+1}) & & \\ & & \ddots & \\ 0 & & & E(\theta_{k+r}) \end{pmatrix} \end{aligned}\right\} \qquad (2.15)$$

Here Q and $E(\theta)$ are defined by (2.3) and I is a $k \times k$ unit matrix. Now those columns of Z that correspond to the diagonal matrices D_1 and D_2 will be taken care of. To make the discussion less abstract let us assume that $\dim(M_i) = m_i \quad i = 1,2$. The projection P_{M_i} is given as $Q_i Q_i^H$ where Q_i, i=1,2 are $n \times m_i$ matrices. The matrix X defined by (2.9) is an $m_1 \times (k+r)$-matrix. If m_1 is greater than $(k+r)$ then it is easy to find a $m_1 \times (m_1 - (k+r))$-matrix X' such that (X,X') is a unitary matrix. Take $Z_2' = Q_1X'$. It follows that $P_{M_1} Z_2' = Q_1Q_1{}^H Q_1 X' = Q_1X' = Z_2'$ and $P_{M_2} Z_2' = Q_2Q_2{}^H Q_1X' = Q_2 Y\Sigma \underbrace{X^H X'}_{=0} = 0$. Similarly, if m_2 is greater than $k+r$ then Y can be augmented to a unitary matrix (Y,Y'') and if $Z_2'' = Q_2Y''$ we get $P_{M_1} Z_2'' = 0; P_{M_2} Z_2'' = Z_2''$. The matrix (Z_1, Z_2', Z_2'') has $k + 2r + m_1 - (k+r) + m_2 - (k+r) = m_1+m_2-k$ orthogonal columns. If m_1+m_2-k is less than n then a matrix Z_2''' can easily be found so that $(Z_1, Z_2', Z_2'', Z_2''')$ is unitary. Since the column of Z''' are orthogonal to those of Q_1 and Q_2 we get

$$P_{M_1} Z''' = 0, \quad P_{M_2} Z''' = 0.$$

Take $Z_2 = (Z_2', Z_2'', Z_2''')$. Obviously

$$P_{M_i} Z_2 = D_i Z_2 , \quad i = 1,2 \tag{2.16}$$

where the matrices D_i, $i = 1,2$ have the desired properties. The theorem now follows from (2.15) and (2.16). A note on the geometry of the representation can be found in appendix 2.

Corollary

$$||P_{M_2} - P_{M_1}|| = \begin{cases} \sin\theta_{k+r} & \text{if } D_1 = D_2 = 0 \text{ or nonexistent} \\ 1 \text{ if } D_1 \neq 0 \text{ or } D_2 \neq 0 \end{cases}$$

If $\dim(M_1) \neq \dim(M_2)$ then $||P_{M_2} - P_{M_1}|| = 1$

If $\dim(M_1) = \dim(M_2)$ then

$||(I-P_{M_1})P_{M_2}|| = ||(I-P_{M_2})P_{M_1}|| = ||P_{M_2} - P_{M_1}||$

If $\dim(M_2) > \dim(M_1)$ then $||(I-P_{M_1})P_{M_2}|| = 1$

Proof Use the representation (2.2) and proceed as in the 2-dimensional case.

3. On metrics for angles between subspaces

Up to now only angles between one dimensional subspaces have been studied. Let us introduce the angle between a vector x and a subspace M.

Definition $\sphericalangle(x,M) = \inf_{y\in M} \sphericalangle(x,y) = \inf_{y\in M} \cos^{-1} \frac{|x^H y|}{||x||\,||y||}$

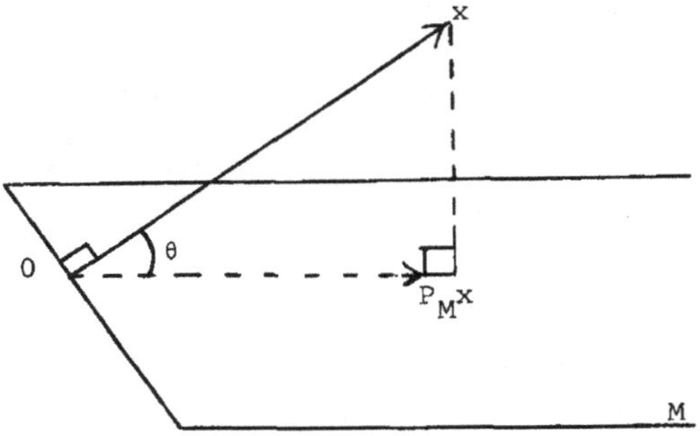

$\theta = \sphericalangle(x,M)$

Figure 4

As in three dimensional geometry it is easily proved that

$$\sphericalangle(x,M) = \sphericalangle(x,P_M x) = \cos^{-1}\frac{||P_M x||}{||x||} = \sin^{-1}\frac{||(I-P_M)x||}{||x||}$$

Hence

$$\sup_{x\in M_1} \sphericalangle(x,M_2) = \sin^{-1}||(I-P_{M_2})P_{M_1}||$$

Take

$$\left.\begin{aligned} \alpha_1 &= \sup_{x\in M_1} \sphericalangle(x,M_2) = \sin^{-1}||(I-P_{M_2})P_{M_1}|| \\ \alpha_2 &= \sup_{y\in M_2} \sphericalangle(y,M_1) = \sin^{-1}||(I-P_{M_1})P_{M_2}|| \end{aligned}\right\} \quad (3.1)$$

From the corollary of the representation theorem it is seen that

i) $\dim(M_1) = \dim(M_2) \Rightarrow \alpha_1 = \alpha_2$

ii) $\dim(M_1) < \dim(M_2) \Rightarrow \alpha_2 = \pi/2$

The function $\operatorname{dist}(M_1,M_2) = ||P_{M_2}-P_{M_1}||$ used in Kato [10] can certainly be used to measure the angle between subspaces M_1 and M_2 of the same dimension. But for subspaces of different dimensions it is completely useless since

(iii) $\operatorname{dist}(M_1,M_2) \equiv ||P_{M_2}-P_{M_1}|| = \max(\sin\alpha_1,\sin\alpha_2)$

(iv) $\dim(M_1) \neq \dim(M_2) \Rightarrow \operatorname{dist}(M_1,M_2) = 1$

Note that $\operatorname{dist}(x,M) = 1$ for the vector x and the two dimensional subspace M of figure 4 even if the angle θ between x and P_Mx goes to zero. The following definition of the angle between two subspaces also works for subspaces of different dimensions.

Definition

$$\measuredangle_2(M_1,M_2) = \text{minimum}(\alpha_1,\alpha_2)$$

where α_1 and α_2 are defined by (3.1).

Note that

$$\dim(M_1) = \dim(M_2) \rightarrow \operatorname{dist}(M_1,M_2) = \sin \measuredangle_2(M_1,M_2).$$

The angle function $\measuredangle_2(M_1,M_2)$ has the following metric properties:

1. $\measuredangle(M_1,M_2) \geq 0$ with equality iff $M_1 \supset M_2$ or $M_2 \supset M_1$
2. $\measuredangle(M_1,M_2) = \measuredangle(M_2,M_1)$
3. $\measuredangle(M_1,M_3) \leq \measuredangle(M_1,M_2) + \measuredangle(M_2,M_3)$

   if $\dim(M_1) \leq \dim(M_2) \leq \dim(M_3)$

   or $\dim(M_1) \geq \dim(M_2) \geq \dim(M_3)$
4. $\measuredangle(UM_1,UM_2) = \measuredangle(M_1,M_2)$ when U is unitary.

Property 1,2 and 4 are trivially true.

3. follows from the corresponding inequality for angles between vectors $\sphericalangle(u,w) \leq \sphericalangle(u,v) + \sphericalangle(v,w)$

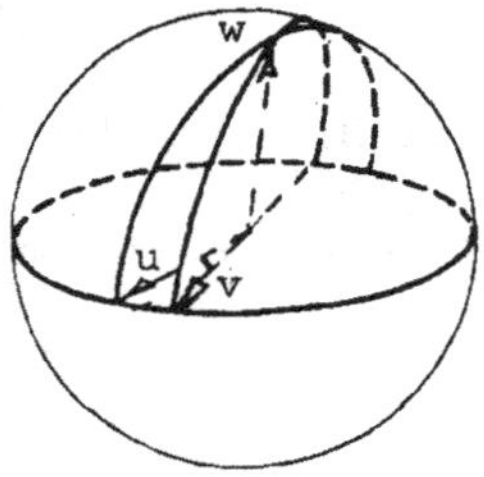

"Δ-inequality on the unit sphere"

Figure 5.

Every function $\sphericalangle(M_1,M_2)$ that satisfies the metrical properties 1-4 is said to be an angle function. Perhaps the most practical choice is the chordal metric used successfully here by Sun Ji-guang [9] :

$$\sphericalangle(M_1,M_2) = \sin(\sphericalangle_2(M_1,M_2)).$$

The metric properties for the chordal metric follow immediately from those for the $\sphericalangle_2$-metric. But there are even more subtle ways to define angle functions. Lars Andersson [2] noted that there is a one-one correspondence between vectors of $\Omega^k(V^*)$ = {alternative k-linear functionals of V^*} and subspaces of V.

If

$$\mathrm{Ann}(x) = \{v\in V:\ \underset{\text{wedge product}}{x\wedge v}=0\}$$

then every k-dimensional subspace of V is Ann(x) for some decomposable $x \in\Omega^k(V^*)$. Here a vector $x\in\Omega^k(V^*)$ is said to be decomposable if it can be written $x_1\wedge\ldots\wedge x_k$, $x_i\in V$. On the space Ω^k an innerproduct is canonically defined, corresponding to the geometry of V.
We can define the angle function like this

Definition

Let $m_1,m_2 \in\Omega^k(R^{n*})$, m_1,m_2 decomposable

$\mathrm{Ann}(m_i) = M_i\subset R^n$, $\dim(M_i) = k$

Take

$$\measuredangle(M_1,M_2) = \cos^{-1} \frac{|<m_1,m_2>|}{||m_1|| \ ||m_2||}$$

where $<m_1,m_2>$ is the innerproduct of m_1 and m_2.

It can be proved that the metric satisfies

$$\measuredangle(M_1,M_2) = \cos^{-1}(\cos\theta_1 \cdot \ \ldots \ \cdot \cos\theta_k)$$

where $\theta_1,\ldots,\theta_k$ are the principal angles of (M_1,M_2).

During the Pencil-conference it came to our knowledge that this metric was introduced by Lu [11] in 1963 and is used by Sun Ji Guang in his paper at this conference.

A triangular inequality for principal angles is derived in Wedin[17]. From the representation theorem and property 4 of the angle function it is known that every angle function depends only on the principal angles. It would be nice (but not very useful) to be able to characterize those functions of the principal angles that are angle functions. A similar characterization by von Neumann [7] says that a matrix norm is unitarily invariant if and only if it is a symmetric gauge function of the singular values of the matrix. The characterization of the angle function as a function of the principal angles is still an open question.

4. Perturbation identity + metric ⇒ perturbation bounds.

In section 4 the angle $\measuredangle(M_1,M_2)$ was defined for subspaces M_1 and M_2. Most often a subspace is defined as the column space of a matrix. To make the formulas look less clumsy let us change the notation slightly.

<u>Definition</u> Let A and B be nxp and nxq matrices, respectively. Take

$$\measuredangle (A,B) = \measuredangle (R(A),R(B))$$

and

$$P_A = P_{R(A)}$$

From the Δ-inequality 3) for angles between subspaces it is seen that

if rank(A) = rank(A+δA) and rank(B) = rank(B+δB)

then

$$|\measuredangle (A+\delta A,B+\delta B) - \measuredangle (A,B)| \leq \measuredangle (A+\delta A,A) + \measuredangle (B+\delta B,B) \tag{4.1}$$

In the rest of this paper take $\measuredangle(\cdot,\cdot) = \measuredangle_2(\cdot,\cdot)$.

How do we estimate $\measuredangle(A+\delta A,A)$?

The following perturbation identity is the clue:

$$(I-P_{A+\delta A})\underbrace{(A+\delta A-A)}_{=\delta A} = -(I-P_{A+\delta A})\underbrace{P_A\ A}_{=A} \tag{4.2}$$

Multiply the identity(4.2) from the right with the pseudo-inverse A^+ and use that $AA^+=P_A$ to get

$$(I-P_{A+\delta A})\delta A \cdot A^+ = -(I-P_{A+\delta A})P_A \tag{4.3}$$

Hence

$$\sin\measuredangle (A,A+\delta A) \leq ||(I-P_{A+\delta A})P_A|| \leq ||\delta A||\ ||A^+|| \tag{4.4}$$

The bound (4.4) can be attained for any matrix A and holds even if the rank of A+δA is different from that of A.

Let us now turn to a more difficult problem, that of studying how a subset of singular vectors of A are perturbed when the whole matrix A is perturbed. Let A_1 be that part of A that corresponds to these singular vectors, i.e. the matrix A is split up like this

$$A = A_0 + A_1 \text{ with } A_0^H A_1 = 0,\quad A_1 A_0^H = 0 \tag{4.5A}$$

Let the perturbation B of A have the corresponding decomposition

$$B = B_0 + B_1 \quad \text{with} \quad B_0^H B_1 = 0, \quad B_1 B_0^H = 0 \tag{4.5B}$$

If the perturbation $B_1 - A_1$ is known it is easy to use the identity (4.2) to estimate $\measuredangle(A_1,B_1)$. With the new notations the identity (4.2) can be written

$$(I-P_{B_1})(B_1-A_1) = -(I-P_{B_1})P_{A_1}A_1 \tag{4.6}$$

If only the complete perturbation $B - A$ is known it is not that simple.

Slightly different perturbation identities can be used to estimate the angle $\theta=\measuredangle(A_1,B_1)$.

The estimate of $\sin 2\theta$ in [5] is derived from the natural decomposition of $(I-P_{A_1})(B-A)P_{A_1^H}$ while the $\sin\theta$ theorem is derived from the decomposition of $(I-P_{B_1})(B-A)P_{A_1^H}$. These very small differences are of no importance at all since these different estimates of $|\delta\theta|/||\delta A||$ approach the same limit as $\delta A \to 0$. It is more important to study the first order perturbation identities and the related estimates to be able to find the relevant condition numbers as has been done by Stewart in several fundamental paper of which [14] is the most easily accessible. Before turning to perturbation identities relevant for the singular value decomposition let us point out that these identities can easily be reformulated using orthogonal matrices F_1, G_1 whose range are $R(A_1)$ and $R(A_1^H)$ respectively. Unfortunately this practically more useful formulation (see [15]) is beyond the scope of this paper. Now to the identities!

Take

$$X = (I-P_{A_1})P_{B_1} \quad \text{and} \quad Y = (I-P_{B_1^H})P_{A_1^H} \tag{4.7}$$

It is easily seen that

$$E_{21} \equiv (I-P_{B_1})(B-A)P_{A_1^H} = B_0 Y - X A_1 \tag{4.8}$$

Let us assume that there is a gap δ between the singular values of B_0 and A_1, i.e.

$$\delta = \sigma_{min}(A_1) - \sigma_{max}(B_0) > 0. \tag{4.9}$$

and that

$$\text{rank}(A_1) \leq \text{rank}(B_1) \tag{4.10}$$

Make the assumption

$$\text{rank}(A_1) \leq \text{rank}(B_1)$$

to get

$$||X|| = \sin \measuredangle (A_1, B_1); \quad ||Y|| = \sin \measuredangle (A_1^H, B_1^H) \tag{4.11}$$

From the identity (4.8) it is seen that

$$||X|| \leq (||E_{21}|| + ||B_0|| \; ||Y||)/\sigma_{min}(A_1) \tag{4.12}$$

From a similar identity

$$E_{12} \equiv P_{A_1}(B-A)(I-P_{B_1^H}) = X^H B_0 - A_1 Y^H \tag{4.13}$$

it is seen that

$$||Y|| \leq (||E_{12}|| + ||B_0|| \; ||X||)/\sigma_{min}(A_1) \tag{4.14}$$

The inequality

$$\max(||X||, ||Y||) \leq \frac{\max(||E_{12}||, ||E_{21}||)}{\delta} \tag{4.15}$$

is an immediate consequence of the two inequalities (4.12) and (4.14). When A and B are Hermitian, the inequality (4.15) becomes the $\sin\theta$ - theorem of Davis and Kahan [6]. The gap assumption (4.9) can be relaxed when the most convenient norm, the Frobenius norm, is used to define the metric for the angles i.e. $\measuredangle (A_1, B_1) = (\text{trace}(X^H X))^{\frac{1}{2}}$. In that case

$$\delta = \max_{i,k} |\sigma_i(A_1) - \sigma_k(B_0)| \tag{4.16}$$

can be used as the gap. But it is probably still an open question (see question 10.1 p 44 in [6]) whether the $\sin\theta$ -theorem p 225 [13] is as valid for the 2-norm as for the Frobenius norm.

It is my opinion that the central identities (4.8) and (4.11) are more interesting than the inequality (4.13) that was derived from them. The identities can be used actively to get new results and more insight, perhaps even algorithms, while the estimates can only be used passively to estimate how good an approximation is.

It is most important that the few basic identities behind most perturbation bounds are presented separetely and scrutinized in detail.

Acknowledgement. I would like to thank Karin Andersson for her careful typing and updating of the manuscript because of changes installed by me and the editors.

Appendix 1 A perturbation identity for oblique projections

Let S be the oblique projection on M_1 along $M_2^{\perp}$ where M_1 and M_2 are complimentary subspaces of C^n. Let $\tilde{M}_i$, $i=1,2$ be perturbations of M_i, $i=1,2$ and $\tilde{S}$ the oblique projection on $\tilde{M}_1$ along $\tilde{M}_2$. Assume that $\tilde{M}_i$ and M_i, $i=1,2$ have the same dimensions. It is easily seen that

$$S = S\,P_{M_2} = P_{M_1}\,S \tag{A1}$$

$$(I-S) = (I-P_{M_2})(I-S) = (I-S)(I-P_{M_1}) \tag{A2}$$

$$\tilde{S} - S = \tilde{S}(I-S) - (I-\tilde{S})S \tag{A3}$$

Hence

$$\tilde{S} - S = \tilde{S}\,P_{\tilde{M}_2}(I-P_{M_2})(I-S) - (I-\tilde{S})(I-P_{\tilde{M}_1})P_{M_1}S \tag{A4}$$

and

$$||\tilde{S}-S|| \leq \frac{1}{\cos\tilde{\theta}_{max}\cos\theta_{max}}(\sin \sphericalangle (M_1,\tilde{M}_1) + \sin \sphericalangle (M_2,\tilde{M}_2)) \tag{A5}$$

where

θ_{max} is the greatest angle between M_1 and M_2.

and $\tilde{\theta}_{max}$ is the greatest angle between $\tilde{M}_1$ and $\tilde{M}_2$.

It can be shown that the estimate (A5) is fairly sharp. As was said before in section 5 the perturbation identity (A4) is much more interesting than the estimate (A5) that follows from it. The perturbation theory of oblique projections will be treated in detail in a forthcoming report. The results will there be used to study the perturbation of the eigenvalue spectrum of an unsymmetric matrix.

Appendix 2 Geometry of subspaces

$M_1 \cap M_2$	has dimension k and is spanned by $(u_1,\dots,u_k)$
$P_{M_1}M_2$	has dimension k+r and is spanned by $(u_1,\dots,u_{k+r})$
$P_{M_2}M_1$	has dimension k+r and is spanned by $(v_1,\dots,v_{k+r})$
$M_1 \cap M_2^{\perp}$	has dimension $m_1-(k+r)$ and is spanned by the columns of Z_2'
$M_1^{\perp} \cap M_2$	has dimension $m_2-(k+r)$ is spanned by the columns of Z_2''
$M_1^{\perp} \cap M_2^{\perp}$	has dimension $n-(m_1+m_2-k)$ and is spanned by Z_2'''

Together with the columns of Z_2'' and Z_2''' the vectors $w_{k+1},\dots,w_{k+r}$ span $M_1^{\perp}$.

References

1. S.N. Afriat, Orthogonal oblique projectors and the characteristics of pairs of vector spaces, Proc. Cambridge Philos. Soc. v. 53(1957), pp 800-816.

2. L. Andersson, The concepts of angle between subspaces ... unpublished notes, Umeå(1980).

3. Adi Ben-Israel, On the geometry of subspaces in Euclidean n-Space, SIAM J. Appl. Math. Vol 15, (1967), pp 1184-1198.

4. Å. Björck, G.H. Golub, Numerical methods for computing angles between linear subspaces, Math. Comp. 27(1973), pp 579-594.

5. Ch. Davis, W.M. Kahan, Some new bounds on perturbation of subspaces Bulletin of the American Math. Society 75(1969), pp 863-868.

6. Ch. Davis, W.M. Kahan, The rotation of eigenvectors by a perturbation III, SIAM J. Numer. Anal. 7(1970), pp 1-46.

7. J.C. Gohberg, M.G. Krein, Introduction to the theory of linear nonselfadjoint operators, Nauka, Moscow 1965, English transl. Math Monographs vol. 18 Amer. Math. Soc., theorem 3.1 p 78, Providence, R.I.,(1969).

8. A.S. Householder, The theory of matrices in numerical analysis, Blaisdell Publishing Company, New York, 1964.

9. Sun Ji Guang, Perturbation analysis for the generalized singular value problem, to appear in SIAM J. Numer. Anal. (1982).

10. T. Kato, Perturbation theory for linear operators, Springer-Verlag Berlin (1966).

11. Q.K. Lu, The elliptic geometry of extended spaces, Acta Math. Sinica, 13(1963), 49-62; translated as Chinese Math. 4(1963), 54-69.

12. L. Mirsky, Symmetric guage functions and unitarily invariant norms, Quart.J.Math., Oxford(2), 11(1960), pp 50-59.

13. B. Parlett, The symmetric eigenvalue problem, chap 11-7, pp 222-225, Prentice-Hall, N.J.(1980).

14. G.W. Stewart, Error and perturbation bounds for subspaces associate with certain eigenvalue problems, SIAM Rev., 15, pp 727-764.

15. P.Å. Wedin, Perturbation bounds in connection with singular value decomposition, BIT 12(1972), 99-111.

16. P.Å. Wedin, A geometrical approach to angles in finite dimensional inequality for angles, UMINF-66.78, Inst. of Inf. proc., Umeå University (1978).

THE MULTIVARIATE CALIBRATION PROBLEM IN CHEMISTRY SOLVED BY THE PLS METHOD

S. Wold, H. Martens and H. Wold

Institute of Chemistry, Umea University, Umea, Sweden,
Norwegian Food Research Institute, As, Norway, and
Department of Statistics, Uppsala University, Sweden and
Departement d'Econometrie, Université de Genève, Switzerland

Introduction.

In chemistry, slow and specialized "wet chemistry" methods are rapidly substituted by fast and general instrumentalised methods (Kowalski, 1975). One common problem is how to use spectroscopy to determine the concentrations of various constituents in complicated samples. When the constituents don't absorb light in separated frequency regions, one must utilize a combination of many spectral frequencies to estimate the concentrations. The problem of how to optimally combine the absorbtions at several frequencies (or other chemical "sensors") in order to approximate a measured set of concentrations is called the multivariate calibration problem.

In this problem we have found the PLS method with two blocks in mode A (H.Wold, 1982), to be very useful. PLS=Partial Least Squares models in latent variables. In a PLS analysis, the data matrix is divided variable-wise into a number of blocks. These blocks are in Mode A modelled by principal components like models as shown in fig.1 for the two block case ($X=TB+E$ and $Y=UC+F$; E and F denoting matrices of residuals). The blocks are related to each other by relations in the latent variables; here between the column vectors u and t (see fig. 1).

The PLS method has not been analysed in strict numerical detail, nor are its statistical properties well investigated. We have found the PLS method work better in several chemical applications of the multivariate calibration problem than traditional methods, i.e. multiple regression (MR), principal components regression (PCR), and ridge regression (RR). See Draper and Smith (1981) for the details of these methods. Hence we wish to present the PLS method to numerical analysists in order to draw the attention to the method and to the fact that its performance is not well investigated.

As an example, we use the measurement of the contents of protein and water in a sample of grain. Traditionally, one analyses protein according to Kjelldahl and water by weighing fresh and dried samples. In an alternative spectroscopic method, one leads infrared light through the sample and measures the light absorbtion at 12 frequencies. In a prototype instrument, one used 34 samples with varying protein and water content as calibration samples. These samples were first analysed by the traditional "wet" methods to estimate the contents of the two constituents. Another 34 samples were used as test samples to evaluate the calibration.

Notation.

The spectroscopic data with p variables measured on the n samples in the calibration set are collected in the n x p matrix X. The n x r matrix Y contains the concentrations of the r constituents. See further figure 1.

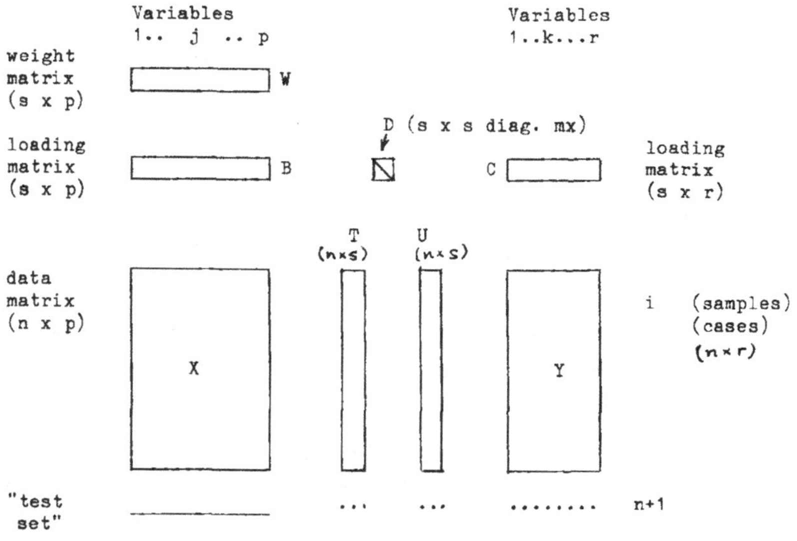

FIGURE 1. Matrices and notation in the PLS two block problem.

Traditional approach.

A linear model in X is formulated for each of the y-variables and estimated by multiple regression (MR). Since the matrix X has less than full rank, however, this regression problem is almost singular and has to be regularized in some way. This is usually done by either selecting a few of the variables in X to get a set which is as little collinear as possible (step-wise regression) or by representing X by its first q principal components (PCR), or by adding a small constant, K, to the diagonal elements in the moment matrix $X'X$ before its inversion, ridge regression (RR).

Numerical example.

To illustrate the details of the methods, we use the small data set of Marquardt (1970) extended with one additional y-variable. The x-values are 3, 4 and 5 divided by $\sqrt{50}$, shown in decimal approximation below. Note that in this example, the variables are used in their non-centered form.

TABLE 1. Data matrix used as numerical example.

i	x_1	x_2	y_1	y_2
1	.424264	.565685	1	1
2	.565685	.424264	2	1.5
3	.707101	.707101	3	2

The regression coefficients in separate MR for y_1 and y_2 are (5.357,-1.714) and (3.107,-.4285) with residual variances 0.364 and 0.0227, respectively.

The PLS approach.

Like in PCR, the matrix X is -- after possible centering and scaling -- modelled by the product of two smaller matrices T and B. In PLS, the Y matrix is modelled in the same way by the product of the two matrices U and C (figure 1). The matrices B and C are calculated so that both $\|X-TB\|$ and $\|Y-UC\|$ are "small" and so that U and T are correlated with each other column-wise more than if they are calculated by separate PC analyses of X and Y. What is meant by "small" depends on the problem; in the present example, one wished to have the norm of the predictive y-residuals, $||Y-TDC||$, be less than 0.25 times $||Y||$.

The main differences between PLS and PCR are that in PCR (1) B and T are calculated only to make $||X-TB||$ "small" and (2) that separate regressions are made for each y-variable.

In PLS a weight matrix W is used in the intermediate calculations, together with a diagonal matrix D. The number of significant dimensions, s, is determined by cross-validation (Stone, 1974, Wahba, 1977, S.Wold, 1978, H.Wold, 1982).

The PLS analysis can be seen as modelling X by TB plus a "noise" matrix of deviations from the model, E. In each dimension, the "latent variable of the Y-block", u, is modelled as u=dt. Hence, at the same time, Y is modelled by TDC except for the residuals F. In this way PLS gives a biased solution to canonical correlation.

For a single y-variable, both PCR and PLS converge towards the multiple regression solution as the dimensionality of the model, s (see fig. 1), goes towards the number of variables in X, p. We believe that a similar situation is at hand with the multivariate Y-case, as seen in the small numerical example, but we have not yet proven this conjecture.

Algorithm.

The PLS method estimates one "dimension" in the model at a time. This dimension consists of one row-vector (b, c, w) in each of the matrices B, C, and W, one column-vector in the matrices T and U (t,u), plus the scalar d which is the corresponding diagonal element in D (see fig. 1). The statistical interpretation of each step is included as a possible clarification. Thus, with PLS, mode A, for two blocks X and Y, the algorithm in the calibration phase is:

(1). Usually, X and Y are centered by subtracting the average for each column. This is not done with the numerical example.

(2). Set the dimension index, s, to zero. Scale each column to variance one. Other scalings may be used if prior information exists about the importance of the variables. In the numerical example, the data are not scaled, but used as shown in table 1, above.

(3). Increment s by one. In each dimension, iterative criss-cross regressions are used to estimate w, t, c and u analogously to the power method.

(3a). Use the first column in Y as starting vector for the Y-block latent variable vector, u. $u_{start}=(1,2,3)'$

(3b). Calculate the X-weight vector w as u'X. w=(.26264,.25254). Thus w_j is the regression coefficient of u in x_j (x-variable j). Normalize w to length one. $w_{norm}=(.72083,.6931)$

(3c). Calculate the X-block latent variable vector t as Xw'. Thus t_i is the regression coefficient of w in x_i (i.th sample x-vector). t=(.6979,.7018,1)'

(3d). Calculate the Y weight and loading vector c as t'Y. c=(2.577,1.895). Thus c_k is the regression coefficient of t in y_k (the k.th y-variable). Normalize c to length one. $c_{norm}=(.8057,.5923)$.

(3e). Calculate a new Y-block latent variable vector u as Yc'. u=(1.398, 2.5, 3.602). Thus u_i is the regression coefficient of c in y_i (the i.th sample y-vector). If convergence, i.e. $\|u-u_{old}\| < 10^2 \mathrm{MACHEP} \|u\|$, then step 4, else back to step 3a. (MACHEP is the precision of the computer, here 10^{-10}). Convergence was reached after 3 iterations, giving the final vectors: w=(.7193,.6947), c=(.8057, 0.5923), t=(.6982,.7016,.9998)' and u=(1.398,2.5,3.602).

(4). Calculate the X-block loading vector b as t'X. b=(.7073,.7071). Use w, t, u, c and b as the s.th vectors in the matrices W, T, U, C and B, respectively. X is now modelled by tb and Y by uc. Since u is modelled as dt by the "inner PLS relation", one can see Y as modelled also by tdc (d is calculated in step 5, below).

(5). Calculate d (the s.th diagonal element in the matrix D) as the regression coefficient of t in u. d=t'u/(t't). $d_1=3.1986$

(6). Form the residuals in X and Y and use them as new X and Y in the next dimension. X:=X-tb and Y:=Y-tdc. If these residuals are sufficiently small or if the last dimension was insignificant according to cross validation (CV), then stop. Else back to step 3.

With CV, parts of the data (here every third row in Y) are kept out of the estimation of one PLS dimension (see below) and the parameter vectors are estimated from the remaining data. Then, the "kept out" y-values are predicted from the corresponding x-values and the estimated model. The sum of squared differences between "observed" and calculated y-values is formed. Then the deletion is rotated to the second part of Y, the calculations repeated, etc., until after a number of rounds (here three) the sum of squares contains one and only one term from each Y-element. If SS is smaller that $||Y||$, the model at the present dimension is judged to have predictive relevance.

With the numerical example, convergence was reached in the second dimension after two iterations, giving the second set of vectors: w=(.6947,-.7193), c=(.8952, .4457), t=(-.1001, .09989, -.00018)´, u=(-.8592, .2478, .4261), b=(.6947, -.7193) and d_2=5.5345. This set of vectors are non-significant according to CV. They are included here to show that in this example the two-dimensional PLS solution coincides with the MR solutions for the individual y-variables. The corresponding regression coefficients of the PLS solution were calculated as described in S.Wold et.al. (1982). They agree with the MR coefficients better than one in the sixth significant figure (calculated on a 8-bit computer with 6 decimals precision).

Phase 2.

In the prediction phase, each "test set" data vector $\underline{x}$ is analysed as follows (after centering and scaling):

(1a). Measure the similarity of the new sample vector x to the calibration set by least squares fitting x to the b-vectors. The residuals from this fitting are $e = x - xB'B/||B||$. If e greatly exceed E from phase 1 (see S.Wold et.al. 1976, 1977), then the sample is dissimilar to the calibration set and the predictions in step 1b are unreliable.

(1b). Calculate predictions of the vector y for the test sample as: (i) set the predicted y-vector to zero: y= 0. (ii) for each dimension, h=1 to s, calculate the t-value: $t = xw_h'$, update y as: $y := y + d_h t_h c_h$ and then update x as: $x := x - t_h b_h$. Consider as an example the "test set" x-vector (1,1). It fits the one dimensional PLS model with the residuals e=(-.00018, .00018). Their standard deviation is considerably smaller than the standard deviation of the residuals of X after one dimension, 0.14. Hence, we conclude that the fit is OK and calculate the predicted Y-values as described above to be (3.644, 2.678).

The predicted values of y are now obtained in scaled and centered form and should hence be transformed back to the original coordinates by applying the reverse of the centering and scaling calculated in the calibration phase.

Example.

The chemical data set described above was scaled to variance one in both y-variables (protein and water content). The models gave the following fit to the 34 cases in the calibration set, expressed as residual standard deviation of Y: MR=0.14. The MR solution was nearly singular and gave "wild" results with the ordinary Gauss elimination routine for matrix inversion. The conjugate gradient method gave stable results, however. The other methods gave: RR=0.16, PCR=0.20 (six dimensions), PLS=0.15 (eight dimensions). The predicted Y-values in the evaluation set differed from the actual values with the following standard deviations: MR=0.28, RR=0.32, PCR=0.36 , PLS=0.28. This result was surprising to us; the fact that the "bad predictive properties" of MR can be cured by proper computations is often overlooked in the statistical literature. Without outliers (see below) PLS=0.23.

Discussion.

The PLS method solves the multivariate calibration problem by an approach similar to principal components regression. In contrast to PCR, however, the PLS solution is calculated using the information in Y. If X contains structure which has a predictive relevance for Y (e.g. according to cross-validation), this structure may be found by the PLS solution even if it is contained in singular vectors corresponding to small singular values in the ordinary singular value decomposition of X. This may be the reason why PLS often gives better results than RR and PCR.

In the present example, the number of cases in the calibration set (n=34) was larger than the number of x-variables (p=12). Hence MR could still be used with good results. In many cases, however, p exceeds n, and MR cannot be used at all. PLS and PCR still work well; in fact they give more precise results the larger the number of x-variables (H.Wold, 1982).

An interesting feature with PLS is that the matrix B can be used to classify new samples (with the data in row vectors x) as similar to the calibration set or not (S.Wold et al, 1976, 1977). Thus if the fit of a new sample to the X-part of the model leaves large residuals, i.e. $||x-xB'B||$ are statistically large compared to $||X-XB'B||$ from the calibration phase, this is an indication that the new sample is dissimilar to the calibration set and therefore the predicted y-values are unreliable. In the example, eight of the "test" samples fitted the calibration model badly according to an F-test. These samples also were the ones obtaining the worst predictions for Y. When removed, the PLS predictions for the remaining samples improved by about 20%.

References.

N.R. DRAPER and H. SMITH (1981). Applied regression analysis, 2.ed. Wiley, New York.

B.R. KOWALSKI (1975). Chemometrics: Views and propositions. J.Chem.Info.Comput. Syst. 15, 201.

D.W. MARQUARDT (1970). Generalized inverses, ridge regression, biased linear and non-linear estimation. Technometrics 12, 591-612.

M. STONE (1974). Cross-validatory choice and assessment of statistical predictions. J.Roy.Statist.Soc. B36, 111.

G. WAHBA (1977). Practical approximate solution to linear operator equations when the data are noisy. SIAM J.Numer.Anal. 14, 651.

H. WOLD (1982) in Systems under indirect observation. Causality, structure, prediction (K.G.Jöreskog and H.Wold, Ed.s). North-Holland, Amsterdam.

S. WOLD (1976). Pattern recognition by means of disjoint principal components models. Pattern Recognition 8, 127.

S. WOLD and M. SJÖSTRÖM (1977). SIMCA, a method for analyzing chemical data in terms of similarity and analogy. In Chemometrics: Theory and application (B.R. Kowalski, Ed.). Amer.Chem.Soc.Symp.Ser. no. 52.

S. WOLD (1978). Crossvalidatory estimation of the number of components in factor and principal components analysis. Technometrics 20, 397.

S. WOLD, H. WOLD, W.J. DUNN, A. RUHE (1982). The collinearity problem in linear regression. The partial least squares (PLS) approach to generalized inverses. Report UMINF-83.80, version 3. Umeå University, Dept. Information Processing.

Vol. 817: L. Gerritzen, M. van der Put, Schottky Groups and Mumford Curves. VIII, 317 pages. 1980.

Vol. 818: S. Montgomery, Fixed Rings of Finite Automorphism Groups of Associative Rings. VII, 126 pages. 1980.

Vol. 819: Global Theory of Dynamical Systems. Proceedings, 1979. Edited by Z. Nitecki and C. Robinson. IX, 499 pages. 1980.

Vol. 820: W. Abikoff, The Real Analytic Theory of Teichmüller Space. VII, 144 pages. 1980.

Vol. 821: Statistique non Paramétrique Asymptotique. Proceedings, 1979. Edited by J.-P. Raoult. VII, 175 pages. 1980.

Vol. 822: Séminaire Pierre Lelong–Henri Skoda, (Analyse) Années 1978/79. Proceedings. Edited by P. Lelong et H. Skoda. VIII, 356 pages, 1980.

Vol. 823: J. Král, Integral Operators in Potential Theory. III, 171 pages. 1980.

Vol. 824: D. Frank Hsu, Cyclic Neofields and Combinatorial Designs. VI, 230 pages. 1980.

Vol. 825: Ring Theory, Antwerp 1980. Proceedings. Edited by F. van Oystaeyen. VII, 209 pages. 1980.

Vol. 826: Ph. G. Ciarlet et P. Rabier, Les Equations de von Kármán. VI, 181 pages. 1980.

Vol. 827: Ordinary and Partial Differential Equations. Proceedings, 1978. Edited by W. N. Everitt. XVI, 271 pages. 1980.

Vol. 828: Probability Theory on Vector Spaces II. Proceedings, 1979. Edited by A. Weron. XIII, 324 pages. 1980.

Vol. 829: Combinatorial Mathematics VII. Proceedings, 1979. Edited by R. W. Robinson et al.. X, 256 pages. 1980.

Vol. 830: J. A. Green, Polynomial Representations of GL_n. VI, 118 pages. 1980.

Vol. 831: Representation Theory I. Proceedings, 1979. Edited by V. Dlab and P. Gabriel. XIV, 373 pages. 1980.

Vol. 832: Representation Theory II. Proceedings, 1979. Edited by V. Dlab and P. Gabriel. XIV, 673 pages. 1980.

Vol. 833: Th. Jeulin, Semi-Martingales et Grossissement d'une Filtration. IX, 142 Seiten. 1980.

Vol. 834: Model Theory of Algebra and Arithmetic. Proceedings, 1979. Edited by L. Pacholski, J. Wierzejewski, and A. J. Wilkie. VI, 410 pages. 1980.

Vol. 835: H. Zieschang, E. Vogt and H.-D. Coldewey, Surfaces and Planar Discontinuous Groups. X, 334 pages. 1980.

Vol. 836: Differential Geometrical Methods in Mathematical Physics. Proceedings, 1979. Edited by P. L. García, A. Pérez-Rendón, and J. M. Souriau. XII, 538 pages. 1980.

Vol. 837: J. Meixner, F. W. Schäfke and G. Wolf, Mathieu Functions and Spheroidal Functions and their Mathematical Foundations Further Studies. VII, 126 pages. 1980.

Vol. 838: Global Differential Geometry and Global Analysis. Proceedings 1979. Edited by D. Ferus et al. XI, 299 pages. 1981.

Vol. 839: Cabal Seminar 77 – 79. Proceedings. Edited by A. S. Kechris, D. A. Martin and Y. N. Moschovakis. V, 274 pages. 1981.

Vol. 840: D. Henry, Geometric Theory of Semilinear Parabolic Equations. IV, 348 pages. 1981.

Vol. 841: A. Haraux, Nonlinear Evolution Equations- Global Behaviour of Solutions. XII, 313 pages. 1981.

Vol. 842: Séminaire Bourbaki vol. 1979/80. Exposés 543–560. IV, 317 pages. 1981.

Vol. 843: Functional Analysis, Holomorphy, and Approximation Theory. Proceedings. Edited by S. Machado. VI, 636 pages. 1981.

Vol. 844: Groupe de Brauer. Proceedings. Edited by M. Kervaire and M. Ojanguren. VII, 274 pages. 1981.

Vol. 845: A. Tannenbaum, Invariance and System Theory: Algebraic and Geometric Aspects. X, 161 pages. 1981.

Vol. 846: Ordinary and Partial Differential Equations, Proceedings. Edited by W. N. Everitt and B. D. Sleeman. XIV, 384 pages. 1981.

Vol. 847: U. Koschorke, Vector Fields and Other Vector Bundle Morphisms – A Singularity Approach. IV, 304 pages. 1981.

Vol. 848: Algebra, Carbondale 1980. Proceedings. Ed. by R. K. Amayo. VI, 298 pages. 1981.

Vol. 849: P. Major, Multiple Wiener-Itô Integrals. VII, 127 pages. 1981.

Vol. 850: Séminaire de Probabilités XV. 1979/80. Avec table générale des exposés de 1966/67 à 1978/79. Edited by J. Azéma and M. Yor. IV, 704 pages. 1981.

Vol. 851: Stochastic Integrals. Proceedings, 1980. Edited by D. Williams. IX, 540 pages. 1981.

Vol. 852: L. Schwartz, Geometry and Probability in Banach Spaces. X, 101 pages. 1981.

Vol. 853: N. Boboc, G. Bucur, A. Cornea, Order and Convexity in Potential Theory: H-Cones. IV, 286 pages. 1981.

Vol. 854: Algebraic K-Theory. Evanston 1980. Proceedings. Edited by E. M. Friedlander and M. R. Stein. V, 517 pages. 1981.

Vol. 855: Semigroups. Proceedings 1978. Edited by H. Jürgensen, M. Petrich and H. J. Weinert. V, 221 pages. 1981.

Vol. 856: R. Lascar, Propagation des Singularités des Solutions d'Equations Pseudo-Différentielles à Caractéristiques de Multiplicités Variables. VIII, 237 pages. 1981.

Vol. 857: M. Miyanishi. Non-complete Algebraic Surfaces. XVIII, 244 pages. 1981.

Vol. 858: E. A. Coddington, H. S. V. de Snoo: Regular Boundary Value Problems Associated with Pairs of Ordinary Differential Expressions. V, 225 pages. 1981.

Vol. 859: Logic Year 1979–80. Proceedings. Edited by M. Lerman, J. Schmerl and R. Soare. VIII, 326 pages. 1981.

Vol. 860: Probability in Banach Spaces III. Proceedings, 1980. Edited by A. Beck. VI, 329 pages. 1981.

Vol. 861: Analytical Methods in Probability Theory. Proceedings 1980. Edited by D. Dugué, E. Lukacs, V. K. Rohatgi. X, 183 pages. 1981.

Vol. 862: Algebraic Geometry. Proceedings 1980. Edited by A. Libgober and P. Wagreich. V, 281 pages. 1981.

Vol. 863: Processus Aléatoires à Deux Indices. Proceedings, 1980. Edited by H. Korezlioglu, G. Mazziotto and J. Szpirglas. V, 274 pages. 1981.

Vol. 864: Complex Analysis and Spectral Theory. Proceedings, 1979/80. Edited by V. P. Havin and N. K. Nikol'skii, VI, 480 pages. 1981.

Vol. 865: R. W. Bruggeman, Fourier Coefficients of Automorphic Forms. III, 201 pages. 1981.

Vol. 866: J.-M. Bismut, Mécanique Aléatoire. XVI, 563 pages. 1981.

Vol. 867: Séminaire d'Algèbre Paul Dubreil et Marie-Paule Malliavin. Proceedings, 1980. Edited by M.-P. Malliavin. V, 476 pages. 1981.

Vol. 868: Surfaces Algébriques. Proceedings 1976–78. Edited by J. Giraud, L. Illusie et M. Raynaud. V, 314 pages. 1981.

Vol. 869: A. V. Zelevinsky, Representations of Finite Classical Groups. IV, 184 pages. 1981.

Vol. 870: Shape Theory and Geometric Topology. Proceedings, 1981. Edited by S. Mardešić and J. Segal. V, 265 pages. 1981.

Vol. 871: Continuous Lattices. Proceedings, 1979. Edited by B. Banaschewski and R.-E. Hoffmann. X, 413 pages. 1981.

Vol. 872: Set Theory and Model Theory. Proceedings, 1979. Edited by R. B. Jensen and A. Prestel. V, 174 pages. 1981.

Vol. 873: Constructive Mathematics, Proceedings, 1980. Edited by F. Richman. VII, 347 pages. 1981.

Vol. 874: Abelian Group Theory. Proceedings, 1981. Edited by R. Göbel and E. Walker. XXI, 447 pages. 1981.

Vol. 875: H. Zieschang, Finite Groups of Mapping Classes of Surfaces. VIII, 340 pages. 1981.

Vol. 876: J. P. Bickel, N. El Karoui and M. Yor. Ecole d'Eté de Probabilités de Saint-Flour IX – 1979. Edited by P. L. Hennequin. XI, 280 pages. 1981.

Vol. 877: J. Erven, B.-J. Falkowski, Low Order Cohomology and Applications. VI, 126 pages. 1981.

Vol. 878: Numerical Solution of Nonlinear Equations. Proceedings, 1980. Edited by E. L. Allgower, K. Glashoff, and H.-O. Peitgen. XIV, 440 pages. 1981.

Vol. 879: V. V. Sazonov, Normal Approximation – Some Recent Advances. VII, 105 pages. 1981.

Vol. 880: Non Commutative Harmonic Analysis and Lie Groups. Proceedings, 1980. Edited by J. Carmona and M. Vergne. IV, 553 pages. 1981.

Vol. 881: R. Lutz, M. Goze, Nonstandard Analysis. XIV, 261 pages. 1981.

Vol. 882: Integral Representations and Applications. Proceedings, 1980. Edited by K. Roggenkamp. XII, 479 pages. 1981.

Vol. 883: Cylindric Set Algebras. By L. Henkin, J. D. Monk, A. Tarski, H. Andréka, and I. Németi. VII, 323 pages. 1981.

Vol. 884: Combinatorial Mathematics VIII. Proceedings, 1980. Edited by K. L. McAvaney. XIII, 359 pages. 1981.

Vol. 885: Combinatorics and Graph Theory. Edited by S. B. Rao. Proceedings, 1980. VII, 500 pages. 1981.

Vol. 886: Fixed Point Theory. Proceedings, 1980. Edited by E. Fadell and G. Fournier. XII, 511 pages. 1981.

Vol. 887: F. van Oystaeyen, A. Verschoren, Non-commutative Algebraic Geometry. VI, 404 pages. 1981.

Vol. 888: Padé Approximation and its Applications. Proceedings, 1980. Edited by M. G. de Bruin and H. van Rossum. VI, 383 pages. 1981.

Vol. 889: J. Bourgain, New Classes of $\mathcal{L}^p$-Spaces. V, 143 pages. 1981.

Vol. 890: Model Theory and Arithmetic. Proceedings, 1979/80. Edited by C. Berline, K. McAloon, and J.-P. Ressayre. VI, 306 pages. 1981.

Vol. 891: Logic Symposia, Hakone, 1979, 1980. Proceedings, 1979, 1980. Edited by G. H. Müller, G. Takeuti, and T. Tugué. XI, 394 pages. 1981.

Vol. 892: H. Cajar, Billingsley Dimension in Probability Spaces. III, 106 pages. 1981.

Vol. 893: Geometries and Groups. Proceedings. Edited by M. Aigner and D. Jungnickel. X, 250 pages. 1981.

Vol. 894: Geometry Symposium. Utrecht 1980, Proceedings. Edited by E. Looijenga, D. Siersma, and F. Takens. V, 153 pages. 1981.

Vol. 895: J.A. Hillman, Alexander Ideals of Links. V, 178 pages. 1981.

Vol. 896: B. Angéniol, Familles de Cycles Algébriques – Schéma de Chow. VI, 140 pages. 1981.

Vol. 897: W. Buchholz, S. Feferman, W. Pohlers, W. Sieg, Iterated Inductive Definitions and Subsystems of Analysis: Recent Proof-Theoretical Studies. V, 383 pages. 1981.

Vol. 898: Dynamical Systems and Turbulence, Warwick, 1980. Proceedings. Edited by D. Rand and L.-S. Young. VI, 390 pages. 1981.

Vol. 899: Analytic Number Theory. Proceedings, 1980. Edited by M.I. Knopp. X, 478 pages. 1981.

Vol. 900: P. Deligne, J. S. Milne, A. Ogus, and K.-Y. Shih, Hodge Cycles, Motives, and Shimura Varieties. V, 414 pages. 1982.

Vol. 901: Séminaire Bourbaki vol. 1980/81 Exposés 561–578. III, 299 pages. 1981.

Vol. 902: F. Dumortier, P.R. Rodrigues, and R. Roussarie, Germs of Diffeomorphisms in the Plane. IV, 197 pages. 1981.

Vol. 903: Representations of Algebras. Proceedings, 1980. Edited by M. Auslander and E. Lluis. XV, 371 pages. 1981.

Vol. 904: K. Donner, Extension of Positive Operators and Korovkin Theorems. XII, 182 pages. 1982.

Vol. 905: Differential Geometric Methods in Mathematical Physics. Proceedings, 1980. Edited by H.-D. Doebner, S.J. Andersson, and H.R. Petry. VI, 309 pages. 1982.

Vol. 906: Séminaire de Théorie du Potentiel, Paris, No. 6. Proceedings. Edité par F. Hirsch et G. Mokobodzki. IV, 328 pages. 1982.

Vol. 907: P. Schenzel, Dualisierende Komplexe in der lokalen Algebra und Buchsbaum-Ringe. VII, 161 Seiten. 1982.

Vol. 908: Harmonic Analysis. Proceedings, 1981. Edited by F. Ricci and G. Weiss. V, 325 pages. 1982.

Vol. 909: Numerical Analysis. Proceedings, 1981. Edited by J.P. Hennart. VII, 247 pages. 1982.

Vol. 910: S.S. Abhyankar, Weighted Expansions for Canonical Desingularization. VII, 236 pages. 1982.

Vol. 911: O.G. Jørsboe, L. Mejlbro, The Carleson-Hunt Theorem on Fourier Series. IV, 123 pages. 1982.

Vol. 912: Numerical Analysis. Proceedings, 1981. Edited by G. A. Watson. XIII, 245 pages. 1982.

Vol. 913: O. Tammi, Extremum Problems for Bounded Univalent Functions II. VI, 168 pages. 1982.

Vol. 914: M. L. Warshauer, The Witt Group of Degree k Maps and Asymmetric Inner Product Spaces. IV, 269 pages. 1982.

Vol. 915: Categorical Aspects of Topology and Analysis. Proceedings, 1981. Edited by B. Banaschewski. XI, 385 pages. 1982.

Vol. 916: K.-U. Grusa, Zweidimensionale, interpolierende Lg-Splines und ihre Anwendungen. VIII, 238 Seiten. 1982.

Vol. 917: Brauer Groups in Ring Theory and Algebraic Geometry. Proceedings, 1981. Edited by F. van Oystaeyen and A. Verschoren. VIII, 300 pages. 1982.

Vol. 918: Z. Semadeni, Schauder Bases in Banach Spaces of Continuous Functions. V, 136 pages. 1982.

Vol. 919: Séminaire Pierre Lelong – Henri Skoda (Analyse) Années 1980/81 et Colloque de Wimereux, Mai 1981. Proceedings. Edité par P. Lelong et H. Skoda. VII, 383 pages. 1982.

Vol. 920: Séminaire de Probabilités XVI, 1980/81. Proceedings. Edité par J. Azéma et M. Yor. V, 622 pages. 1982.

Vol. 921: Séminaire de Probabilités XVI, 1980/81. Supplément: Géométrie Différentielle Stochastique. Proceedings. Edité par J. Azéma et M. Yor. III, 285 pages. 1982.

Vol. 922: B. Dacorogna, Weak Continuity and Weak Lower Semicontinuity of Non-Linear Functionals. V, 120 pages. 1982.

Vol. 923: Functional Analysis in Markov Processes. Proceedings, 1981. Edited by M. Fukushima. V, 307 pages. 1982.

Vol. 924: Séminaire d'Algèbre Paul Dubreil et Marie-Paule Malliavin. Proceedings, 1981. Edité par M.-P. Malliavin. V, 461 pages. 1982.

Vol. 925: The Riemann Problem, Complete Integrability and Arithmetic Applications. Proceedings, 1979-1980. Edited by D. Chudnovsky and G. Chudnovsky. VI, 373 pages. 1982.

Vol. 926: Geometric Techniques in Gauge Theories. Proceedings, 1981. Edited by R. Martini and E.M.de Jager. IX, 219 pages. 1982.

Vol. 927: Y. Z. Flicker, The Trace Formula and Base Change for GL (3). XII, 204 pages. 1982.

Vol. 928: Probability Measures on Groups. Proceedings 1981. Edited by H. Heyer. X, 477 pages. 1982.

Vol. 929: Ecole d'Eté de Probabilités de Saint-Flour X – 1980. Proceedings, 1980. Edited by P.L. Hennequin. X, 313 pages. 1982.

Vol. 930: P. Berthelot, L. Breen, et W. Messing, Théorie de Dieudonné Cristalline II. XI, 261 pages. 1982.

Vol. 931: D.M. Arnold, Finite Rank Torsion Free Abelian Groups and Rings. VII, 191 pages. 1982.

Vol. 932: Analytic Theory of Continued Fractions. Proceedings, 1981. Edited by W.B. Jones, W.J. Thron, and H. Waadeland. VI, 240 pages. 1982.

Vol. 933: Lie Algebras and Related Topics. Proceedings, 1981. Edited by D. Winter. VI, 236 pages. 1982.

Vol. 934: M. Sakai, Quadrature Domains. IV, 133 pages. 1982.

Vol. 935: R. Sot, Simple Morphisms in Algebraic Geometry. IV, 146 pages. 1982.

Vol. 936: S.M. Khaleelulla, Counterexamples in Topological Vector Spaces. XXI, 179 pages. 1982.

Vol. 937: E. Combet, Intégrales Exponentielles. VIII, 114 pages. 1982.

Vol. 938: Number Theory. Proceedings, 1981. Edited by K. Alladi. IX, 177 pages. 1982.

Vol. 939: Martingale Theory in Harmonic Analysis and Banach Spaces. Proceedings, 1981. Edited by J.-A. Chao and W.A. Woyczyński. VIII, 225 pages. 1982.

Vol. 940: S. Shelah, Proper Forcing. XXIX, 496 pages. 1982.

Vol. 941: A. Legrand, Homotopie des Espaces de Sections. VII, 132 pages. 1982.

Vol. 942: Theory and Applications of Singular Perturbations. Proceedings, 1981. Edited by W. Eckhaus and E.M. de Jager. V, 363 pages. 1982.

Vol. 943: V. Ancona, G. Tomassini, Modifications Analytiques. IV, 120 pages. 1982.

Vol. 944: Representations of Algebras. Workshop Proceedings, 1980. Edited by M. Auslander and E. Lluis. V, 258 pages. 1982.

Vol. 945: Measure Theory. Oberwolfach 1981, Proceedings. Edited by D. Kölzow and D. Maharam-Stone. XV, 431 pages. 1982.

Vol. 946: N. Spaltenstein, Classes Unipotentes et Sous-groupes de Borel. IX, 259 pages. 1982.

Vol. 947: Algebraic Threefolds. Proceedings, 1981. Edited by A. Conte. VII, 315 pages. 1982.

Vol. 948: Functional Analysis. Proceedings, 1981. Edited by D. Butković, H. Kraljević, and S. Kurepa. X, 239 pages. 1982.

Vol. 949: Harmonic Maps. Proceedings, 1980. Edited by R.J. Knill, M. Kalka and H.C.J. Sealey. V, 158 pages. 1982.

Vol. 950: Complex Analysis. Proceedings, 1980. Edited by J. Eells. IV, 428 pages. 1982.

Vol. 951: Advances in Non-Commutative Ring Theory. Proceedings, 1981. Edited by P.J. Fleury. V, 142 pages. 1982.

Vol. 952: Combinatorial Mathematics IX. Proceedings, 1981. Edited by E. Billington, S. Oates-Williams, and A.P. Street. XI, 443 pages. 1982.

Vol. 953: Iterative Solution of Nonlinear Systems of Equations. Proceedings, 1982. Edited by R. Ansorge, Th. Meis, and W. Törnig. VII, 202 pages. 1982.

Vol. 954: S.G. Pandit, S.G. Deo, Differential Systems Involving Impulses. VII, 102 pages. 1982.

Vol. 955: G. Gierz, Bundles of Topological Vector Spaces and Their Duality. IV, 296 pages. 1982.

Vol. 956: Group Actions and Vector Fields. Proceedings, 1981. Edited by J.B. Carrell. V, 144 pages. 1982.

Vol. 957: Differential Equations. Proceedings, 1981. Edited by D.G. de Figueiredo. VIII, 301 pages. 1982.

Vol. 958: F.R. Beyl, J. Tappe, Group Extensions, Representations, and the Schur Multiplicator. IV, 278 pages. 1982.

Vol. 959: Géométrie Algébrique Réelle et Formes Quadratiques, Proceedings, 1981. Edité par J.-L. Colliot-Thélène, M. Coste, L. Mahé, et M.-F. Roy. X, 458 pages. 1982.

Vol. 960: Multigrid Methods. Proceedings, 1981. Edited by W. Hackbusch and U. Trottenberg. VII, 652 pages. 1982.

Vol. 961: Algebraic Geometry. Proceedings, 1981. Edited by J.M. Aroca, R. Buchweitz, M. Giusti, and M. Merle. X, 500 pages. 1982.

Vol. 962: Category Theory. Proceedings, 1981. Edited by K.H. Kamps, D. Pumplün, and W. Tholen, XV, 322 pages. 1982.

Vol. 963: R. Nottrot, Optimal Processes on Manifolds. VI, 124 pages. 1982.

Vol. 964: Ordinary and Partial Differential Equations. Proceedings, 1982. Edited by W.N. Everitt and B.D. Sleeman. XVIII, 726 pages. 1982.

Vol. 965: Topics in Numerical Analysis. Proceedings, 1981. Edited by P.R. Turner. IX, 202 pages. 1982.

Vol. 966: Algebraic K-Theory. Proceedings, 1980, Part I. Edited by R.K. Dennis. VIII, 407 pages. 1982.

Vol. 967: Algebraic K-Theory. Proceedings, 1980. Part II. VIII, 409 pages. 1982.

Vol. 968: Numerical Integration of Differential Equations and Large Linear Systems. Proceedings, 1980. Edited by J. Hinze. VI, 412 pages. 1982.

Vol. 969: Combinatorial Theory. Proceedings, 1982. Edited by D. Jungnickel and K. Vedder. V, 326 pages. 1982.

Vol. 970: Twistor Geometry and Non-Linear Systems. Proceedings, 1980. Edited by H.-D. Doebner and T.D. Palev. V, 216 pages. 1982.

Vol. 971: Kleinian Groups and Related Topics. Proceedings, 1981. Edited by D.M. Gallo and R.M. Porter. V, 117 pages. 1983.

Vol. 972: Nonlinear Filtering and Stochastic Control. Proceedings, 1981. Edited by S.K. Mitter and A. Moro. VIII, 297 pages. 1983.

Vol. 973: Matrix Pencils. Proceedings, 1982. Edited by B. Kågström and A. Ruhe. XI, 293 pages. 1983.